Development and Neurobiology of *Drosophila*

BASIC LIFE SCIENCES
Alexander Hollaender, General Editor
Associated Universities, Inc.
Washington, D.C.

Volume 1 • GENE EXPRESSION AND ITS REGULATION
Edited by F. T. Kenney, B. A. Hamkalo, G. Favelukes,
and J. T. August

Volume 2 • GENES, ENZYMES, AND POPULATIONS
Edited by A. M. Srb

Volume 3 • CONTROL OF TRANSCRIPTION
Edited by B. B. Biswas, R. K. Mandal, A. Stevens,
and W. E. Cohn

Volume 4 • PHYSIOLOGY AND GENETICS OF REPRODUCTION
(Parts A and B)
Edited by E. M. Coutinho and F. Fuchs

Volume 5 • MOLECULAR MECHANISMS FOR REPAIR OF DNA
(Parts A and B)
Edited by P. C. Hanawalt and R. B. Setlow

Volume 6 • ENZYME INDUCTION
Edited by D. V. Parke

Volume 7 • NUTRITION AND AGRICULTURAL DEVELOPMENT
Edited by N. Scrimshaw and M. Béhar

Volume 8 • GENETIC DIVERSITY IN PLANTS
Edited by Amir Muhammed, Rustem Aksel, and R. C. von Borstel

Volume 9 • GENETIC ENGINEERING FOR NITROGEN FIXATION
Edited by Alexander Hollaender, R. H. Burris, P. R. Day,
R. W. F. Hardy, D. R. Helinski, M. R. Lamborg, L. Owens,
and R. C. Valentine

Volume 10 • LIMITATIONS AND POTENTIALS FOR
BIOLOGICAL NITROGEN FIXATION IN THE TROPICS
Edited by Johanna Döbereiner, Robert H. Burris,
Alexander Hollaender, Avilio A. Franco, Carlos A. Neyra,
and David Barry Scott

Volume 11 • PHOTOSYNTHETIC CARBON ASSIMILATION
Edited by Harold W. Siegelman and Geoffrey Hind

Volume 12 • GENETIC MOSAICS AND CHIMERAS IN MAMMALS
Edited by Liane B. Russell

Volume 13 • POLYPLOIDY: Biological Relevance
Edited by Walter H. Lewis

Volume 14 • GENETIC ENGINEERING OF OSMOREGULATION: Impact on
Plant Productivity for Food, Chemicals, and Energy
Edited by D. W. Rains, R. C. Valentine, and Alexander Hollaender

Volume 15 • DNA REPAIR AND MUTAGENESIS IN EUKARYOTES
Edited by W. M. Generoso, M.D. Shelby, and F. J. deSerres

Volume 16 • DEVELOPMENT AND NEUROBIOLOGY OF *DROSOPHILA*
Edited by O. Siddiqi, P. Babu, Linda M. Hall, and Jeffrey C. Hall

A Continuation Order Plan is available for this series. A continuation order will bring delivery of each new volume immediately upon publication. Volumes are billed only upon actual shipment. For further information please contact the publisher.

Development and Neurobiology of *Drosophila*

EDITED BY

O. SIDDIQI
P. BABU

Tata Institute of Fundamental Research
Bombay, India

LINDA M. HALL

Albert Einstein College of Medicine
The Bronx, New York

AND

JEFFREY C. HALL

Brandeis University
Waltham, Massachusetts

PLENUM PRESS · NEW YORK AND LONDON

Library of Congress Cataloging in Publication Data

Conference on Development and Behavior of Drosophila, Tata Institute of Fundamental Research, 1979
Development and neurobiology of Drosophila.

(Basic life sciences; v. 16)
Includes index.
1. Drosophila melanogaster—Development—Congresses. 2. Drosophila melanogaster—Physiology—Congresses. 3. Nervous system—Insects—Congresses. 4. Insects—Development—Congresses. 5. Insects—Physiology—Congresses. I. Siddiqi, Obaid, 1932- II. Title.
QL537.D76C66 1979   595.77'4   80-19900
ISBN 0-306-40559-8

Proceedings of the International Conference on Development and
Behavior of *Drosophila melanogaster,* held at the Tata Institute
of Fundamental Research, Bombay, India, December 19—22, 1979.

© 1980 Plenum Press, New York
A Division of Plenum Publishing Corporation
227 West 17th Street, New York, N.Y. 10011

All rights reserved

No part of this book may be reproduced, stored in a retrieval system, or transmitted,
in any form or by any means, electronic, mechanical, photocopying, microfilming,
recording, or otherwise, without written permission from the Publisher

Printed in the United States of America

FOREWORD

There is no multicellular animal whose genetics is so well understood as *Drosophila melanogaster*. An increasing number of biologists have, therefore, turned to the fruitfly in pursuit of such diverse areas as the molecular biology of eukaryotic cells, development and neurobiology. Indeed there are signs that *Drosophila* may soon become the most central organism in biology for genetic analysis of complex problems.

The papers in this collection were presented at a conference on Development and Behavior of *Drosophila* held at the Tata Institute of Fundamental Research from 19th to 22nd December, 1979. The volume reflects the commonly shared belief of the participants that *Drosophila* has as much to contribute to biology in the future as it has in the past. We hope it will be of interest not merely to *Drosophilists* but to all biologists.

We thank Chetan Premani, Anil Gupta, K.S. Krishnan, Veronica Rodrigues, Hemant Chikermane and K.Vijay Raghavan for help with recording and transcription of the proceedings and Vrinda Nabar and K.V. Hareesh for editorial assistance. We thank Samuel Richman, Thomas Schmidt-Glenewinkel and T.R. Venkatesh for their valuable assistance in proofreading the manuscripts, and we also thank Patricia Rank for her excellent effort in the preparation of the final manuscripts. The conference was supported by a grant from Sir Dorabji Tata Trust.

> O. Siddiqi
> P. Babu
> Linda M. Hall
> Jeffrey C. Hall
>
> Tata Institute of Fundamental Research
> Bombay, January, 1980

CONTENTS

Opening Remarks   A. *Garcia-Bellido*   1

I.   GENE ORGANIZATION AND EXPRESSION

Structural and Functional Organization of a Gene in *Dro-
    sophila melanogaster*   A. *Chovnick, M. McCarron,
    S.H. Clark, A.J. Hilliker* and *C.A. Rushlow*   3

Genetics of Minute Locus in *Drosophila melanogaster*
    A.K. *Duttagupta* and D.L. *Shellenbarger*   25

Effect of zeste on white Complementation   P. *Babu*
    and S. *Bhat*   35

The Heat Shock Response: A Model System for the Study
    of Gene Regulation in *Drosophila*   M.L. *Pardue,
    M.P. Scott, R.V. Storti* and *J.A. Lengyel*   41

Regulation of DNA Replication in *Drosophila*   A.S.
    *Mukherjee, A.K. Duttagupta, S.N. Chatterjee,
    R.N. Chatterjee, D. Majumdar, C. Chatterjee,
    M. Ghosh, P.M. Achary, A. Dey* and *I. Banerjee*   57

II.   GENETICS OF DEVELOPMENT

A Combined Genetic and Mosaic Approach to the Study
    of Oogenesis in *Drosophila*   E. *Wieschaus*   85

Developmental Analysis of *fs(1)1867*, an Egg Resorp-
    tion Mutation of *Drosophila melanogaster*   J.
    Szabad and J. *Szidonya*   95

Genetics of Pattern Formation   P.J. *Bryant* and
    J.R. *Girton*   109

The Control of Growth in the Imaginal Discs of Drosophila   P. Simpson and P. Morata — 129

An Analysis of the Expressivity of Some bithorax Transformations   G. Morata and S. Kerridge — 141

What is the Normal Function of Genes Which Give Rise to Homeotic Mutations?   A. Shearn — 155

Genetic and Developmental Analysis of Mutants in an Early Ecdysone-inducible Puffing Region in Drosophila melanogaster   I. Kiss, J. Szabad, E.S. Belyaeva, I.F. Zhimulev and J. Major — 163

Developmental and Genetic Studies of the Indirect Flight Muscles of Drosophila melanogaster   I.I. Deak, A. Rahmi, P.R. Bellamy, M. Bienz, A. Blumer, E. Fenner, M. Gollin, T. Ramp, C. Reinhardt, A. Dubendorfer and B. Cotton — 183

Monoclonal Antibodies in the Analysis of Drosophila Development   M. Wilcox, D. Brower and R.J. Smith — 193

III. DEVELOPMENT AND WIRING OF THE NERVOUS SYSTEM

The Effect of X-chromosome Deficiencies on Neurogenesis in Drosophila   J.A. Campos-Ortega and F. Jiménez — 201

Formation of Central Patterns by Receptor Cell Axons in Drosophila   J. Palka and M. Schubiger — 223

Sensory Pathways in Drosophila Central Nervous System   A. Ghysen and R. Janson — 247

Peripheral and Central Nervous System Projections in Normal and Mutant (bithorax) Drosophila melanogaster   N.J. Strausfeld and R.N. Singh — 267

IV: CHEMISTRY OF THE NERVOUS SYSTEM

Use of Neurotoxins for Biochemical and Genetic Analysis of Membrane Proteins Involved in Cell Excitability   L.M. Hall — 293

CONTENTS

The Acetylcholinesterase from *Drosophila melanogaster*
    S. Zingde and K.S. Krishnan ... 305

Isolation and Characterization of Membranes from *Drosophila melanogaster*   T.R. Venkatesh, S. Zingde and K.S. Krishnan ... 313

Phosphorylated Proteins in *Drosophila* Membranes   P. Thammana ... 323

V.   SENSORY AND MOTOR BEHAVIOR

Photoreceptor Function   W.L. Pak, S.K. Conrad, N.E. Kremer, D.C. Larrivee, R.H. Schinz and F. Wong ... 331

Genetic Analysis of a Complex Chemoreceptor   O. Siddiqi and V. Rodrigues ... 347

Olfactory Behavior of *Drosophila melanogaster*   V. Rodrigues ... 361

Mutants of Brain Structure and Function: What is the Significance of the Mushroom Bodies for Behavior?
    M. Heisenberg ... 373

Visual Guidance in *Drosophila*   K.G. Götz ... 391

VI.   COMPLEX BEHAVIOR

Effects of a Clock Mutation on the Subjective Day -- Implications for a Membrane Model of the *Drosophila* Circadian Clock   R. Konopka and D. Orr ... 409

Apparent Absence of a Separate B-Oscillator in Phasing the Circadian Rhythm of Eclosion in *Drosophila pseudoobscura*   M.K. Chandrashekaran ... 417

Higher Behavior in *Drosophila* with Mutants that Disrupt the Structure and Function of the Nervous System   J.C. Hall, L. Tompkins, C.P. Kyriacou, R.W. Siegel, F. von Schilcher and R. Greenspan ... 425

Attractants in the Courtship Behavior of *Drosophila melanogaster*   R. Venard ... 457

A Review of the Behavior and Biochemistry of dunce,
    a Mutation of Learning in *Drosophila*   D. Byers   467

Concluding Remarks   *L.M. Hall, J.C. Hall, O. Siddiqi
    and P. Babu*   475

Conference Participants

Index

OPENING REMARKS

A. Garcia-Bellido

Centro de Biologia Molecular
Universidad Autonoma de Madrid
Madrid, Spain

I think we had the best introduction to this meeting yesterday evening when we went to a restaurant and had our first taste of Indian food. Looking at the program, I have the same feeling as I had when looking through that menu card. There is abundant food of every kind, food for thought. This is the most ambitious program I have ever seen; the most ambitious conference I have attended. It seems to be one piece because of *Drosophila* alone. When we contemplate this diversity, we might be inclined to react like the biologists of the forties, before the advent of molecular biology. There were many things not known that had to be described before attempting to understand them. You could just about take any problem and go in any direction: development, behavior, ecology and so on. To contemplate any small parcel of this diversity was socially justified because there was no hope of understanding the rules at any level of biological complexity.

The approach of molecular biology succeeded because it could show that there are rules and principles that explain a variety of molecular processes. We now feel that we can understand the logic behind diversity. When we contemplate the complexity of an organism made with different cells and tissues, which performs a series of behavioral responses, and which can evolve into other types of organisms, can we ever hope to explain it in molecular terms, or even attempt a complete molecular description? The answer must be no because complexity does not result from simple addition of elements at the lower level but from combinations of certain subsets of these elements. We can explain DNA in terms of nucleotides but we cannot predict which four of the possible nucleotides are actually found in the DNA. From the DNA we cannot predict recombination or Mendel's laws, although, no doubt, these depend on the very nature of the DNA. Complexity results from interactions and the rules at any level have to be found at that level. They might be explained in terms of elements of a lower level and eventually in terms of molecules. If this conclusion is correct, the practical consequences would be to go back to the biological realm and keep observing, describing and experimenting, back to the scientific attitude of the biologists of the forties. It is a pessimistic conclusion. If we believe that there are rules in each level of complexity, the first question is by which methods can we hope to abstract these rules.

In my view, the best instrument to see order in this jungle is genetics. Genetics is precisely a science of interaction. It is effectively discovering the elements of a set, and establishing their hierarchical relationships; it is a science of grammar rather than of phonetics. It could help us as the Ariadna's thread to not get lost in the labyrinth. Genetics provides a very different approach from that of classical biochemistry. In a biochemical approach, we break things into pieces and put the pieces together *in vitro* to reconstruct the natural state. By induction, we then infer the properties of the original system. The geneticists' approach to complex systems, to a developing egg, to a population, to ecology or to evolution is somewhat different. He tries to keep the norms of the system intact and introduces small changes at a time. He then observes the effects on the whole and draws conclusions about the interaction rules of the system.

Genetics is, however, not a formal science. Genes are active in cells and may well determine how cells differentiate and consequently organize into tissues or make the cellular connections which underly neural responses and behavior. Genes, after all, are elements which are preserved throughout evolution. They are also responsible for making species different. The intellectual challenge is to find out to what extent these discrete properties of different levels of complexity are coded in the DNA; what are the instructions to construct them. And this, precisely, is our justification for choosing *Drosophila*. About the genes of *Drosophila* we have a great deal of information. If genetics is going to be the Ariadna's thread that I have talked about, I would urge everyone, including myself, to contemplate the beautiful Ariadna, but care even more for her thread. I want to end these opening remarks paraphrasing the last words of Geothe: Genetics, more Genetics.

STRUCTURAL AND FUNCTIONAL ORGANIZATION OF A GENE IN *DROSOPHILA MELANOGASTER*

        A. Chovnick, M. McCarron, S.H. Clark, A.J. Hilliker and C.A. Rushlow

        Genetics and Cell Biology Section
        Biological Sciences Group
        The University of Connecticut
        Storrs, Connecticut 06268 U.S.A.

INTRODUCTION

    In recent years, considerable interest and research effort has been directed towards an understanding of the mechanisms underlying the control of gene expression in higher organisms. Several experimental approaches have been employed to this end, and the results of such studies have been the subject of recent review (Axel et al., 1979). Major emphasis in this laboratory has been directed towards the development of an experimental system in *Drosophila melanogaster* involving the genetic dissection of a specific genetic unit. The end-product of this effort will be a gene whose DNA is marked and mapped to permit eventual identification of the various sequences that control its expression.

THE GENETIC SYSTEM

    The rosy locus in *Drosophila melanogaster* (*ry*:3-52.0) is a genetic unit concerned with xanthine dehydrogenase (XDH) activity, located on the right arm of chromosome 3 (Figure 1) within polytene chromosome section 87D. The locus was originally defined by a set of brownish eye color mutants deficient in drosopterin pigment. Such mutants were shown subsequently to exhibit no detectable XDH activity (Glassman and Mitchell, 1959). Figure 2 summarizes reactions used in this laboratory to assay *Drosophila* XDH (Forrest et al., 1956; Glassman and Mitchell, 1959). Zygotes possessing little or no XDH activity are unable to complete development and die before eclosion on standard *Drosophila* culture medium supplemented with purine. This fact serves as the basis for a nutritional selective procedure which

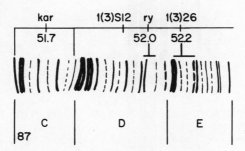

Figure 1: Polytene chromosome map of the rosy region of chromosome three in *Drosophila melanogaster* (Bridges, 1941).

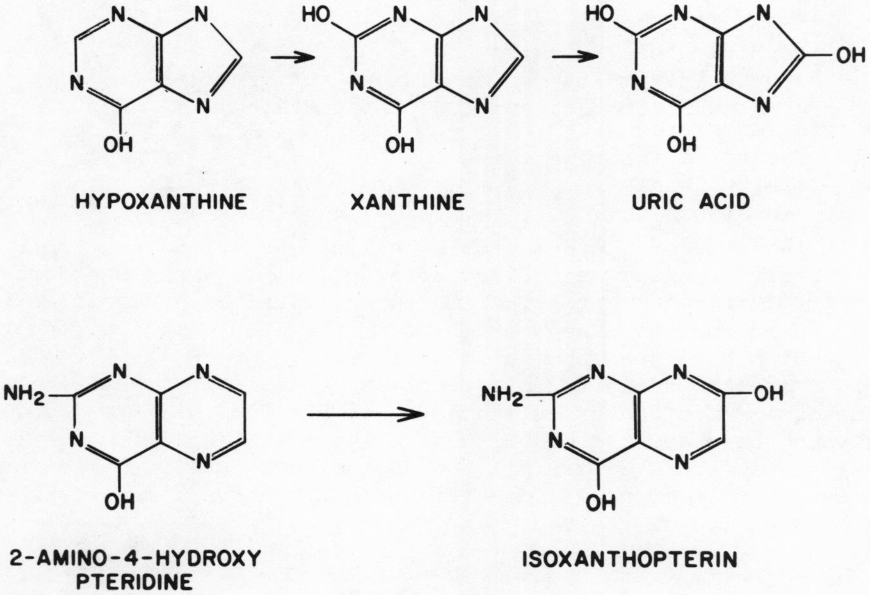

Figure 2: Reactions of *Drosophila* xanthine dehydrogenase.

has made large scale fine structure mapping within the rosy locus a routine laboratory exercise (Chovnick et al., 1971).

It is now clear that *Drosophila* XDH is a homodimer (Gelbart et al., 1974), with a subunit molecular weight of 150,000 daltons (Edwards et al., 1977). Two observations serve to demonstrate that the structural information for XDH is encoded by the rosy locus. (A) Variation in dosage of ry^+ alleles, from 0-3 doses, appears to be the limiting factor in determining level of XDH activity/fly in otherwise wild-type flies (Grell, 1962; Glassman et al., 1962). (B) The genetic basis for variation in XDH electrophoretic mobility seen in wild-type strains, maps to sites within the rosy locus that is defined by rosy eye color mutants which are XDH⁻ (Gelbart et al., 1974).

NOMENCLATURE

Electrophoretic mobility variants of XDH are readily isolated from laboratory stocks and natural populations of *Drosophila melanogaster*. From these sources, we have established a number of wild-type isoalleles of the rosy locus. These are maintained as stable lines that possess XDH molecules with distinctive electrophoretic mobilities and thermal properties. Moreover, the XDH enzyme activity level associated with each of these wild-type alleles is also a distinctive, stable phenotypic character. Table 1 summarizes our present array of ry^+ isoalleles. The XDH produced by ry^{+0} serves as a mobility standard, and is designated $XDH^{1.00}$. All variants XDHs that are slower are designated by relative mobility values < 1.00, while faster XDHs are designated by superscripts > 1.00. The ry^{+4} allele is associated with sharply higher XDH activity than all others and is classified in Table 1 as high (H), while ry^{+10} exhibits much lower activity than all others and is classified as low (L). The activity levels of the remaining alleles are representative of intermediate levels which we presently classify as normal (N).

Enzymatically inactive rosy mutants are readily selected by virtue of their visibly mutant eye color phenotype. Over a period of many years, X-ray, γ-ray and EMS mutagenesis experiments have provided us with a large number of such mutants. We have adopted the convention of labeling each mutant with a superscript which identifies the ry^+ isoallele from which the mutant was derived. Thus, the XDH⁻, rosy eye color mutant ry^{402}, is the second variant isolated from ry^{+4}, and ry^{1201} is the first mutant derivative of the

Table 1. Wild-type Isoalleles of rosy

ry^+ Alleles	XDH Mobility	XDH Activity
+12, +13	0.90	N
+14	0.94	N
+10	0.97	L
+0, +6, +7, +8	1.00	N
+1, +11, +16	1.02	N
+4	1.02	H
+2	1.03	N
+3, +5	1.05	N

ry^{+12} allele. Another class of rosy locus mutants exhibits normal or near-normal eye color in homozygotes, possesses low levels of XDH activity, and dies on purine supplemented media (Gelbart et al., 1976). These mutants are designated by the prefix *ps* (purine sensitive). Thus, ry^{ps214}, the 14th identified variant of ry^{+2}, is a purine sensitive mutant associated with a low level of XDH activity and a wild-type eye color. Electrophoretic sites are noted with the prefix *e*. The site responsible for the mobility difference between the ry^{+4} product ($XDH^{1.02}$) and the ry^{+0} enzyme ($XDH^{1.00}$) is designated *e408*. Thus, ry^{+0} possesses *e408s* (slow), while ry^{+4} carries the *e408f* (fast) alternative. The difference in intensity or level of XDH activity associated with the ry^{+4} isoallele as compared to other wild-type isoalleles (Table 1) has been localized to a site designated *i409*, with ry^{+4} possessing *i409H* (high) while ry^{+0} and other alleles possess *i409N* (normal).

GENETIC DISSECTION OF THE ROSY REGION

The rosy region of chromosome 3 has been the subject of extensive cytogenetic analysis in our laboratory. The precise segment which has been the major focus of this analysis extends from 87D2-4 to 87E12-F1, a chromosomal section of 23 to 24 polytene chromosome bands (Figure 1). A detailed description of this work will appear elsewhere (Hilliker et al., 1980a, b).

The raw material of this study is a group of 153 non-rosy, lethal and semi-lethal, apparent site mutants, induced largely with EMS, but including some radiation-induced mutants. These were selected, originally, as being lethal with one or another of several chromosomes possessing deletions of part or all of this chromosomal segment. Then from a series of experiments involving *inter se* complementation and tests for lethality with chromosomes carrying smaller overlapping deletions of this region, we have been able to identify and order a total of 20 lethal complemention groups within this segment of 3R (Figure 1). Thus, with the addition of the rosy locus, a total of 21 complementation groups have been identified in this segment of 23 to 24 polytene bands, a result entirely consistent with the one gene:one chromomere hypothesis (Bridges, 1935; Judd et al., 1972; Hochman, 1973).

In addition to defining the cytogenetic limits of the rosy locus, another purpose of this work was to establish the limits of the genetic entity in the rosy region that controls XDH. That the rosy locus itself plays a central role in the control of XDH is well established. Of interest is the possibility that the rosy locus may be part of a larger genetic complex in control of XDH, perhaps including genes immediately adjacent to rosy. This question was examined in two ways. One approach was to seek single site rosy eye color mutations that did not complement the vital functions associated with the adjoining non-rosy complementation groups. From experiments carried out earlier (Schalet et al., 1964) and the recent study (Hilliker et al., 1980a,b) a total of 32 lethal rosy alleles were recovered. All were found to be associated with rearrangements, largely deletions, and the lethality in each case involved functional deficiency for one or more of the vital genetic units surrounding the rosy locus.

The second approach was to ask if the recessive lethal lesions in the genetic units immediately adjacent to the rosy locus could be shown to influence rosy locus function. Mutants of the left flanking *1(3)S12* locus and the right flanking *pic* (piccolo) locus were examined for effects upon XDH production without success. Taken together, these experiments strongly support the contention that the rosy locus is an isolated genetic unit concerned with XDH activity,

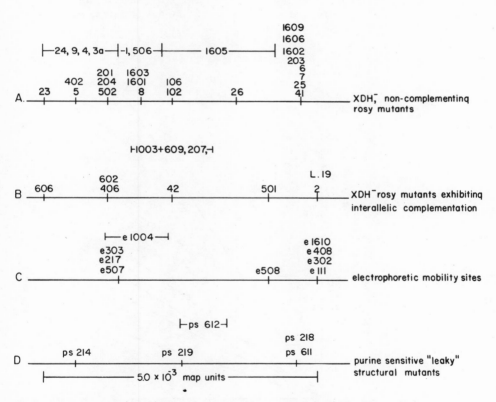

Figure 3: Genetic fine structure maps of rosy locus sites. Map locations of unambiguous structural element variants (B, C and D) are positioned relative to map of XDH⁻ non-complementing mutants (A).

and that all adjacent genetic units are functionally as well as spatially distinct from the rosy gene.

ORGANIZATION OF THE ROSY LOCUS

Rosy locus variants have been the subject of an intensive and continuing intragenic mapping analysis. Figure 3 summarizes our progress in this effort. Figure 3A presents a map of XDH⁻, non-complementing, rosy eye color mutant sites. Estimation of the boundaries of the XDH coding element is provided by the maps of three classes of unambiguous coding element site variants presented in Figure 3B (XDH⁻, allele-complementing, rosy eye color mutant sites), Figure 3C (electrophoretic mobility sites), and Figure 3D (purine sensitive "leaky" structural mutant sites). The left boundary is set by the leftmost allele-complementing site mutant, ry^{606}. On the basis of comparative recombination data and the failure of large-scale tests with ry^{606} to produce recombinants, the non-complementing site mutant, ry^{23}, must also mark the left border. At the

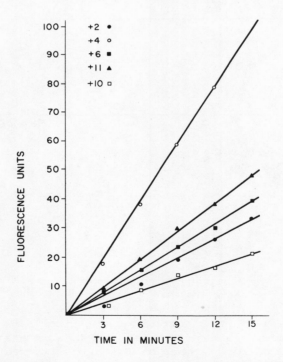

Figure 4: Fluorimetric assay of XDH activities of matched extracts of strains homozygous for the indicated wild-type isoalleles.

ROSY GENE ORGANIZATION

right end of the map, several electrophoretic sites and the complementing mutants, ry^2 and $ry^{L.19}$, identify the right boundary of the coding element, with no known XDH variants beyond them. The maps of Figure 3 position 51 sites to the right of our present left boundary of the XDH structural element. Moreover, an additional ten sites, not indicated, map in the structural element.

VARIATION IN INTENSITY OF XDH ACTIVITY

In addition to electrophoretic mobility differences, we have already noted that the various ry^+ isoalleles are associated with variation in level of XDH activity, which also behaves as a stable phenotypic character. Consider the ry^{+4} and ry^{+10} lines which exhibit much greater and much less activity, respectively, than all of our other wild-type lines (Table 1). These differences are observed by following either the purine or pteridine reaction (Figure 2), and are readily classified in cuvette assays (spectrophotometry or fluorimetry) or upon gel electrophoresis. Figure 4 illustrates typical fluorimetric assays of XDH activity in matched, partially purified extracts of several ry^+ isoallelic stocks including both ry^{+4} and ry^{+10}. Measurements of XDH activity/mg protein, activity/fly and activity/mg wet weight invariably yield similar relationships. Activity levels associated with ry^{+2}, ry^{+6} and ry^{+11} (Figure 4) are typical of wild-type alleles presently classified as normal (N) in Table 1, and never overlap the ry^{+4} and ry^{+10} extremes. Indeed, ry^{+10} homozygotes exhibit such low activity, that they may be distinguished from other wild-type strains by virtue of their purine

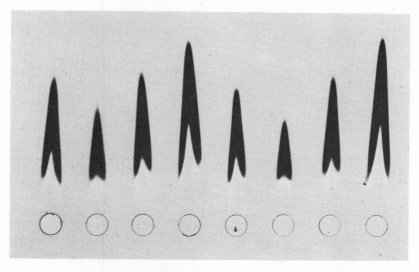

Figure 5: "Rocket" electropherogram. Matched extracts of homozygotes run against anti-XDH serum. From l. to r., ry^{+11}, ry^{+10}, ry^{+0} and ry^{+4} repeated in that order across the gel.

sensitivity. Detailed analyses of the genetic bases for the ry^{+4} and ry^{+10} phenotypes are presented elsewhere (Chovnick et al., 1976; McCarron et al., 1979). Together, these studies identify a cis-acting control element located adjacent to the left (centromere proximal) side of the XDH structural element. The following sections outline the experimental basis for this conclusion.

FURTHER CHARACTERIZATION OF THE XDH ACTIVITY LEVEL VARIANTS

The variation in level of XDH activity (Table 1 and Figure 4) is a property of the rosy locus derived from standard mapping experiments. However, the task of distinguishing whether the genetic bases for the level of activity differences reside in the XDH structural element, or possibly in a control element is not as simply resolved. Based upon fine structure recombination analysis, some of our ry^+ isoalleles differ by as many as five or six structural element sites. Independent studies (Coyne, 1976; Singh et al., 1976) of XDH structural gene polymorphism in natural populations additionally support the view that any two isolated lines will very likely possess structural element differences. Obviously, the genetic polymorphism exhibited by the XDH structural element might also extend into an adjacent control element. Certainly, level of enzyme activity is a superficial phenotypic character that might reflect either structural or control element variation. Thus, we are compelled to define further this character, and to carry out high resolution fine structure mapping experiments.

We have considered the possibility that there are structural differences between the XDH molecules produced by ry^{+4} and those produced by other ry^+ isoalleles that are responsible for the sharply increased activity of ry^{+4} preparations. However, we have been unable to associate this activity difference in any systematic manner with electrophoretic, thermolability or kinetic differences (Chovnick et al., 1976; Edwards et al., 1977). Similar comparisons involving XDH molecules produced by ry^{+10} and those produced by other ry^+ isoalleles also have failed (McCarron et al., 1979).

On the other hand, immunological experiments clearly support the notion that the activity differences reflect differences in number of XDH molecules (Chovnick et al., 1976; McCarron et al., 1979). Figure 5 presents a typical immunoelectrophoresis experiment utilizing the method of "rocket electrophoresis" (Laurell, 1966; Weeke, 1973) to compare the relative numbers of molecules of XDH in matched extracts of ry^{+11}, ry^{+10}, ry^{+0} and ry^{+4} homozygotes. Quantitative analysis of such gels (Chovnick et al., 1976) reveals that the rocket heights parallel quite closely the XDH activity levels associated with these isoalleles.

The time course of appearance of enzyme during development has been determined for a number of our wild-type isoallelic lines, as well as for the activity level variants. Synchronous populations from the various lines are raised to the appropriate developmental stages, at which time extracts are examined for XDH activity, total protein and response to XDH antibody (CRM). Thus, we are able to examine and compare XDH activity/individual, XDH activity/mg protein or activity in terms of CRM levels throughout development. No gross differences have been detected among the various strains. While ry^{+4} exhibits relatively increased amounts of enzyme throughout development and ry^{+10} is associated with reduced enzyme, their developmental profiles are otherwise normal. Similarly, tissue distribution studies have failed to identify qualitative differences that might be associated with the activity level variants.

GENETIC FINE STRUCTURE EXPERIMENTS

Let us now consider localization of the genetic bases for the level of enzyme activity differences described above. Fine structure recombination experiments were carried out that were designed to pursue the difference between the ry^{+4} allele (associated with high activity) and other wild-type alleles which exhibit normal levels of activity (Chovnick et al., 1976). These studies led to the identification of a site, *i409*, with ry^{+4} possessing *i409H* (associated with high activity) while such alleles as ry^{+0}, ry^{+2}, ry^{+6} and ry^{+11} carry *i409N* (associated with normal levels of activity). Moreover, these experiments localized this site to the left, but definitely outside of the genetic boundaries of the XDH structural element. Figure 6 illustrates this localization of *i409* between the left end of the XDH structural element and *l(3)S12*, a site mutant in the very next genetic unit to the left of rosy.

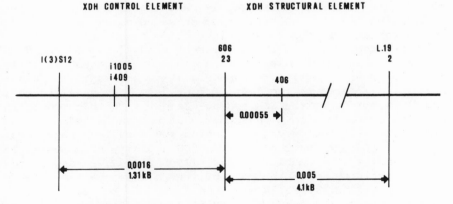

Figure 6: The rosy locus. Size estimates of the structural and control elements.

Now let us turn to the low level of XDH activity associated with the ry^{+10} isoallele. A series of fine structure recombination experiments were carried out that parallel, in many respects, the ry^{+4} experiments (McCarron et al., 1979). From these studies we identify a site, designated $i1005$, with ry^{+10} carrying $i1005L$ (Low), and our normal wild-type alleles carrying $i1005N$ (Normal). Additionally, these crosses position $i1005$ to the right of $l(3)S12$ and to the left of the XDH structural element. Clearly, these experiments localize $i1005$ to the same region as $i409$ (Figure 6).

At this juncture, one is drawn to the possibility that the observed variation in level of XDH activity (Low, Normal and High) characterizing our various ry^+ alleles (Table 1) may reflect alternatives that map to the same site. Might $i1005$ and $i409$ be synonyms designating the same site, or do they mark separable sites? Heterozygotes of the composition $\frac{+\ \ L\ \ +}{kar\ H\ ry^{406}}$ were mated to appropriate tester males following a selective protocol designed to kill L and ry^{406} bearing progeny. (There were additional markers in the cross to facilitate diagnosis of all survivors.) If $i1005L$ and $i409H$ are located at the same site, then all crossovers should exhibit the high activity associated with $i409H$. Similarly, conversions of $i1005L$ should yield $i409H$. In fact, neither of these expectations are fulfilled. In a total of 1.3×10^6 progeny sampled, there were 16 surviving ry^+ recombinants. Three were conversions of ry^{406} and exhibited the <u>high</u> level of XDH activity associated with $i409H$. One was a conversion of $i1005L$ and possessed a <u>normal</u> level of XDH activity. Finally, there were twelve ry^+ progeny associated with exchange for the flanking markers. Nine of these possessed high activity, while the remaining three exhibited <u>normal</u> XDH activity. The results of this cross indicate quite clearly that $i1005$ and $i409$ mark separable sites.

Unfortunately, this experiment is unable to determine the relative positions of the two sites because we do not know what effect the double variant, $i1005L\ i409H$, might have on XDH activity. Figure 7 illustrates the ambiguity in our present situation. If the double variant is associated with less than normal levels of XDH activity, then the map order illustrated in Figure 7A is correct. In this case, $i1005$ is located to the left of $i409$, and thus still further from the left border of the XDH coding element. On this model, the 12 recovered crossovers involve exchange at the indicated positions. Crossover class (1) represents the 9 high XDH activity crossovers, while class (2) describes the 3 normal XDH activity crossovers. On the other hand, if the double variant, $i1005L\ i409H$, is associated with normal activity, then the possibility remains the the reverse order illustrated in Figure 7B might be correct. Resolution of this ambiguity requires further experiments which are in progress.

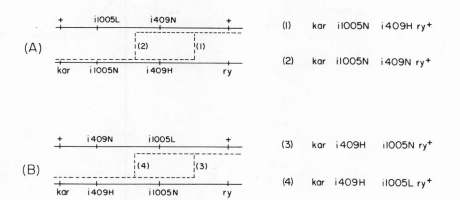

Figure 7: Relative map positions of *i1005* and *i409*. Ambiguity inherent in random strand mapping experiment.

RELATIONSHIP BETWEEN *i409*, *i1005* AND THE STRUCTURAL ELEMENT

Might *i409H* represent a tandem duplication of the XDH structural element? On this notion the ry^{+4} allele would be considered to possess two functional XDH structural elements in tandem, presumably resulting from an unequal exchange event. Such a model is precluded on several counts:

1. EMS mutagenesis of ry^{+4} has produced rosy eye color mutations at a frequency that does not distinguish this allele from other ry^{+} isoalleles.

2. The ry^{+4} allele is associated with a single XDH electrophoretic band of the mobility class, $XDH^{1.02}$ (Table 1). XDH is a homodimer, and the presence of two electrophoretically distinct structural elements will produce individuals possessing three XDH moieties. The tandem duplication model then requires that the ry^{+4} allele possess two XDH structural elements whose peptide products are indistinguishable, and of the mobility class $XDH^{1.02}$. Thus, *i409H* should be associated with an $XDH^{1.02}$. On this point, the tandem duplication model fails. In all experiments which recombine *i409H* with other electrophoretically distinct structural elements there is no evidence of the production of an $XDH^{1.02}$ moiety.

3. Tandem duplications are characterized by instability in homozygotes due to increased incidence of unequal exchange events. The ry^{+4} stock has been quite stable. Moreover, fine structure recombination experiments involving tests of ry^{400} series mutants against other XDH⁻ mutants have been characterized by regular exchange events, and the complete absence of unequal crossing over.

4. Cytological examination of polytene chromosomes reveals no apparent tandem duplication.

Now let us consider *i1005L*. While the association of *i1005L* with a reduction in the number of XDH molecules/fly does not seemingly lend itself to the tandem duplication model, precedence for such consideration exists in the case of the Bar duplication in *Drosophila melanogaster* and its associated position effect (Sturtevant, 1925; Muller et al., 1936; Bridges, 1936; Muller, 1936). On such a notion, the ry^{+10} isoallele would possess two XDH structural elements in tandem. Moreover, by virtue of the resulting change in position of each member of the duplex relative to some adjacent genetic element(s) a disturbed function of both XDH structural elements results. Such a model is precluded by the same arguments described above for *i409H*.

RELATIONSHIP BETWEEN *i409*, *i1005* AND A CONTROL ELEMENT

Having eliminated the possibility that either *i409H* or *i1005L* might be associated with a tandem duplication of the XDH structural element, we are drawn, at this point, to the notion that these sites mark one or more genetic elements that serve to regulate XDH. We believe that the following points provide a compelling argument.

1. We now possess stocks carrying *i409N* and *i409H* with structural elements producing $XDH^{0.97}$, $XDH^{1.00}$, $XDH^{1.02}$, $XDH^{1.03}$ and $XDH^{1.05}$. Examination of these lines has failed to associate *i409* with any XDH structural characteristic.

2. On a more limited scale, we have similarly produced stocks carrying *i1005N* and *i1005L* with structural elements producing $XDH^{0.97}$, $XDH^{1.00}$ and $XDH^{1.05}$. Similarly, we are unable to associate *i1005* with any structural characteristic.

3. As noted above, the rosy locus genetic maps of Figure 3 establish the genetic boundaries of the XDH coding element. As a result of extensive mapping experiments over a number of years, these boundaries enclose a region that today includes more than 60 sites representing all of the mapped XDH⁻ rosy eye color mutant sites, sites of electrophoretic variation and "leaky" structural mutant sites. Eventually, we hope to relate these genetic boundaries to the amino and carboxy termini of the XDH peptide. However, the fact that these boundaries have not changed in recent years (Gelbart et al., 1974; McCarron et al., 1979), coupled with the extensive data upon which

they are based, strongly suggests that these boundaries may well approximate the amino and carboxy termini. Only two mapped sites fall outside of the coding element boundaries (*i409* and *i1005*).

4. While we have failed to associate *i409* and *i1005* variants with XDH structural variation, both are associated with phenotypic effects that conform precisely with expectation for control element variants.

On the basis of evidence thus far presented, a broad array of regulatory roles are possible, and further specification would be premature. However, we are able to describe one key feature of their regulatory function(s). Under one class of regulatory roles, dominance-recessiveness, or incomplete-dominance in heterozygotes would obtain (*trans*-regulation). In still another class of roles, the function of a given regulatory element would be restricted to the specific XDH structural element adjoining it on the chromosome (*cis*-regulation). Figure 8 presents an electropherogram demonstrating the *cis*-acting nature of *i409* and *i1005* variation. The photograph is printed as a negative in order to enhance the contrast between XDH bands and the acrylamide slab gel. Allele designations are presented in Figure 8 in terms of electrophoretic mobility and level of XDH activity. Thus, *1.05H* refers to a recombinant ry^+ allele exhibiting the fast mobility of the ry^{+5} allele (Table 1), and the high XDH activity level of the ry^{+4} allele. In terms of intensity sites, this allele should be considered as *i1005N, i409H*. Those alleles with activity levels designated as N are *i1005N, i409N*, while L refers to *i1005L, i409N*. The heterozygote marked $\frac{1.05N}{0.97N}$ illustrates the characteristic 1:2:1 three banded pattern of XDH dimers to be found in a heterozygote possessing approximately equal proportions of the fast and slow monomers. Intensity site substitutions, as indicated in Figure 8, produce heterozygotes whose XDH electrophoretic

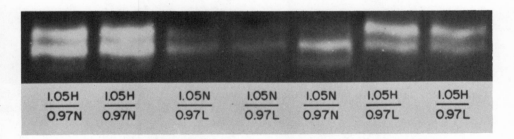

Figure 8: XDH electropherogram indicating the relative amounts of $XDH^{1.05}$, $XDH^{1.01}$ and $XDH^{0.97}$ present in matched extracts of flies heterozygous for the indicated combinations of structural and control elements.

gel patterns indicate that the *i1005* and *i409* sites are *cis*-acting at the level of XDH monomers, to determine the number of monomers available for dimer formation.

Thus, we may eliminate all regulatory models which associate *i1005* and *i409* with the synthesis of a negative or positive acting diffusible regulatory molecule (Dickson et al., 1975; Gilbert and Müller-Hill, 1970). Rather, we are drawn to the possibility that *i1005* and *i409* mark the 5' control element of the rosy locus. Although the present state of this analysis is reminiscent of early stages in the analysis of the *lac* region promoter in *E. coli* (Ippen et al., 1968; Arditti et al., 1968), we caution the reader that the present data do not, as yet, permit specification of the control function involved in either *i409* or *i1005* variation. Certainly, it is reasonable to expect that *cis*-acting variants of the 5' control element might involve alterations in DNA sequences which serve as binding sites for regulatory signal(s), sites for RNA polymerase binding and initiation of transcription, transcript processing sites, ribosome binding and initiation of translation. In view of the significant recombination between *i1005* and *i409*, it is most likely that these sites mark different DNA sequences in the control element, and thus involve variation in different control functions.

STRUCTURAL AND CONTROL ELEMENT SIZE ESTIMATES

Figure 6 summarizes our present "best" estimates of map lengths for the rosy locus structural and control elements. Boundaries for the structural element are described in an earlier section. The XDH structural element map distance of 5×10^{-3} map units emerges from an enormous data base (Figure 3), and is not likely to change. However, control element mapping data is limited, and size estimates should be considered as quite tentative. Indeed, the present estimate (Figure 6) represents a significant downward revision of the size of the control element from an earlier estimate (Chovnick et al., 1976), and is based largely upon recent additional mapping data (McCarron et al., 1979). While the map of Figure 6 indicates ambiguity in the positions of *i1005* and *i409* as described above, the crossovers separate these sites with a map distance of 0.46×10^{-3} standard map units.

Translation of map distance estimates into DNA base lengths proceeds from the XDH peptide molecular weight of 150,000 daltons. Assuming an average amino acid molecular weight (adjusted for peptide linkage) to be 110, then the number of nucleotides in the structural element required to code for such a peptide is approximately 4.1 kB (150,000 x 3/110). Then from the recombination map length of the structural element (0.005 map units), we relate map length to number of bases (0.01 map unit - 8.2 kB), which we apply directly to the adjacent control element (Figure 6). Thus, our present recombination data lead to a total size estimate for the rosy locus of 5.4 kb. It should be appreciated that the extrapolation from recombination map distance to DNA base length assumes that the XDH structural element is not interrupted by one or more non-translated intervening sequences that have been found in the coding elements of a number of higher organism genes. If such an insert were present in the coding element of the rosy locus DNA, and if it were to participate in recombination, then we would have underestimated the size of the coding element, the relationship of nucleotide bases to map distance and consequently have underestimated the size of the XDH control element, as well.

Nevertheless, our size estimate for the rosy locus of 5.4 kB is entirely consistent with the cytogenetic localization of rosy to the very fine polytene chromosome bands 87D12-13, and cytochemical estimate of the DNA content of such bands. Figure 9 presents a photograph of polytene chromosome segment 87D provided by Dr. George Lefevre. Estimate may be made of the haploid DNA content of bands 87D12-13 by relating them to the DNA content of bands measured by Rudkin (1965). The bands that might be associated with the rosy locus are much finer than the average sized bands measured by Rudkin (30 kB), and may well approach the finest bands (5 kB) that he measured.

HETEROCHROMATIC POSITION EFFECT: A SPECIAL CASE OF GENE REGULATION

This section summarizes the results of preliminary experiments that demonstrate a heterochromatic position effect upon rosy locus function. Position effect in *Drosophila* is a long known (Sturtevant, 1925), extensively studied and reviewed (Lewis, 1950; Baker, 1968; Spofford, 1976), but little understood group of phenomena.

Gamma irradiation of flies bearing kar^2 ry^{+11} chromosomes, followed by a mating and selection protocol designed to recover "leaky" structural mutants, "underproducer" control variants and "null" enzyme mutants succeeded in providing a number of interesting mutations for study. Among these were two purine sensitive alleles of rosy, ry^{Ps1149} and $ry^{Ps11136}$, that comprise the subject of this section. Subsequent genetic and cytological analysis revealed that

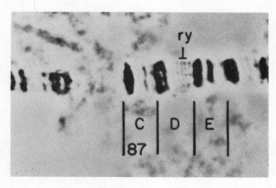

Figure 9: The rosy region of section 87D of polytene chromosome 3R.

these mutants were associated with rearrangements placing the rosy region of polytene chromosome segment 87D in association with centromeric heterochromatin.

That the purine sensitivity of these mutants reflects reduced XDH activity levels was confirmed by direct fluorimetric enzyme assays. Efforts to produce mutant homozygotes or hemizygotes with rosy region deficiencies revealed that ry^{ps1149} is pic^- (the locus immediately to the right of rosy), while $ry^{ps11136}$ is $l(3)S12^-$ (the locus immediately to the left of rosy). Lethal mutants representative of all other rosy region vital genes appear to be complemented by these rosy mutant bearing chromosomes. Cytological study revealed that ry^{ps1149} was associated with a translocation involving a break in chromosome 4 heterochromatin as well as break in 87D. On the other hand, $ry^{ps11136}$ appears to be associated with an inversion involving section 87D and the centromeric heterochromatin. In the paragraphs that follow, experiments are described that characterize ry^{ps1149}. Parallel observations have been made for $ry^{ps11136}$.

Figure 10 illustrates experiments that demonstrate two facts about ry^{ps1149}. The experiments involve the method of "rocket" immunoelectrophoresis (see earlier section) to compare the number of XDH molecules in matched extracts of $kar^2\ ry^{+11}/Tp(3)MKRS,\ M(3)S34\ kar\ ry^2\ Sb$ and $kar^2\ ry^{ps1149}/Tp(3)\ MKRS,M(3)S34\ kar\ ry^2\ Sb$. The MKRS chromosome is a rosy region balancer that carries the ry^2 mutant, long known as a CRM^- mutant in homozygotes (Glassman and Mitchell, 1959). The upper gel of Figure 10 compares extracts of females with two X chromosomes and no Y chromosomes. Clearly, the purine sensitivity and reduced XDH activity associated with ry^{ps1149} is also associated with a considerable reduction in the number of molecules of XDH.

The lower gel of Figure 10 compares extracts of females whose third chromosomes are identical to those of the upper gel. However, the sex chromosome composition of the lower gel females are X/Y^S $X\cdot Y^L$. Certainly, the restoration of approximately normal levels of XDH production in the Y bearing, ry^{ps1149} females is strong indication of a heterochromatic position effect. Similar observations (not shown) are seen in males. Thus XO males carrying ry^{ps1149} produce exceedingly low levels of XDH while comparable XY males have increased, but less than normal levels of XDH.

Further support for the notion of a heterochromatic position effect is seen in the electropherogram illustrated in Figure 11. Matched extracts of heterozygotes of ry^{+11}/ry^{+13} and ry^{ps1149}/ry^{+13} are run on acrylamide slab gels, and developed as described elsewhere (McCarron et al., 1979). While ry^{+11} and ry^{+13} are associated with normal levels of XDH production, the ry^{+11} enzyme exhibits a faster mobility (1.02) than the ry^{+13} enzyme (0.90) (Table 1). The heterozygote, ry^{+11}/ry^{+13}, exhibits the characteristic 1:2:1 three

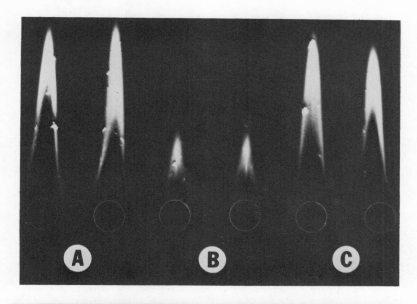

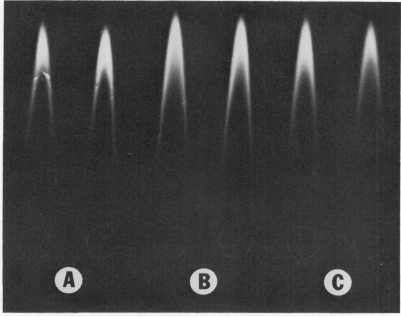

Figure 10: "Rocket" electropherogram. Upper gel consists of matched extracts, run in duplicate, of X/X females with the following third chromosomes: (A) $kar^2\ ry^{+11}/MKRS$; (B) $kar^2\ ry^{ps1149}/MKRS$; (C) $kar\ ry^{+11}/MKRS$. Lower gel consists of matched extracts, run in duplicate, of $X/Y^S X \cdot Y^L$ females with third chromosome genotypes as in the upper gel.

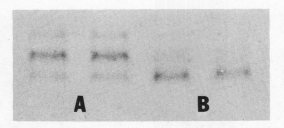

Figure 11: XDH electropherogram indicating the relative amounts of $XDH^{1.02}$, $XDH^{0.96}$ and $XDH^{0.90}$ present in matched extracts, run in duplicate, of (A) kar^2 ry^{+11}/ry^{+13}, and (B) kar^2 ry^{ps1149}/ry^{+13}.

banded pattern of XDH dimers to be found in a heterozygote possessing approximately equal proportions of the fast and slow monomers. Now consider the pattern of XDH dimers to be found in extracts of ry^{ps1149}/ry^{+13}. Clearly, there is a very low production of the ry^{ps1149} associated product. Nevertheless, there are traces of dimer production at positions that indicate some production of a 1.02 mobility product (the ry^{+11} enzyme) and a hybrid dimer with approximately intermediate mobility. Certainly, the strong band at 0.90 mobility indicates that most of the ry^{+13} product is present in the form of homozygous dimers, reflecting a low level of available faster monomers with which to form hybrid dimers. These observations indicate a cis-acting reduction of XDH production associated with ry^{ps1149}.

Finally, the results illustrated in Figures 10 and 11 further suggest that the protein product of ry^{ps1149} is a normal ry^{+11} monomer. It is clearly capable of associating in homozygous and hybrid dimers with expected mobilities (Figure 11). Moreover, the homozygous dimer production revealed in the immunoelectrophoresis experiments (Figure 10) indicate only quantitative variation in parallel with enzyme activity levels.

Taken together, these observations indicate that we are dealing with a Y-suppressed, cis-acting heterochromatic position effect upon rosy locus function. Additionally, the data suggest that the defect does not involve the XDH coding element of the rosy locus, but rather involves an effect upon one or more steps in transcription.

In view of the extensive information already available about the rosy locus, and the chromosomal region about it, we view this system as an excellent model with which to pursue questions concerning the mechanism of position effect.

ACKNOWLEDGEMENTS

This investigation has been supported by research grants, GM-09886, from the Public Health Service, and BMS74-19628 from the National Science Foundation. Additionally, S.H.C. was a postdoctoral fellow of the National Institutes of Health, and C.A.R. is a predoctoral trainee supported by a training grant from the National Institutes of Health, GM-07584.

REFERENCES

Arditti, R.R., Scaife, J.G., and Beckwith, J.R., 1968, The nature of mutants in the *lac* promoter region, J. Mol. Biol., 38:421.

Axel, R., Maniatis, T., and Fox, C.F., 1979, Eucaryotic gene regulation (ICN-UCLA Symposia on Molecular and Cellular Biology). In press.

Baker, W.K., 1968, Position effect variegation, Adv. in Genetics, 14:133.

Bridges, C.B., 1935, Salivary chromosome maps. With a key to the banding of the chromosomes of *Drosophila melanogaster*, J. Hered., 26:60.

Bridges, C.B., 1936, The Bar "gene" a duplication, Science, 83:210.

Bridges, P.N., 1941, A revision of the salivary gland 3R-chromosome map of *Drosophila melanogaster*, J. Hered., 32:299.

Chovnick, A., Ballantyne, G.H., and Holm, D.B., 1971, Studies on gene conversion and its relationship to linked exchange in *Drosophila melanogaster*, Genetics, 69:179.

Chovnick, A., Gelbart, W., McCarron, M., Osmond, B., Candido, E.P.M., and Baillie, D.L., 1976, Organization of the rosy locus in *Drosophila melanogaster*: Evidence for a control element adjacent to the xanthine dehydrogenase structural element, Genetics, 84:233.

Coyne, J.A., 1976, Lack of genic similarity between two sibling species of Drosophila as revealed by varied techniques, Genetics, 84:593.

Dickson, R.C., Abelson, J., Barnes, W.M., and Reznikoff, W.S., 1975, Genetic regulation: the Lac control region, Science, 187:27.

Edwards, T.C.R., Candido, E.P.M. and Chovnick, A., 1977, Xanthine dehydrogenase from *Drosophila melanogaster*. A comparison of the kinetic parameters of the pure enzyme from two wild-type isoalleles differing at a putative regulatory site, Molec. Gen. Genet., 154:1.

Forrest, H.S., Glassman, E., and Mitchell, H.K., 1956, Conversion of 2-amino-4-hydroxypteridine to isoxanthopterin in *D. melanogaster*, Science, 124:725.

Gelbart, W.M., McCarron, M., Pandey, J., and Chovnick, A., 1974, Genetic limits of the xanthine dehydrogenase structural element within the rosy locus in *Drosophila melanogaster*, Genetics, 78:869.

Gelbart, W., McCarron, M., and Chovnick, A., 1976, Extension of the limits of the XDH structural element in *Drosophila melanogaster*, Genetics, 84:211.

Gilbert, W., and Müller-Hill, B., 1970, The lactose repressor, in: "The Lactose Operon", J.R. Beckwith and D. Zipser, eds., Cold Spring Harbor Laboratory, Cold Spring Harbor, New York.

Glassman, E., and Mitchell, H.K., 1959, Mutants of *Drosophila melanogaster* deficient in xanthine dehydrogenase, Genetics, 44:153.

Glassman, E., Karam, J.D., and Keller, E.C., 1962, Differential response to gene dosage experiments involving the two loci which control xanthine dehydrogenase of *Drosophila melanogaster*, Z. Vererb., 93:399.

Grell, E.H., 1962, The dose effect of $ma-1^+$ and ry^+ on xanthine dehydrogenase activity in *Drosophila melanogaster*, Z. Vererb., 93:371.

Hilliker, A.J., Clark, S.H., Chovnick, A., and Gelbart, W.M., 1980a, Cytogenetic analysis of the chromosomal region immediately adjacent to the rosy locus in *Drosophila melanogaster,* Genetics, In press.

Hilliker, A.J., Clark, S.H., Gelbart, W.M., and Chovnick, A., 1980b, Cytogenetic analysis of the rosy micro-region, polytene chromosome interval 87D2-4; 87E12-F1, of *Drosophila melanogaster*, Drosophila Inform. Serv., 56: In press.

Hochman, B., 1973, Analysis of a whole chromosome in Drosophila, Cold Spring Harbor Symp. Quant. Biol., 38:581.

Ippen, K., Miller, J.H., Scaife, J.G., and Beckwith, J.R., 1968, New controlling element in the *Lac* operon of *E. coli*, Nature, 217:825.

Judd, B.H., Shen, M.W., and Kaufman, T.C., 1972, The anatomy and function of a segment of the X chromosome of *Drosophila melanogaster*, Genetics, 71:139.

Laurell, C.-B., 1966, Quantitative estimation of proteins by electrophoresis in agarose gel containing antibodies, Anal. Biochem., 15:45.

Lewis, E.B., 1950, The phenomenon of position effect, Adv. in Genetics, 3:73.

McCarron, M., O'Donnell, J., Chovnick, A., Bhullar, B.S., Hewitt, J., and Candido, E.P.M., 1979, Organization of the rosy locus in *Drosophila melanogaster*: Further evidence in support of a *cis*-acting control element adjacent to the xanthine dehydrogenase structural element, Genetics, 79:275.

Muller, H.J., 1936, Bar duplication, Science, 83:528.

Muller, H.J., Prokofyeva-Belgovskaya, A.A., and Kossikov, K.V., 1936, Unequal crossing-over in the Bar mutant as a result of duplication of a minute chromosome section, C.R. (Dokl.) Acad. Sci. U.R.S.S., N.S., 1(10):87.

Rudkin, G.T., 1965, The relative mutabilities of DNA in regions of the X chromosome of *Drosophila melanogaster*, Genetics, 52:665.

Schalet, A., Kernaghan, R.P., and Chovnick, A., 1964, Structural and phenotypic definition of the rosy cistron in *Drosophila melanogaster*, Genetics, 50:1261.

Singh, R.S., Lewontin, R.C., and Felton, A.A., 1976, Genetic heterogeneity within electrophoretic "alleles" of xanthine dehydrogenase in *Drosophila pseudoobscura*, Genetics, 84:609.

Spofford, J.B., 1976, Position effect variegation in Drosophila, in: "The Genetics and Biology of Drosophila, Vol. 1C", M. Ashburner and E. Novitski, eds., Academic Press, London.

Sturtevant, A.H., 1925, The effects of unequal crossing-over at the Bar locus in Drosophila, Genetics, 10:117.

Weeke, B., 1973, Rocket immunoelectrophoresis, in: "A Manual of Quantitative Immunoelectrophoresis: Methods and Applications", N.H. Axelson, J. Kroll and B. Weeke, eds., Scand. J. Immunol., 2 Suppl. 1:37.

DISCUSSION

J.C. Hall: Is there any experimental method to determine whether a cis-acting locus is affecting transcription or translation? Although the cis action you have shown implies that there is no diffusible product specified by your regulatory sites, these alleles could be acting cytoplasmically by being transcribed normally and then having defects at the translational level.

A Chovnick: I am rather reluctant to think of action at the translational level, but there are any number of possibilities, as we have noted in the publications on these mutants.

GENETICS OF MINUTE LOCUS IN *DROSOPHILA MELANOGASTER*

Ashish K. Duttagupta[*] and David L. Shellenbarger[**]

Genetics Laboratory
Department of Zoology
University of British Columbia, Canada

INTRODUCTION

Minutes are a class of dominant mutants that are characterized by their recessive lethality, increased developmental time, smaller body with short fine bristles in the heterozygotes and with varying degrees of reduced survival and fertility (Lindsley et al., 1972). This class of mutants was first described by Bridges and Morgan (1923) and as a group presents many interesting genetical features. These include: autonomy of the rate of the mitotic division in some Minutes (Morata and Ripoll, 1975), induction of non-disjunction in the X-chromosome (Miklos, 1970), cell autonomy of some Minutes in chaetae differentiation, and cell lethality in genetic mosaics (Stern and Tokunaga, 1971), and their high rate of induction by ethylmethanesulfonate (Huang and Baker, 1976).

While some of the Minutes show no cytologically detectable alterations, there are at least 42 "haplo abnormal" loci which produce Minute phenotype (Lindsley et al., 1972). On the basis of complementation patterns between different Minute mutants and the dosage effect of Minute loci, Schultz (1929) concluded that the similar pleiotropic effect caused by different Minutes may be the result of common "phenogenetic" failure and that the qualitatively different gene products of the different Minutes may perform very similar functions. Atwood's hypothesis (see Lindsley and Grell, 1968) that

---

[*] Present address: Genetics Laboratory, Department of Zoology, University of Calcutta, Calcutta (India)
[**] Present address: Department of Biochemistry, University of Washington, Seattle, Washington (USA)

Minute loci correspond to the sites of the redundant structural genes for tRNA is one of the most interesting suggestions made in respect to such functions. The similarity of the Minute phenotype with those for the partial deficiencies for the ribosomal DNA (Ritossa et al., 1966) or 5S ribosomal RNA genes (Procunier and Tartof, 1975) added force to the hypothesis. It was of interest therefore to look into the genetic organization of a Minute locus and to know how the different alleles within the locus behave in their expression and complementation patterns.

MATERIALS AND METHODS

The locus chosen for the present study was $M(2)173$ (location 2-92.3) on the distal end of the right arm of second chromosome (56F-57B, see Lindsley et al., 1972). $M(2)173$ is a mutation leading to moderate Minute phenotype that is cytologically normal (Figure 1a). Two other Minutes, $M(2)O17$ and $M(2)U$ were found to non-complement each other as well as $M(2)173$. Of these, $M(2)O17$ was found to carry a small deletion in the 56F5-15 region (Figures 1b and c). Since it was viable when heterozygous with the aneuploid deficiency extending from 56E to 56F, it was concluded that $M(2)O17$ is not deficient for 5S ribosomal RNA genes, which are located in the 56E to 56F segments (Procunier and Tartof, 1975).

For screening of the $M(2)173$ alleles, newly eclosed b pr cn males were fed with 0.025 M EMS (Lewis and Bacher, 1968) and were mated with twice the number of virgins carrying a Cy balancer (SM1, SM5 or CyO DTS-513). Treated second chromosomes were recovered over the Cy balancer in the F_1 and single b pr cn*/Cy males (where * denotes the mutagenized chromosome) were mated to 3-5 virgins of $M(2)173$ or $M(2)O17$ females in glass vials and two or three days after they were transferred to fresh vials. In the F_2, the putative lethal alleles of $M(2)173$ or $M(2)O17$ were those cultures where straight-winged non-Cy (b pr cn*/M) flies were absent. The lethal was collected over Cy in the surviving b pr cn*/Cy progeny and a stock was established from these flies. The F_1 b pr cn*/Cy was also screened for the dominant "Minute" phenotype and all of them were crossed with $M(2)173$ for allelism tests.

All the putative alleles of $M(2)173$ were also crossed with both $M(2)173$ and $M(2)O17$ and could be classified into three groups: (1) those that were lethal only with $M(2)173$; (2) those lethal only when heterozygous with $M(2)O17$; and (3) those that were lethal when heterozygous with either $M(2)173$ or $M(2)O17$. Lethals in each group were then crossed with each other for complementation tests, and the progeny were scored not only for the complementing classes but also for any new phenotype. In addition, all lethals were tested in the presence of an extra dose of $M(2)173^+$ in combination with $M(2)173$ or $M(2)O17$. For this a $T(Y;2)$ translocation $T(Y;2)J64$ carrying a duplication for 56F-57A was used; and lethal $l/M/M^+$ flies were generated

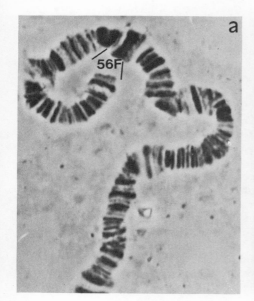
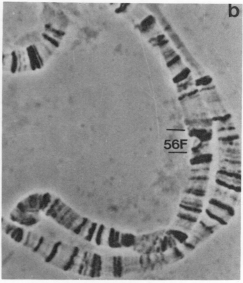
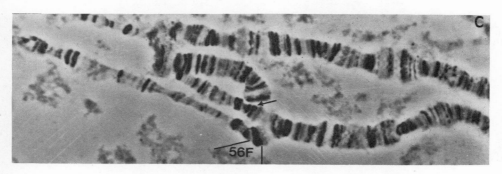

Figure 1: Photomicrographs showing the 56F region of the salivary gland chromosomes of (a) *M(2)173/b pr cn*, (b) *M(2)017/+* and (c) *M(2)017/Cy* larvae. The *M(2)173* chromosome is normal, while the *M(2)017* shows a small deletion. This is clear in (c) where due to partial asynapsis of the balancer the smaller width of the deleted 56F1--9 region (arrow) is clearly revealed.

from the cross $1/Cy$ ♀♀ × $\hat{XY}/y^+Y$ B^S Dp $J64$; nw^D $M(2)173/+$ ♂♂ and $1/Cy$ ♀♀ × $\hat{XY}/y^+Y$ B^S Dp $J64$; Tft $M(2)017/+$ ♂♂.

For fine structure mapping, second chromosome dominant markers on either side of the $M(2)173$ were crossed onto the lethals as needed and the following crosses were made: (1) $nw^D\ l_1/l_2\ Pu^2$ ♀♀ X $M(2)173/Cy$ ♂♂ and (2) $nw^D\ l_1\ Pu^2/l_2$ X $M(2)173/Cy$. The Cy balancer had a dominant temperature-sensitive mutation which at 29°C allowed only the $l_1^+\ l_2^+$ recombinant classes to survive. The symbols l_1 and l_2 designate any two complementary recessive lethals.

RESULTS

A total of 7750 mutagenized second chromosomes were tested in combination with the two Minutes, viz., $M(2)173$ and $M(2)017$; 38 lethals were collected. Within them 14 lethals were collected against the deficiency $M(2)017$ and were labelled with the prefix $D-$. The number of chromosomes tested in this experimental set up was 2150. The remaining 24 lethals isolated from 5600 pair matings were found to be allelic to $M(2)173$. But one of them, 1362, which initially behaved as a lethal when tested with $M(2)173$, was subsequently found to allow survival and enhance the bristle phenotype of $M(2)173$. Only 13 mutants, however, were found to be non-complementing with both $M(2)173$ and $M(2)017$. Two others, 12 and 2362 were semilethal with both the Minutes and the survivors exhibited an enhanced bristle phenotype. Mutants 29, 47, $M(2)U$ exhibited a dominant small bristle phenotype and the mutant 51 survived as a homozygote.

In our effort to score the dominant short bristle phenotype in the $Cy/b\ pr\ cn^*\ F_1$ flies we could isolate 69 such flies out of 2231 progeny. Of these 25 transmitted the phenotype to their progeny, and of the 18 mutants tested for allelism with $M(2)173$ none was confirmed as an allele.

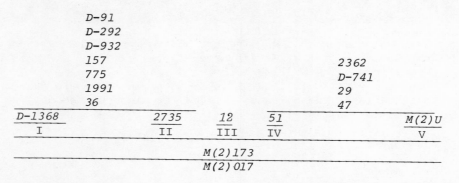

Figure 2: Final complementation map of alleles at the $M(2)173$ locus.

Table 1: Results of complementation analysis of the different alleles of *M(2)173* at 22°C*

	M(2)173	M(2)017	D-91	D-292	D-932	157	775	1991	36	D-1368	2735	12	M(2)U	47	29	D-741	2362	51
D-91	O	O																
D-292	O	O	O															
D-932	O	O	O	O														
157	O	O	O	O	O													
775	O	O	O	O	O	O												
1991	O	O	O	O	O	O	O											
36	O	O	O	O	O	O	O	O										
D-1368	O	O	O	O	O	O	O	O	+									
2735	O	O	O	O	O	O	O	O	+	O								
12	M	M	M	M	M	M	M	M	+	O	+							
M(2)U	O	O	O	O	O	O	O	O	+	O	+	O						
47	O	O	O	O	O	O	O	O	+	M	M	+	O					
29	O	O	O	O	O	O	O	O	+	M	M	+	M	M				
D-741	O	O	O	O	O	O	O	O	+	M	M	+	M	M	M			
2362	M	M	M	M	M	M	M	M	+	O	M	+	O	M	M	M		
51	O	O	O	O	O	O	O	O	M	O	M	+	M	M	M	M	M	

* O, non-complementing for survival, greater than 100 heterozygotes expected, none observed; M, surviving heterozygotes have more extreme Minute phenotype than either mutant alone, survival highly variable; +, complementing, survival of heterozygotes is greater than 30% and survivors have no stronger (if any) Minute phenotype than either mutant alone.

Results of our complementation tests are presented in Table 1. Of the 16 mutants that were tested, 14 were lethal to both $M(2)173$ and $M(2)017$ and two, viz., 12 and 2362 expressed as Minutes. Five complementation groups could be identified (Figure 2) and the mutants could approximately be divided into two groups. Seven alleles fail to complement $D-1368$ and 2735, while 4 fail to complement 51 and $M(2)U$. Interestingly all these 11 alleles fail to complement mutant 12. The mutant 12 and all the mutants in the complementation groups IV and V are of considerable importance since all these mutants have Minute phenotype or they interact with the other to cause a Minute phenotype. Groups I and II on the other hand, have no dominant short bristle phenotype and appear to be related with the groups IV and V, only because the mutant 12 interacts with mutations of all the others in a number of ways. While 12 complements $D-1368$, 2735, $M(2)U$ and 51, it interacts with all the other mutants to produce short bristle phenotype, accompanied sometimes with many other developmental anomalies. For example, it will interact with groups I and II mutants to produce not only short bristles, but also may exhibit missing bristles, rough eyes, bifurcation of scutellum, broad wings and reduced viability.

Table 2: Results of allelism test in the presence of the Y-linked M^+ duplication.

Lethal	$X/\hat{XY}$ ♀♀		$\hat{XY}/y^+$ Y B^S DpJ64 ♂♂ (Critical class)	
	l/M	Cy/M	l/M	Cy/M
$M(2)173$	0	80	48*	62
$M(2)017$	0	62	38*	53
$D-91$	0	73	0	58
$D-292$	0	56	44	37
$D-932$	0	55	53	47
775	0	93	68	76
1991	0	86	64	72
$D-1368$	0	35	55	24
2735	0	78	55	50
12	0	28	24	31
$M(2)U$	0	47	44*	67
47	0	52	60*	45
29	0	61	40	64
$D-741$	0	86	47	44
51	0	41	49	50

* Adults express the Minute small bristle phenotype.

Results of our experiments to test the lethals in the presence of the M^+ duplications have been summarized in Table 2. It was observed that the lethals in heterozygous condition with either of the Minutes, i.e., $M(2)173$ or $M(2)017$ survive in the presence of the duplication; an exception is the lethal $D-91$. It was further observed that for the expression of Minute phenotype at least two doses of the Minute was required. For example, the Minute phenotype is not expressed when the fly is of the genotype $X/y^+ \ Y \ B^S \ DpJ64 \ M^+; \ Cy/nw^D \ M(2)173$, but is expressed in the genotype $X/y^+ \ Y \ B^S \ DpJ64 \ M^+; \ M(2)173/nw^D \ M(2)173$ (Table 2). The exception however was $M(2)U$, where two doses of the M^+ region could not suppress the dominant small bristle phenotype; instead bristles of variable and intermediate size appeared in the $X/y^+ \ Y \ B^S \ DpJ64 \ M^+; \ M(2)U/+$ flies.

Our results on the fine structure recombination mapping are summarized briefly. By the selective marker system, described in Materials and Methods, we could map the different lethals. It was found that the lethals, $157,775,1991,2735$ and $D-1368$ map to the left of $D-741$ and that 775 maps to the left of $M(2)U$. Recombination frequencies varied from 0.004 to 0.012 map units, with respect to different pairwise combinations from among the mutations in this cluster.

DISCUSSION

An array of phenotypic diversity was exhibited by the different alleles of $M(2)173$. While some are recessive lethals, others are not and still others survive as homozygotes and exhibit a dominant small bristle phenotype. In the complementation tests there has been a clear separation of two classes of alleles; while one of them (Groups III and IV) includes the mutants which have a dominant small bristle phenotype occupying one end of the locus, the other class includes the recessive lethals (Groups I and II) occupying the other end. The distance between the groups of mutants varies roughly between 1.2×10^{-2} to 4×10^{-3} map units, a distance nearly similar to length of the rosy gene (Gelbart et al., 1976). Since $M(2)173$ fails to complement both the groups, one may surmise it to be a multigenic deletion. The normal cytology, the point mutant nature of the $M(2)173$ and the close linkage between the major groups of mutants however makes the idea more complex. The failure to recover a mutant which like $M(2)173$ would have failed to complement all the remaining mutants, may be due to the fact that mutagens like EMS usually induce point mutations.

The manifestation of the dominant bristle phenotype of the $M(2)U$ in the presence of two doses of M^+ may be due to the production of a substance which interferes with the normal gene product. On the other hand the survival of the homozygous $M(2)173$ in presence of a single M^+ added as a duplication, is a situation which is probably different from the two doses of Minute in a triploid, where they are lethal (Schultz, 1929).

It is difficult to comment on the relationship that may exist between the Minute mutants and protein synthesis. Recent studies indicate that a heterozygous deficiency in a tRNA locus that reduces the amount of tRNA made is not Minute (Dunn, 1977). It would be however interesting to note whether or not there is any change in the tRNA profile in the different alleles of *M(2)173* in the presence of the Y linked M^+ duplication.

ACKNOWLEDGEMENTS

The authors are grateful to D.T. Suzuki for suggesting the problem and to T.C. Kaufmann, D.P. Cross, T.A. Grigliatti and D.T. Suzuki for their enlightening discussion and helpful criticism.

REFERENCES

Bridges, C.B., and Morgan, T.H., 1923, The third chromosome group of mutant characters of *Drosophila melanogaster*, Carnegie Inst. Wash.Publ., 327.

Dunn, R.J., 1977, Studies on transfer RNA and transfer RNA genes in *Drosophila melanogaster*, Ph.D. Thesis, The University of British Columbia.

Gelbart, W., McCarron, M., and Chovnick, A., 1976, Extension of the limits of the XDH structural element in *Drosophila melanogaster*, Genetics, 84:211.

Huang, S.L., and Baker, B.S., 1976, The mutability of the Minute loci of *Drosophila melanogaster* with ethylmethanesulfonate, Mutation Res., 34:407.

Lewis, E.B., and Bacher, F., 1968, Method of feeding ethylmethanesulfonate (EMS) to *Drosophila* males, Drosophila Inform. Serv., 43:193.

Lindsley, D.L., and Grell, E.H., 1968, Genetic variations of *Drosophila melanogaster*, Carnegie Institution of Washington, Publication No. 627.

Lindsley, D.L., Sandler, L., Baker, B.S., Carpenter, A.T.C., Denell, R.E., Hall, J.C., Jacobs, P.A., Miklos, G.L.G., Davis, B.K., Gethmann, R.C., Hardy, R.W., Hessler, A., Miller, S.M., Nozawa, H:, Perry, D.M., and Gould-Somero, M., 1972, Segmental aneuploidy and the genetic gross structure of the *Drosophila* genome, Genetics, 71:157.

Miklos, G.L.G., 1972, The effects of Minute loci and the possible involvement of transfer-RNA in sex chromosome non-disjuction in the *Drosophila melanogaster* male, Molec. Gen. Genet., 115:289.

Morata, G., and Ripoll, P., 1975, Minutes: mutants of *Drosophila* autonomously affecting cell division rate, Devel. Biol., 42:211.

Procunier, J.D., and Tartof, K., 1975, Genetic analysis of the 5S RNA genes in *Drosophila melanogaster*, Genetics, 81:515.

Ritossa, F.M., Atwood, K.C., and Spiegelman, S., 1966, A molecular explanation of the bobbed mutants of *Drosophila* as partial deficiencies of 'ribosomal' DNA, Genetics, 54:819.

Schultz, J., 1929, The Minute reaction in the development of *Drosophila melanogaster*, Genetics, 14:366.

Stern, C., and Tokunaga, 1971, On cell lethals in *Drosophila*, Proc. Natl. Acad. Sci. USA, 68:329.

EFFECT OF ZESTE ON WHITE COMPLEMENTATION

P. Babu and S. Bhat

Molecular Biology Unit
Tata Institute of Fundamental Research
Bombay, India

The locus in which we are most interested is the white locus; it is on the first chromosome and maps at 1.5 map units. There are five recombinationally separable sites known in this locus (Judd, 1964). The site which maps right-most is called white-spotted (w^{sp}); the white (w) allele which has a pure white eye color as well as white-cherry (w^{ch}) which has a translucent pink eye phenotype also map at the next right-most site. Basically, white forms a single complementation group with the exception of w^{sp}; i.e. all white mutants fail to complement each other with the exception of w^{sp}. The phenotype of w^{sp} itself is unusual compared to other white mutants; w^{sp} males and females have spotted eyes of yellowish to brown pigment facets in a light color background. We are here interested in the fact that the w^{sp} mutation partially complements all other white mutations. This is clearly seen in the case of w/w^{sp} and w^{ch}/w^{sp} flies; both these trans-heterozygotes have uniformly pigmented brownish red eyes approaching the wild type. The new observation we have made is that this complementation between w^{sp} and other white alleles depends on the allelic conditions of the zeste locus, where zeste is also a mutation on the first chromosome at 1.0 map units. Before we describe our observation, we will briefly summarize what is known about mutations at the zeste locus.

The mutation zeste (z) discovered by Gans (1953) has the unusual property of causing lemon yellow eye color in female flies; this mutation has no effect on the eye color of males. Gans also found that it is the number of functional copies of the white locus which determines the eye color; males with two copies of w^+ show the mutant phenotype whereas female flies with one copy of w^+ are wild type. In her study of the zeste-white interaction Gans found a null allele of the z^+ locus which she called z^a. The mutation z^a

behaves like the deficiency of zeste locus; i.e. z/z^a has the same phenotype as z homozygote whereas z^+/z^a is like wild type. By itself z^a has no noticeable effect on the eye color of males or females.

Recently, Jack and Judd (1979) have made an important observation about the expression of the zeste mutation. They have found that not only are two copies of w^+ needed for the expression of the zeste phenotype, it is also necessary for the two w^+ to pair with each other for the mutant phenotype to be expressed. In other words, z flies with two copies of w^+ which are so placed in the genome that they presumably cannot pair with each other have wild type eye color. Thus, they show that zeste mutant expression shows transvection of the w^+ locus. Transvection can be described as a situation where pairing of the two alleles of a gene plays a role in the function of that gene.

The oldest known example of transvection, in the bithorax complex studied by Lewis (1954), is also of interest to us. This transvection phenomenon is best seen between bx^{34e}, a weak allele of bithorax, and Ultrabithorax *(Ubx)* mutations of the bithorax complex which map at 58.8 map units, on the third chromosome. Wild type flies have no observable dorsal metathorax (metanotum); bx^{34e}/bx^{34e} flies have hairy bristled metanotal tissue between the mesonotum and the first abdominal segment. Even though *Ubx* is homozygous lethal, by generating gynanders with *Ubx/Ubx* male tissue, it is observed that the *Ubx/Ubx* genotype leads to a well developed metanotum which looks like a normal mesonotum (Lewis, 1963). The bx^{34e}/Ubx flies have no metanotum and thus for this phenotype the two mutants complement each other. Lewis discovered that this complementation is dependent on allelic pairing; if one causes rearrangements in the right arm of the third chromosome so that pairing between the two alleles is disrupted, one gets development of metanotum. It was found that different degrees of metanotal development are seen with different rearrangements, presumably because these rearrangements cause different degrees of disruption of pairing between bx^{34e} and *Ubx*.

Lewis (1959) discovered another mutation, enhancer of bithorax *(en(bx)*; 1-1.0) which enhanced the mutant phenotype of bx^{34e}/Ubx flies. Thus, *en(bx)*; bx^{34e}/Ubx flies have metanotal development comparable to flies of genotype bx^{34e}/Ubx which also carry an extreme rearrangement which disrupts the pairing of bx^{34e} and *Ubx*. Kauffman, Tasaka and Suzuki (1973) later found that *en(bx)* is a z^a-like allele of zeste and that z^a has the same effect as *en(bx)*. The relationship between the transvection and the enhancement produced by z^a in the trans-combination bx^{34e}/Ubx has not been studied so far. But in light of our recent observation that z^a abolishes complementation between white alleles, it is tempting to suggest that the complementation dependence on allelic pairing seen in

bx^{34e}/Ubx flies needs the normal function of z locus.

Our observation of the effect of z^a on the complementation of white alleles is summarized in Table 1. z^a by itself has no noticeable effect on the eye phenotype of w^{sp}, w or wch flies, either in male or female; this was observed by Kauffman, Tasaka and Suzuki (1973). We find that in the presence of z^a, w^{sp} does not complement w or w^{ch}. $z^a w^{sp}/z w$ flies have a spotted eye very similar to w^{sp} (or $z^a w^{sp}$) males in contrast to w^{sp}/w flies which have uniform brownish red pigmentation. $z^a w^{sp}/z^a w^{ch}$ flies also have a mottled eye distinctly lighter than w^{sp}/wch flies. Thus, the complementation of w^{sp} with other white alleles is eliminated when z^+ is replaced by z^a. It is worth noting that when z^+ is replaced by z, this complementation is unaffected.

Table 1. Effect of z^a on white Complementation.

No.	Genotype	Eye Color
1)	$w^{sp}; z^a w^{sp}$	light with yellow to brownish spots
2)	$w^{sp}/w^{sp}; z^a w^{sp}/z^a w^{sp}$	light mottled pigmentation; lighter than (1)
3)	$w^{ch}; z^a w^{ch}$	uniform translucent pink
4)	$w^{ch}/w^{ch}; z^a w^{ch}/z^a w^{ch}$	uniform translucent pink; darker than (3)
5)	$w; z^a w; w/w; z^a w/z^a w$	pure white
6)	w^{sp}/w	uniform brownish red, much darker than w^{sp}
7)	$z^a w^{sp}/z^a w$	light spotted; similar to w^{sp}
8)	w^{sp}/w^{ch}	uniform brownish red; darker than (6)
9)	$z^a w^{sp}/z^a w^{ch}$	lighter than (8) and mottled.

We do not yet know whether the complementation observed between w^{sp} and other w alleles exhibits transvection or not; nor do we know whether z^+ is directly involved in the transvection observed in bx^{34e}/Ubx flies. But with the scanty evidence that we have we want to build a model of complementation dependent on pairing which applies both to the bithorax complex and the white locus. We assume that z^+ gene product is needed for complementation in both these loci. The model is rather specific and thus provides a framework for our ideas.

Our model of complementation is schematically represented in Fig. 1. We assume that the z^+ gene product can bind simultaneously to two sites (A and B) on the DNA and bring them together such that

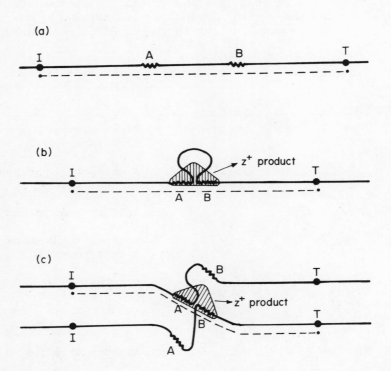

Figure 1: Schematic representation of the zeste product binding to DNA. I and T represent the initiation and termination sites of the RNA transcript. A and B are the sites on the DNA to which the zeste product can bind simultaneously and bring them close to each other so that the transcription can jump across from A to B.
(a) transcription in the absence of z^+ product; the transcript now includes a copy of the region between A and B.
(b) z^+ product binds to A and B and the region between them is "looped out" in the transcription process.
(c) z^+ product binds to the sites A and B on opposite paired chromosomes leading to a hybrid transcript including copies of regions from both chromosomes.

the region between A and B is "looped out" in the transcription process. That is, the RNA polymerase jumps across from A to B and the RNA transcript does not contain a copy of the looped out region. When the two homologus chromosomes are paired so that the sites A and B of the two chromosomes are close to each other, we assume that with some probability the z^+ product can bind to sites A and B on the opposite paired chromosomes. This leads to transcription jumping from one chromosome to the paired chromosome at this point between sites A and B. The transcript thus produced contains a copy of one of the chromosomes to the left of A and a copy of the other paired chromosome to the right of B. It is obvious that the production of such a hybrid transcript would lead to partial complementation between mutations of the complex which map on opposite sides of sites A and B.

There is some operational similarity between our model of production of hybrid RNA transcript and the RNA splicing models considered long ago; but the mechanism of hybrid RNA production in our model requires the jumping of the polymerase across paired chromosomes and thus is expected to lead to transvection. No such transvection is expected in RNA splicing models where the splicing is specially separated from the process of transcription.

Since the z^a homozygote is viable, we are forced to assume that the binding of z^+ product is not essential; this implies that the transcript which includes the region between A and B can also be utilized by the organism, presumably with a lower efficiency. Thus the coding region of the transcript has to be restricted to one side of the sites A and B. There are several other consequences of this model which lead to testable predictions. These will be discussed in detail elsewhere.

In summary, we have observed that the complementation seen in the white locus is abolished by z^a mutation. This observation, together with the enhancement of the mutant phenotype of bx^{34e}/Ubx by z^a observed by Kauffman, Tasaka and Suzuki (1973) leads us to suggest that a similar mechanism of complementation dependent on z^+ gene product operates in both cases. We present an explicit model which incorporates these features. This model in its essence assumes that the z^+ product binds simultaneously to paired homologous chromosomes and facilitates the jumping of RNA polymerase from one chromosome to its paired homologue, leading to the production of a hybrid RNA transcript.

REFERENCES

Gans, M., 1953, Étude génétique et physiologique der mutant z de
    Drosophila melanogaster, Bull. Biol. France. Belg., (Suppl.),
    38:1.

Jack, J.W., and Judd, B.H., 1979, Allelic pairing and gene regulation: A model for zeste-white interaction in *Drosophila melanogaster,* Proc. Natl. Acad. Sci. U.S.A., 76:1368.

Judd, B.H., 1964, The structure of interlocus duplication and deficiency produced by recombination in *Drosophila melanogaster*, with evidence for polarized pairing, Genetics, 49:253.

Kauffman, T.C., Tasaka, S.E., and Suzuki, D.T., 1973, zeste-bithorax interactions, Genetics, 75:299.

Lewis, E.B., 1954, The theory and application of a new method of detecting chromosomal rearrangements in *Drosophila melanogaster,* Am. Naturalist, 88:225.

Lewis, E.B., 1959, New mutant report, Drosophila Inform. Serv., 33:96.

Lewis, E.B., 1963, Genes and developmental pathways, Am. Zoologist, 3:33.

DISCUSSION

I. Kiss: If zeste plays such an important role as you suggest, how is it that the z^a mutation is not lethal? Wouldn't you expect z^a to be pleiotropic?

P. Babu: I do not really have an answer to your question.

A. Chovnick: You suggest that the degree of pairing is very high in diploid cells. Is there any evidence for such tight pairing except in polytene cells?

P. Babu: I presume that you are asking about the degree of pairing in diploid interphase cells. The evidence from mitotic recombination by A. Garcia-Bellido and F. Wandosell (Mol. Gen. Genet., 161:1978) with chromosomes containing inversions indicates that except near the breakpoints, there is a high degree of pairing.

THE HEAT SHOCK RESPONSE: A MODEL SYSTEM FOR THE STUDY OF GENE

REGULATION IN *DROSOPHILA*

M.L. Pardue[1], M.P. Scott[1], R.V. Storti[2] and J.A. Lengyel[3]

[1]Biology Department
Massachusetts Institute of Technology
Boston, Massachusetts 02139 U.S.A.

[2]Department of Biological Chemistry
University of Illinois Medical Center
Chicago, Illinois 60612 U.S.A.

[3]Department of Biology
University of California
Los Angeles, California 90024 U.S.A.

INTRODUCTION

It is generally believed that development in higher organisms is controlled by sets of coordinately regulated genes, sometimes referred to as "batteries of genes". Such gene batteries would provide the most economical explanation for cases where a single developmental stimulus, such as a hormone, seems to induce a complex but well-regulated differentiation response.

In spite of biological evidence suggesting the existence of sets of coordinated genes acting during development, we still know little about the make-up of these putative gene sets beyond a small number of genes coding for abundant proteins. We know even less about how the coordination of the genes might be effected. In fact there is still very little information about how much eukaryotic genetic activity is controlled at the level of gene transcription, how much is controlled at the level of RNA processing and/or transport to the cytoplasm, or how much is controlled at the level of translation. There is even less information about the molecular mechanisms involved in control at any level.

Recently, there has begun to be much interest in the heat shock response of *Drosophila* as a model system for study of problems of genetic control of development (Ritossa, 1962; Ashburner and Bonner, 1979). During the heat shock a small number of apparently coordinated genes are induced while most other genes become inactive. The heat shock response is not itself a developmental response; it can be induced in all types of *Drosophila* cells, including cultured cell lines, and is reversed when the inducing conditions are removed. Nevertheless, the small number of genes involved, the apparent coordinate control of these genes, and the ease of experimental manipulation make the heat shock response a promising system for gaining insights into mechanisms which might be used in development. In addition, the heat shock response is of interest in its own right. The evidence that has accumulated strongly suggests that the response may be a homeostatic mechanism enabling the organism to cope with a variety of stresses.

THE BIOLOGY OF THE HEAT SHOCK RESPONSE

Although now known to be induced by a variety of agents, the response was named the heat shock response because it was first induced by a temperature shift. This effect was first reported by Ritossa (1962) who found that when larvae or their salivary glands were moved from their customary temperature of $25^{\circ}C$ to $37^{\circ}C$, nine new puffs appeared on the polytene chromosomes almost immediately. A puff appears to be the uncoiling of a small section of the chromosome and contains a high concentration of newly synthesized RNA. Because this morphology is so strikingly that which might be envisioned for an active gene, the induction of puffing has long been taken to indicate the activation of genes. However only recently has it been possible to show a correlated appearance of RNA sequences complementary to DNA of the puff (Spradling et al., 1977; Bonner and Pardue, 1977). At least some of these RNAs code for proteins which appear in the cytoplasm after the induction of puffing (Mirault et al., 1978; McKenzie and Meselson, 1977).

The induction of nine new puffs in Ritossa's experiments implied that the temperature shift had activated a small set of genes in the salivary gland cells. Further work by Ritossa and others (Ritossa, 1964; Ashburner, 1970; Leenders and Berendes, 1972; Lakhotia, 1971) showed that exactly the same set of puffs could be induced by a wide variety of environmental factors, including recovery from anaerobiosis, inhibitors of electron transport, uncouplers of oxidative phosphorylation, and inhibitors of various cellular functions. The finding that such a varied list of inducers can induce the same small set of genes argues that these genes can respond to a common control although the evidence has not yet permitted the identification of the molecular basis of this control. The known list of inducing agents has suggested several possible mechanisms of induction, but there is little evidence to support any of these postulated mechanisms.

Although the inductive mechanism has not yet been identified, the known list of environmental inducers suggests that the heat shock response is of physiological importance. A *Drosophila* might be expected to encounter some of the known inducers, especially heat and anoxia, as stresses in its everyday life. A reaction enabling the insect to cope with these stresses would clearly be advantageous.

As might be expected of a homeostatic response, the heat shock response is neither tissue- nor stage-specific. Each *Drosophila* tissue studied has shown the same response. All polytene tissues show the same set of induced puffs. Non-polytene tissues produce the same set of heat shock-induced RNAs and polypeptides. It is important to note that *Drosophila* can recover totally after rather long periods of the altered metabolic state of the heat shock response. We have kept cultured cells for as long as ten hours at $37^{\circ}C$ and seen recovery of essentially all of the cells after the culture was returned to $25^{\circ}C$. Experiments by Lindsley and Poodry (1977) showed that pupae could undergo reversible periods of developmental arrest of at least six hours if placed under conditions of heat shock or anoxia. During this developmental arrest the animals were immune to a second environmental insult which would have produced developmental abnormalities if given under normal conditions.

Thus, what we now know about the heat shock response is consistent with the idea that it may be important to the insect as a means of surviving stress. Further attempts to confirm this hypothesis will require identification of the functions of the proteins synthesized during the response, an endeavor which has not yet been successful.

THE HEAT SHOCK RESPONSE AS A MODEL SYSTEM FOR CONTROL AT THE LEVELS OF RNA TRANSCRIPTION, PROCESSING AND TRANSPORT

The early studies on polytene puffing gave the first indication that heat shock (and the other inducers) produced changes in the pattern of transcription. Not only were nine new puffs induced but many of the pre-existing puffs disappeared (Ashburner, 1970), suggesting increased transcription at some loci and decreased transcription at many others.

We have used another approach to measure transcription during heat shock. Cells are placed under heat shock conditions, ^3H-uridine is added to the medium, and incubation is continued for one hour. Nuclear RNA isolated from the cells is analyzed by hybridization *in situ* to polytene chromosomes. Only RNA transcribed during the heat shock will be radioactive. The *in situ* hybridization makes it possible to identify the chromosomal sites complementary to the radioactive RNA and to measure the relative amount of ^3H-RNA in the nuclear RNA which is complementary to each locus.

Most of our analysis of nuclear RNA has been done on cells of the continuously cultured Schneider 2-L line (J.A. Lengyel, L.J. Ransom, M.L. Graham and M.L. Pardue, in preparation). Comparison of the pattern of *in situ* hybridization of nuclear RNA from heat shocked cells with the pattern of nuclear RNA from control cells shows clear differences between the two populations of transcripts (Pardue et al., 1977). Nuclear RNA from heat shocked cells contains sequences complementary to DNA in the heat shock puffs as well as to DNA in additional chromosomal sites. However the pattern of hybridization by heat shock nuclear RNA is significantly less complex than the pattern of hybridization of nuclear RNA from control cells (Figure 1).

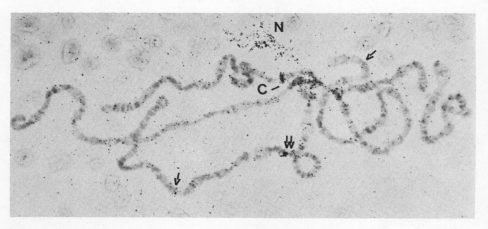

Figure 1: Autoradiograph of salivary gland chromosomes showing hybridization of ^{3}H-nuclear RNA from Schneider 2-L cells labelled for 1 hr at 37°C. Hybridization is to heat shock puff loci (some indicated by arrows) and a number of other sites. Hybridization at the site marked C is due to an internal standard added to the hybridization mix. The cluster of grains marked n is over the nucleolus. Giemsa stain; X400; exposure 44d.

A more detailed analysis of transcripts complementary to the DNA of the three major heat shock loci, 87A, 87C, and 93D, shows that transcripts from these three loci are metabolized in significantly different ways, despite the apparent coordinate induction of the puffs by heat shock (J.A. Lengyel, L.J. Ransom, M.L. Graham and M.L. Pardue, in preparation). The cells used in this set of experiments were heat shocked at 35°C since the rate of synthesis of heat shock RNA in the Schneider 2-L line increases with temperature up to 35°C, but falls off precipitously above this temperature. Furthermore, at least one major species of heat shock-induced poly-(A)$^+$ RNA does not appear in the cytoplasm at temperatures above 35°C in this cell line (Spradling et al., 1977). Cells were put at 35°C for ten minutes, ^{3}H-uridine and ^{3}H-guanosine were added, and incubation was continued for one hour. RNA was then prepared from both nuclear and cytoplasmic fractions of the cells, separated to give poly(A)$^+$ and poly(A)$^-$ species by oligo(dT)-cellulose chromatography, and analyzed by *in situ* hybridization to polytene chromosomes from larval salivary glands. Quantitative analysis of the amount of hybrid formed at each of the loci as a function of the amount of radioactive RNA added allowed us to measure the relative amounts of RNA complementary to each locus in the different RNA fractions (Figure 2).

In earlier experiments (Spradling et al., 1977) we had found two different cytoplasmic RNAs, both of which were complementary to, and presumably transcribed from, polytene band 87C1. One of these RNAs has been shown to code for the 70K heat shock protein (Mirault et al., 1978). The function of the other RNA is unknown. More recently several groups have obtained recombinant DNA molecules carrying sequences complementary to one or the other of these two cytoplasmic RNAs (Lis et al., 1978; Livak et al., 1978; Artavanis-Tsakonas et al., 1978). Both types of recombinant DNA hybridize to band 87C1 where the sequences are close but distinct at the molecular level (Ish-Horowicz et al., 1980). In our *in situ* experiments with nuclear RNA (J.A. Lengyel, L.J. Ransom, M.L. Graham and M.L. Pardue, in preparation), the kinetics of hybridization were consistent with the presence of at least two RNA species complementary to band 87C1 in the nuclear RNA.

We have analyzed sequences complementary to 87C1 further by filter hybridization to recombinant DNA from plasmid 229.1, which contains sequences coding for the 70K protein (Livak et al., 1978), and cDm 703, which codes for RNA of unknown function (Lis et al., 1978). Although the concentrations of RNA sequences complementary to these two plasmids were nearly equal in the nuclear RNA, the cytoplasmic concentrations differed significantly, the RNA coding for the 70K protein being at least five-fold more abundant in the cytoplasm than the RNA species complementary to cDm 703.

Puff 93D differed from the puffs at 87A and 87C in that both saturation and competition experiments showed the cytoplasmic RNA to have only 41-55% of sequence length of the nuclear transcript complementary to 93D. The analyses of hybridization to both 87A and 87C had been consistent with all of the sequences of both these loci being represented in cytoplasmic RNA. The transcript at 93D is also unusual in that, although nuclear RNA contains approximately equal concentrations of poly(A)$^+$ and poly(A)$^-$ transcripts, the sequences in cytoplasmic RNA complementary to 93D are predominantly poly(A)$^-$, suggesting preferential retention of poly(A)$^+$ sequences in the nucleus. Cytoplasmic RNA complementary to 87A and 87C is more nearly evenly divided between poly(A)$^+$ and poly(A)$^-$ fractions.

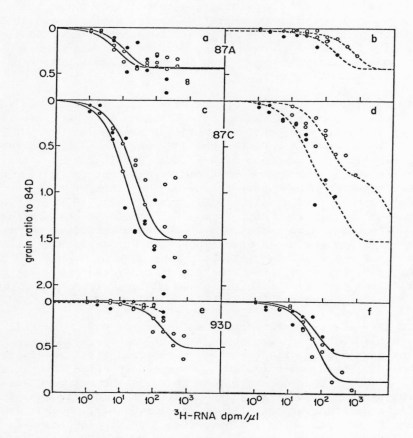

Figure 2: Saturation hybridization of heat shock sites 87A, 87C and 93D, with various RNA fractions. ^{3}H-nuclear and cytoplasmic poly-(A)$^+$ and poly(A)$^-$ RNA was prepared from Schneider 2-L cells labelled for 1 hour at 35°C. The RNA was hybridized *in situ* to salivary gland polytene chromosomes. A constant amount of ^{3}H-cRNA to plasmid pDm2 (complementary to site 84D) was included in each hybridization mixture as an internal standard. After autoradiography (9-14 days), grains over sites 84D, 87A, 87C and 93D were counted in 10-20 fields for each slide, without knowledge of the identity of the slide. The data are plotted as the ratio of grains over the

Figure 2 (continued):

particular site to the grains over 84D, and were fit to pseudo-first order reaction kinetics by a non-linear least squares computer program (Pearson et al., 1977). A solid line (_____) has been drawn through data which reached a saturation plateau, and a dotted line (------) through data which did not.

(a) Hybridization of poly(A)$^+$ (●_____●) and poly(A)$^-$ (o_____o) cytoplasmic RNA to 87A. Because the plateaus of the best one-component fits of poly(A)$^+$ and poly(A)$^-$ RNA were similar (.48 and .42, respectively), an average plateau (.44) was determined by fitting the combined sets of data (after normalizing for the difference in rate constants). Forcing the fits to this plateau increased the rms error of each fit (over the "best fit") by less than 3%.

(b) Hybridization of poly(A)$^+$ (●-----●) and poly(A)$^-$ (o-----o) nuclear RNA to 87A. These hybridization data were fit to the same plateau as that for the cytoplasmic RNA (.44), resulting in rms error increases of less than 2% over the rms errors of the best fits.

(c) Hybridization of poly(A)$^+$ (●_____●) and poly(A)$^-$ (o_____o) cytoplasmic RNA to 87C. The plateaus of the best one-component fits were very similar (1.6 and 1.4, respectively). The hybridizations were therefore fit to an average of these plateaus (1.5). The fits shown here are two-component fits, as there is independent evidence that there are two different RNA species derived from 87C (see text). The two components of the fit were each fixed to 50% of the plateau value, as this is the approximate proportion in which the two different sequences are present at 87C and does not bias the fit toward either component. The rms errors of these two component fits are the same as those for one component fits forced to a plateau of 1.5.

(d) Hybridization of poly(A)$^+$ (●-----●) and poly(A)$^-$ (o-----o) nuclear RNA to 87C. As was done for the cytoplasmic RNA (see c) the data are fit to two components with the plateau fixed at 1.5. For poly(A)$^+$ RNA the rms error is 27% less, and for poly(A)$^-$ RNA 5% less, than the rms error for a one-component fit forced to a plateau of 1.5. The rate constant for the second component of the poly(A)$^-$ nuclear RNA was fixed at .0001 (even though lower rate constants gave a somewhat better fit) because the concentration of species hybridizing to 87C in poly(A)$^-$ nuclear RNA is at least one tenth of that in poly(A)$^+$ nuclear RNA, for which the second component rate constant is .0027.

(e) Hybridization of poly(A)$^+$ (●-----●) and poly(A)$^-$ (o-----o) cytoplasmic RNA to 93D. The poly(A)$^+$ cytoplasmic RNA hybridization was not significantly above background; it was therefore not fit by computer. The curve through the poly(A)$^-$ RNA hybridization data is the best one-component fit.

(f) Hybridization of poly(A)$^+$ (●_____●) and poly(A)$^-$ (o_____o) nuclear RNA to 93D. The curves for both poly(A)$^-$ and poly(A)$^+$ RNA are the best one-component solutions.

Although the data on hybridization at 93D might indicate that there is a large nuclear precursor which is processed to yield the cytoplasmic RNA, it is also possible that there are two or more transcripts of the locus with quite different rates of transport to the cytoplasm and/or different life-times. The question can best be resolved with the aid of recombinant DNA, carrying sequences from 93D; this is the focus of current experiments.

This analysis of the three major heat shock loci showed that, although the transcription of these puffs is apparently coordinately controlled under heat shock conditions, the further history of the transcripts varies from locus to locus. Differences in the fate of the transcripts exist not only between puffs but also within puffs. At least one, and probably two, of the loci codes for more than one RNA species. The multiple species are distinguishable by their relative distribution in the nucleus and in the cytoplasm. This implies that these RNAs differ from each other in the rate of export from the nucleus and/or their rate of turn-over in the cytoplasm. Thus, a group of coordinately transcribed RNAs may be separately regulated at a post-transcriptional level.

THE HEAT SHOCK RESPONSE AS A MODEL SYSTEM FOR CONTROL AT THE LEVEL OF TRANSLATION

The heat shock response is also a system in which there is control of gene expression at the level of translation. In addition to preferential transcription of a small set of RNAs there is preferential translation of these RNAs. This control at the level of translation removes the pre-existing mRNAs from the polysomes rapidly and so increases the abruptness of the physiological responses.

The rapid decline in translation of pre-existing mRNAs is not due to degradation of these RNAs. If cells under heat shock conditions are given actinomycin D and then returned to 25°C, they will

---

Figure 3: Autoradiographs of polyacrylamide gels of proteins synthesized *in vivo* by L-2 cultured cells and *in vitro* by rabbit reticulocyte lysates, demonstrating the presence of many 25° mRNAs in heat shocked cells.
  a) Proteins labelled *in vivo* during a one hour pulse with ^{35}S-methionine. Cells at 25°.
  b) Proteins labelled *in vivo* during a one hour pulse with ^{35}S-methionine. Cells at 36°.
  c) Protein synthesized *in vitro* in lysates programmed by poly(A)-containing RNA from cells heat-shocked to 36° for one hour.
  Proteins from whole cells (a,b) or *in vitro* translation reactions (c) were separated on two-dimensional polyacrylamide gels as described by O'Farrell (1975). *In vitro* translation procedures were as described by Pelham and Jackson (1976).

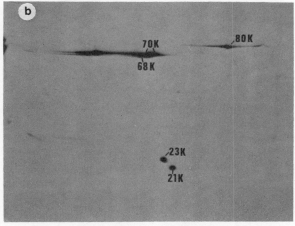
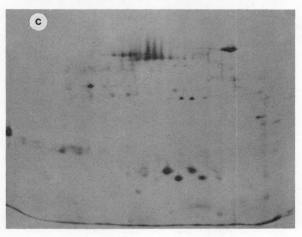

eventually begin to synthesize the proteins normally made at 25°C. The mRNAs which direct this new synthesis must have been made before the heat shock began since the drug inhibited new transcription of nuclear DNA after the cells were returned to the normal growth temperature (McKenzie, 1976; R.V. Storti, M.P. Scott, A. Rich, and M.L. Pardue, in preparation). This implies that the pre-existing mRNA, although not translated during heat shock, was somehow sequestered until normal growth conditions returned.

Functional mRNAs from the set being translated before the heat shock can also be detected by cell-free translation of RNA isolated from heat-shocked cells. Experiments done in several laboratories have shown that mRNA extracted from heat shocked cells can be translated in cell-free translation systems made from rabbit reticulocytes or wheat germ (Figure 3). In every case the RNA directed synthesis of the polypeptides normally made by cells growing at 25°C as well as the heat shock polypeptides (S. Lindquist, in preparation; Mirault et al., 1978; R.V. Storti, M.P. Scott, A. Rich, and M.L. Pardue, in preparation). *In vivo* labelling under the same heat shock conditions showed little except the heat shock proteins being synthesized in the intact cells.

Although cell-free translation systems made from rabbit reticulocytes or wheat germ appear to translate *Drosophila* mRNA accurately, these heterologous systems apparently do not discriminate against the pre-existing mRNA species in RNA prepared from heat shocked cells. In order to investigate further the mechanism of the heat shock translational control we found it necessary to develop a homologous cell-free translation system which could be dissected and reconstituted using defined components. The system which we have described (Scott et al., 1979) is prepared from cells of established *Drosophila* cells lines. (We have used Schneider 2-L cells (Lengyel et al., 1975) and K_C cells (Eschalier and Ohanessian, 1969)). These cell lysates are capable of polypeptide initiation and can be made dependent on exogenous mRNA by digesting endogenous mRNA with micrococcal nuclease which is then inhibited by the removal of Ca^{++} (Pelham and Jackson, 1976). We have used the system successfully to translate a number of viral, invertebrate, and vertebrate mRNAs (Figure 4).

With increasing interest in the molecular biology of *Drosophila* the cell-free translation system should provide a useful homologous system for various types of studies. For our present interests, one of the most useful features of the lysate is its ability to carry out the heat shock-induced translational control. Figure 5 shows that although lysates prepared from control cells translated mRNA from both control and heat shocked cells effectively, lysates prepared from cells growing under heat shock conditions show a clear discrimination in favor of heat shock mRNAs. The data support two conclusions at this point. First, the lysate is discriminating on the basis of the primary sequence of the RNA. The RNA added to each

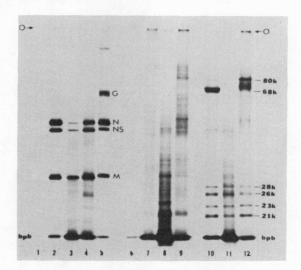

Figure 4: Products of cell-free translation in Drosophila lysates, wheat germ extracts, and rabbit reticulocyte lysates. Electrophoresis was in a 12% SDS-polyacrylamide slab gel.

1) Two microliters (3200 cpm) of a 25 µL reaction mix of Drosophila lysate with no added RNA (endogenous synthesis).
2) Two microliters (32000 cpm) of a 25 µL reaction mix of Drosophila lysate with 192 µg/mL vesicular stomatitis virus (VSV) mRNA added.
3) 32,000 cpm of the products translated from VSV mRNA in wheat germ extract.
4) 33,000 cpm of the products translated from VSV mRNA in rabbit reticulocyte lysate.
5) Proteins labeled in vivo in VSV-infected Chinese hamster ovary (CHO) cells.
6) 5 µL (8000 cpm) of a 25 µL reaction mix of Drosophila lysate without added mRNA (endogenous synthesis).
7) 7 µL (33,000 cpm) of a 25 µL reaction mix of Drosophila lysate containing approximately 320 µg/mL total RNA from Schneider cells grown at 25°.
8) Same RNA as in (7), translated in wheat germ extracts (30,000 cpm).
9) Same RNA as in (7), translated in rabbit reticulocyte lysate (30,000 cpm).
10) 5 µL (33,000 cpm) of a 25 µL reaction mix of Drosophila lysate containing approximately 320 µg/mL total RNA from Schneider cells heat shocked to 36° for two hours.
11) Same RNA as in (10), translated in wheat germ extracts (33,000 cpm).
12) Same RNA as in (10), translated in rabbit reticulocyte lysate (29,000 cpm).

Exposure was for two days. (O) Origin; (bpb) bromphenol blue dye marker.

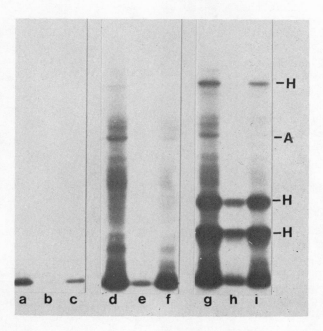

Figure 5: Translation of embryo poly(A)$^+$ RNA in normal and heat shock *Drosophila* lysates.
    a)   25° lysate; no added RNA; 9500 cpm.
    b)   36° lysate; no added RNA, 5400 cpm.
    c)   Mixed lysates; no added RNA; 15,400 cpm.
    d)   25° lysate with 25° mRNA; 36,000 cpm.
    e)   36° lysate with 25° mRNA; 12,500 cpm.
    f)   Mixed lysates with 25° mRNA; 31,000 cpm.
    g)   25° lysate with 36° mRNA; 44,600 cpm.
    h)   36° lysate with 36° mRNA; 19,800 cpm.
    i)   Mixed lysates with 36° mRNA; 26,900 cpm

Lysates were prepared as described by Scott et al. (1979) from cultured cells kept at 25°C, and from cells warmed to 36°C for one hour. Poly(A)$^+$ RNA was prepared from 0.5 g of 12-18 hour *Drosophila* embryos kept at 23°C or at 36°C for one hour. Heat shock proteins (H) and actin (A) are indicated. The autoradiograph of a 10% SDS-polyacrylamide gel exposed for one day is shown. Equal amounts of mRNA were added to each translation reaction, and equal amounts of each reaction were loaded on the gel.

lysate had been treated with protease and extracted with phenol so it is unlikely that the translation selection is based on a protein which might have become associated with the RNA molecule as it was processed in the nucleus. Second, the effect of the heat shock persists for some time after the removal of the inducer. Lysates from both control and heat shocked cells are maximally active at 28°C (a non-heat shock temperature) so the in vitro translation reactions are all performed at that temperature. In spite of this the pattern of translation in the two lysates clearly reflects the state of the cells from which the lysates were prepared. The persistance of the translational control suggests that the heat shock induction may induce molecular changes which have a measurable lifetime under non-heat shock conditions.

CONCLUDING REMARKS

The heat shock response offers an especially favorable experimental situation for studying questions of control of gene expression applied to specific genes. In outline the response is apparently a simple one in which expression of a small number of genes can be induced by a variety of agents. However, the accumulating experimental evidence shows a number of post-transcriptional points, such as processing, transport from the nucleus, and translation, where individual RNA species may be treated differently. Eventually, study of the heat shock response may help to understand the mechanisms of these discriminations and the extent to which they may modulate gene expression.

REFERENCES

Artavanis-Tsakonas, S., Schedl, P., Mirault, M.-E., Moran, L.,and Lis, J., 1979, Genes for the 70,000 dalton heat shock protein in two cloned *D. melanogaster* DNA segments, Cell, 17:9.

Ashburner, M., 1970, Patterns of puffing activity in salivary glands of *Drosophila* V. Responses to environmental treatment, Chromosoma, 31:356.

Ashburner, M., and Bonner, J.J., 1979, The induction of gene activity in *Drosophila* by heat shock, Cell, 17:241.

Bonner, J.J., and Pardue, M.L., 1977, Polytene chromosome puffing and in situ hybridization measure different aspects of RNA metabolism, Cell, 12:227.

Eschalier, G., and Ohanessian, A., 1969, In vitro culture of *Drosophila melanogaster* embryonic cells, In Vitro, 6:162.

Ish-Horowicz, D., Pinchin, S.M., Schedl, P., Artavanis-Tsakonas, S., and Mirault, M.-E., 1980, Genetic and molecular analysis of the 87A7 and 87Cl heat-inducible loci of *Drosophila melanogaster*, Cell, in press.

Lakhotia, S.C., 1971, Benzamide as a tool for gene activity studies in *Drosophila*, Abstract of the Fourth Cell Biol. Conf., New Delhi.

Leenders, H.J., and Berendes, H.D., 1972, The effect of changes in the respiratory metabolism upon genome activity in *Drosophila* I. The induction of gene activity, Chromosoma, 37:433.

Lengyel, J., Spradling, A., and Penman, S., 1975, Methods with insect cells in suspension culture II. *Drosophila melanogaster*, Meth. Cell Biol., 10:195.

Lindsley, D.E., and Poodry, C.A., 1977, A reversible temperature-induced developmental arrest in *Drosophila*, Devel. Biol., 56:213.

Lis, J., Prestidge, L., and Hogness, D.S., 1978, A novel arrangement of tandemly repeated genes at a major heat shock site in *D. melanogaster*, Cell, 14:901.

Livak, K.F., Freund, R., Schweber, M., Wensink, P.C., and Meselson, M., 1978, Sequence organization and transcription at two heat shock loci in *Drosophila*, Proc. Natl. Acad. Sci. USA, 75:5613.

McKenzie, S.L., 1976, Protein and RNA synthesis induced by heat shock in *Drosophila melanogaster* tissue culture cells, Ph.D. Thesis, Harvard University, Cambridge, Massachusetts.

McKenzie, S.L., and Meselson, M., 1977, Translation in vitro of *Drosophila* heat-shock messages, J. Mol. Biol., 117:279.

Mirault, M.-E., Goldschmidt-Clermont, M., Moran, L., Arrigo, A.P., and Tissieres, A., 1978, The effect of heat shock on gene expression in *Drosophila melanogaster*, Cold Spring Harbor Symp. Quant. Biol., 42:819.

O'Farrell, P.H., 1975, High resolution two-dimensional electrophoresis of proteins, J. Biol. Chem., 250:4007.

Pardue, M.L., Bonner, J.J., Lengyel, J.A., and Spradling, A.C., 1977, *Drosophila* salivary gland polytene chromosomes studied by in situ hybridization, in:"International Cell Biology, 1976-1977", B.R. Brinkley and K.R. Porter, eds., Rockefeller Press, New York, p. 509.

Pearson, W.R., Davidson, E.H., and Britten, R.J., 1977, A program for least squares analysis of reassociation and hybridization data, Nucleic Acids Res., 4:1727.

Pelham, H.R.B., and Jackson R.J., 1976, An efficient mRNA-dependent translation system from reticulocyte lysates, Eur. J. Biochem., 67:247.

Ritossa, F.M., 1962, A new puffing pattern induced by heat shock and DNP in *Drosophila*, Experientia, 18:571.

Ritossa, F.M., 1964, Experimental activation of specific loci in polytene chromosomes of *Drosophila*, Exp. Cell Res., 35:601.

Scott, M.R., Storti, R.V., Pardue, M.L., and Rich, A., 1979, Cell-free protein synthesis in lysates of *Drosophila melanogaster* cells, Biochemistry, 18:1588.

Spradling, A.C., Pardue, M.L., and Penman, S., 1977, Messenger RNA in heat-shocked *Drosophila* cells, J. Mol. Biol., 109:559.

DISCUSSION

I. Kiss: I am going to ask a naive question. What happens if you heat shock a lysate made from non-heat shocked cells? Can you induce any change?

M.L. Pardue: I don't think it is a naive question at all. We have been trying to do this. The trouble is that for some reason when we take any lysate above 34°C, the lysate is no longer active. I suspect that there might be some degradation at high temperature. This is a technical problem and it does not say anything about heat shock.

REGULATION OF DNA REPLICATION IN *DROSOPHILA*

A.S. Mukherjee, A.K. Duttagupta, S.N. Chatterjee, R.N. Chatterjee, D. Majumdar, Chandana Chatterjee, Mita Ghosh, P. Mohan Achary, Anup Dey and Indranath Banerjee

Department of Zoology
Calcutta University
35 Ballygunge Circular Road
Calcutta 700 019, India

INTRODUCTION

The biosynthesis of DNA in the eukaryotic system distinguishes itself from that in prokaryotes by many significant points. Although much evidence indicates that in both systems replication proceeds in both directions from the initiation point in the complementary strands (Hubermann and Riggs, 1968; Callan, 1974; Kriegstein and Hogness, 1974; Painter, 1976), the existence of multiple initiation points, the multiple termini, the differences in the size of replication units in different eukaryotic systems, the rate of chain growth, etc., are some of the complex features of eukaryotic systems. The organization of eukaryotic chromatin, the existence of heterochromatin and euchromatin, and the new concept of DNA-histone organization in the nucleosome (Seale, 1978) demand additional assumptions in establishing the regulatory mechanism of DNA replication in eukaryotes.

From the fiber autoradiographic studies of Hubermann and Riggs (1968), and of others (Callan, 1972, 1974; Painter, 1976; Hand and Tamm, 1973), which allow the visualization of replication of molecular DNA with the light microscope, it was clear that: (a) DNA replication in eukaryotes starts at multiple initiation sites arranged in tandem order; (b) replication proceeds from the two growing points on the complementary strands in opposite directions towards the replication forks; (c) the chain growth takes place in both directions from the initiation points, the growing chain meeting at the point called terminus; (d) the points of initiation in a particular tissue system are fixed, but the termini are not fixed sequences; (e) the length of terminus-to-terminus is considered to be one replicon.

There is no definite knowledge regarding the relation between different replication units or replicons with respect to their time of initiation or termination, or with respect to the determination of their chain growth. Callan (1972) suggested that the time and number of initiation sites might be determined either by qualitatively different sets of recognition sites and number of DNA polymerases, or by an inherent difference in the macromolecular configuration over the sequences of the genome. Upon comparison of the DNA replication in somatic and pre-meiotic S phase in *Triturus* groups, Callan (1972) confirmed that each chromomere in an extended chromatin is equivalent to one replicon. It may be recalled that Pelling (1966) proposed that each chromomere in the polytene chromosomes of diptera is a unit of replication.

However, studies on the replication in dividing cells (Stubblefield, 1974) and non-dividing giant cells of diptera (Rodman, 1968; Plaut et al.,1966) suggest that clusters of replicons somehow maintain a synchrony of replication with respect to initiation, chain growth and termination. Such clusters were termed by Stubblefield (1974) as replicon series. Existence of such harmony among groups of replicons was confirmed by Painter (1970, 1976), Blumenthal et al. (1974), and Callan (1974), at the level of DNA; and Mulder et al. (1968), Arcos-Terán (1972), Lakhotia and Mukherjee (1970), Chatterjee, S.N. and Mukherjee (1975), and Chatterjee, R.N. and Mukherjee (1977a) at the level of the chromosome or chromatin.

Arcos-Terán and Beermann(1968) and later Mukherjee and his collaborators (1978) conceived the possible existence of factors which might guide the harmonious chain growth and termination in a replicon series. Chatterjee and Mukharjee (1975) proposed a distinction between the factors responsible for synchronous initiation and termination. This report presents information bearing upon the possible levels of control of *in vivo* DNA replication in *Drosophila* and possible future work that may yield information on the regulation of DNA replication.

PROBLEMS ENCOMPASSING DNA REPLICATION IN EUKARYOTES

DNA replication in eukaryotic systems differs from that in prokaryotes in several important features. These features, although serving the important purpose of organization and differentiation, create special problems with respect to regulation and maintenance of the system. First, DNA replication in eukaryotes starts at multiple initiation points. The problems that ensue are concerned with how the initiation sites are determined, how they are maintained, and whether there are fixed nucleotide sequences in DNA for initiation sites to be recognized by the enzymes. Second, the work of Mueller and Kajiwara (1966) using sequential labelling of synchronized HeLa cells successively with ^{14}C-thymidine and BUdR showed that the same sequences are replicated at least at the beginning of each replication, which implies that initiation sites on the euchromatin may be fixed. This is probably true also for *Drosophila* polytene chromosomes, where it is seen that certain specific puff sites are labelled at the initial phase when given a short pulse of ^{3}H-thymidine (Roy and Lakhotia, 1979; Majumdar and Mukherjee, 1980). Obviously, this implies that there is sequential initiation of DNA replication at many sites on the eukaryotic genome. What accounts for such fixed sequence of initiation of replication?

Third, the number of initiation sites varies in different tissues and in different developmental stages of the same organism (Callan, 1974; Blumenthal et al., 1974). This means that there is a problem of inactivation and reactivation of some initiation sequences in different tissues and at different developmental stages.

Fourth, Hubermann and Riggs (1968), on the basis of fiber autoradiographic patterns, proposed that the eukaryotic DNA consists of tandemly arranged initiation sequences alternating with termini. As questioned by Blumenthal et al. (1974) and Hand (1975), the presumed existence of fixed origin and termini could cause problems not only in regard to their maintenance but also with respect to replication in genetically-altered genomes such as translocations, deficiencies, etc. In actuality, such problems do not occur. This implies that there must be some other means of termination and chain growth signals.

Fifth, the fact that a differential replication pattern of replicating units exists between eu- and heterochromatin, introduces yet another kind of problem both with respect to the phenomenon of regulation, as well as the timing of chain growth (Mulder et al., 1968; Rudkin, 1972).

Last, but not least, is the existence of a synchrony among certain adjacent replicons, which resolves into the formation of a cluster of replicons (Blumenthal et al., 1974), originally termed by Stubblefield (1974) as replicon series. Such clusters of replicons are thought to be activated at approximately the same time and have a close mean spacing between their origin (Blumenthal et al., 1974). At the level of chromatin replication, this idea of the existence of synchrony among clusters of adjacent units has been emphasized by Mukherjee and Chatterjee (1975) and Chatterjee, R.N. and Mukherjee (1977a) in *Drosophila* polytene system. They have described the cluster as a complex replication unit, which consists of a band length from 1 to 8 and has a common time for initiation, chain growth and termination. This aspect of eukaryotic DNA replication raises the problem of existence of regulatory factors for specific initiation and termination at the level of super-replicon organization.

REPLICATION OF *DROSOPHILA* DNA

In the past several years investigations on the replication of the *Drosophila* genome have been carried out both at the level of chromosomes as well as macromolecular DNA. While both cytological and molecular studies, whether ultrastructural, cytochemical, conventional autoradiographic, fiber autoradiographic or biochemical, have provided many findings in common; different approaches have yielded crucial information about events during DNA replication in this important eukaryotic system. Some of the general features, as revealed by the cytological and conventional autoradiographic approaches, are presented first.

General Features

In the early part of investigations on DNA replication of *Drosophila* genome it was shown that the *Drosophila* genome can be divided into a heterochromatic part and a euchromatic part. In many somatic tissues like the larval salivary gland, after a few initial cycles, either does not replicate at all or is under-replicated (Henning and Meer, 1971; Spear and Gall, 1973), or replicates late. The latter includes "DNA segments" which may vary, not only in their time of initiation, but also in their rate of chain elongation and termination of replication. Such DNA segments, to which we shall return later, include one or more replicons.

The replication of both parts has to follow a certain sequence of events during the S phase, and indeed a sequence of replication events is observed among the ^{3}H-thymidine autoradiograms. Such a sequence is derived from the replication of *Drosophila* larval polytene chromosomes. In these chromosomes one can resolve the different segments of DNA by their characteristic cytological configuration, such as pericentric or chromocentric heterochromatin, thick bands, intercalary heterochromatin, thin bands, puffs, interbands, etc. When replication occurs in the presence of radioactive thymidine, a considerable number of nuclei are unlabelled, but in most of those which are labelled, either all chromosomes are labelled, or none at all. A nucleus in which some chromosomes are labelled and others are not is a rare event. The only exception is the X chromosome in the male and, in some cases, certain specific autosomes. They will be discussed later.

From the autoradiographic studies, three to five distinct phases of replication events can be resolved and they can be arranged in a sequential order with a beginning and an end. At the beginning of replication, several sites on every chromosome start replication almost simultaneously. These sites include many interbands, certain puffs and certain thin bands (or the adjacent interband). The heterochromatic part around the centromere as well as that forming the chromocenter (that is, the proximal heterochromatin at the basal part of each arm) remain unlabelled. One such typical nucleus at the initial phase of replication is shown in Figure 1. At this stage, the chromosomes appear discontinuously labelled on certain sites, as mentioned above, and the grains are dispersely distributed on these sites. Such patterns of labelling have been termed disperse discontinuous (or DD). Whether the very initiation of replication starts at puffs, interbands, or thin bands, however, is not yet established. This difficulty has been due to the rather short duration of the onset of replication, and to the overlapping nature of the events following the onset. Mukherjee et al. (1978) proposed that the events of onset of replication take place at certain interbands. Whether such interbands are specific is yet to be established. Roy and Lakhotia (1979) showed that the onset of the initiation event takes place at two specific puffs in *D. kikkawai* polytene chromosomes. In *D. melanogaster*, at least one puff (87 Cl) has been found to show labelling at the very onset, and this is the only site labelled during this period (D. Majumdar, unpublished). However, labelling on a puff does not precisely resolve whether the onset took place on the band involved in the puffing or on the adjacent interband which is also included in the puffing process.

After the primary set of intiation sites, a secondary set of initiation follows on the remaining DNA segments of the chromosomes. The chromosomes labelled with ^{3}H-thymidine appear to be continuously labelled during this phase (Figure 2). The only part that remains unlabelled is the centromeric heterochromatin which cannot be resolved under the light microscope but has been shown by EM autoradiography (Lakhotia and Jacob, 1974). During this stage, the autoradiograms reveal a more or less uniform distribution of silver grains on all the chromosomes (except the X in the male, which will be discussed later), and hence the pattern is termed the continuous or C pattern. At the level of molecular DNA in DNA fiber autoradiography, this should correspond to a pattern in which the length of the labelled segments is mostly in the lower range. Our preliminary results on single-pulse fiber autoradiograms following the FUdR inhibition and release show that the length of the labelled segments mostly lies within the lower range between 1-5 μm (Figure 3). The release from the FUdR block results in a spurt of C patterns within one hour after the removal of the block. These data suggest that all DD and all C patterns are chromosomal representations of the active initiation. This also implies that while DD represents the pattern where some initiation sites are activated first (primary initiation), C represents such time in the sequence of events when all "activable" initiation sites are activated (primary + secondary initiation). After C no new initiation takes place under normal physiological conditions. However, it is likely that during this stage (when the continuous or C type labelling pattern is cytologically realized), the DNA segments (whose initiation took place during the cytological DD pattern) start their chain growth. The data obtained from our preliminary fiber autoradiography of the DNA, replicated following the removal of FUdR block after 24 hrs (Figure 4), corrob-

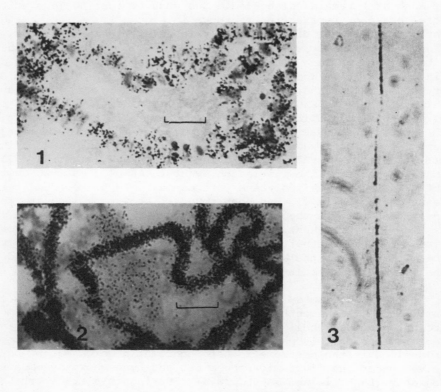

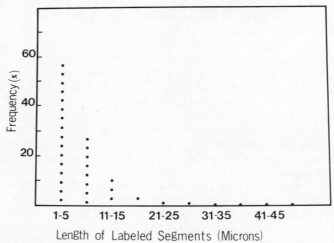

4

Figure 1: An autoradiogram labelled with ^{3}H-TdR showing an initial DD pattern in *D. melanogaster*. (Scale = 10 µm in **all figures**).

Figure 2: A ^{3}H-TdR labelled autoradiogram showing the middle or continuous pattern (3C).

Figure 3: ^{3}H-TdR labelled fiber autoradiogram of DNA from *Drosophila* larval polytene chromosomes treated for 24 hrs with FUdR.

Figure 4: Histogram showing the distribution of the length of labelled segments from fiber-autoradiograms of DNA from FUdR-treated *Drosophila* larval salivary glands.

---

orates this finding. We found that the highest frequencies of labelled DNA segments were of two sizes, 1-5 µm and 6-10 µm; the next longer labelled segments of 11-15 µm were also of considerable frequency.

The third and terminal phase of replication starts soon after the continuous phase. The labelled polytene chromosomes during this part of replication show a varying number of unlabelled gaps (discontinuous or D patterns). The unlabelled gaps represent the regions or DNA sequences whose replication has been terminated. The conclusive evidence in favor of this has been provided by Mulder et al. (1968) using cytophotometric measurement of the unlabelled gaps. They showed that the unlabelled gaps have a high DNA content proportional to the C value of one replication. The number (and length) of unlabelled gaps increases as the replication proceeds toward completion (Figures 5 and 6). The sites, in which labelling continues to be present for a considerable period after the unlabelling starts, are considered to be late replicating sites and probably have repetitive sequences of a higher order (Evgenev et al., 1977). However, this aspect needs to be re-examined and confirmed by fiber autoradiography and *in situ* hybridization.

At the end of one cycle of replication when all the arms of the chromosomes are mostly unlabelled (only one to three sites on each being labelled), the pericentric heterochromatin (i.e. the chromocenter) is seen to be labelled (Figure 6). As a matter of fact, labelling on the chromocenter can be seen from the C stage. The labelling on the chromocenter serves as one of the principal features distinguishing the initial discontinuous (DD) from the terminal discontinuous (D) patterns. Nuclei have also been observed in which only the chromocenter is labelled and all the arms are unlabelled (chromocenter labelled pattern or CL pattern; Figure 7).

From the analysis of the frequencies of labelling of various sites, it appears that the major part of replication of a genome or of the entire DNA molecule of a chromosome is completed within

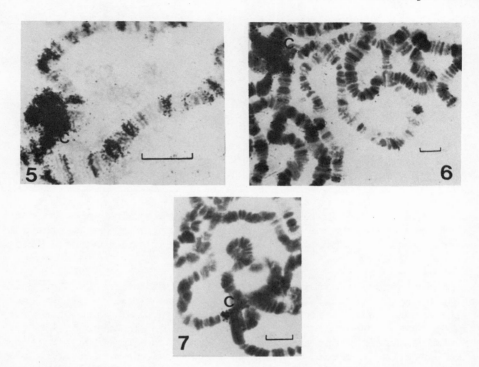

Figures 5-7: Three terminal patterns of ^{3}H-TdR labelled polytene chromosomes in sequence, 2D (Figure 5), 1D (Figure 6) and CL (Figure 7). C = chromocenter.

about two-thirds of the total replication time. This is true for many species of Drosophila examined. However, the precise proportion of genome which finishes replication in the first part, the number or proportion of genomic sites which finishes during the later part of the S, and the percentage of time taken to finish replication by the early and late replicating sites, vary from species to species (Table 1). It appears that D. pseudoobscura shows maximum interchromosomal variation in timing. This cytological finding is at variance with the fiber autoradiographic data of Blumenthal et al. (1974). Whether this variation is due to differences in the systems is to be examined.

Special Features in the Replication Cytology of *Drosophila* Polytene Chromosomes and Problems of Regulation

The ^{3}H-TdR labelling patterns of Drosophila polytene chromosomes clearly distinguish the terminal discontinuous patterns from either the initial disperse discontinuous patterns or the middle continuous patterns. The chromosomes which show the terminal discontinuous labelling never show labelled sites mimicking the initial labelling

Table 1. Variation in the time of replication in different chromosomes and different species

Species	Chromo- somes	Percentage of sites finishing replication in				Source of Data
		First Quarter	Second Quarter	Third Quarter	Fourth Quarter	
D. melan- ogaster	X 2R	30 20	40 40	17.5 20.0	12.5 20.0	Chatterjee,R.N. & Mukherjee,1977a
D. hydei	X 4	34.3 30.8	5.7 15.4	20 7.8	40 46	Chatterjee,S.N. & Mukherjee,1973
D. pseudo- obscura	XL XR 2 3	10 4 18 75	17.5 64.0 28.6 12.5	47.5 12.0 46.4 12.5	25 20 7 0	Mukherjee and Chatterjee,S.N., 1975; Chatterjee et al., 1976

under normal conditions. (See Figures 5 & 6). In other words, no nucleus is found to be labelled simultaneously with initial and terminal patterns. This implies that new initiation remains suppressed until the completion of replication of the whole genome. New initiation can, however, be induced under experimental conditions; for example, by blocking the chain growth, as reported by Blumenthal et al. (1974). Chatterjee et al. (1978) have induced reinitiation before completion by puromycin and other such agents (to be discussed below).

This lack of new initiation before completion of the replication cycle poses a problem of regulation which is set with the biological clock of the cell cycle. Rao et al. (1977) using fused HeLa cells have shown that certain factors or inducers are accumulated gradually through the G_1 until they reach a critical level at G_1-S transition; they attain a peak at early to mid-S and then decline below the critical level when DNA synthesis ceases. According to Rao et al. (1977), these factors are probably protein. An important point that emerges from these works is that the factors that induce initiation of DNA synthesis are probably cytoplasmic or nucleoplasmic (Hand, 1975).

The second important feature in the replication of *Drosophila* chromatin that suggests the existence of other regulation, is the asynchrony of replication. The asynchrony lies at different levels.

First, it has been evident from autoradiographs of polytene chromosomes that replication is initiated at multiple initiation points. Yet, not all sites are simultaneously initiated. A set of about 5% to 20% of the genomic DNA may initiate replication first; the remaining 80% to 95% of the genome quickly follows and initiation is completed within 1/4 to 1/3 of the total S period. This implies that there is a stage of discrimination between sequences with respect to initiation. On the basis of the relative frequency of the initial (DD) and continuous (C) labelling patterns of different sites in the autoradiograms, an estimate of the relative time spent by the sites during initiation can be made. Usually this time is very short, but it varies from species to species. It appears to be species-specific within a given physiological norm. It seems to be shortest in *D. melanogaster* and longest in *D. pseudoobscura*, as reported by Chatterjee and Mukherjee (1975); it may be even longer in *D. kikkawai* as reported by Roy and Lakhotia (1979).

Second, there seems to be a secondary level of asynchrony during the termination of the replication series or super-replication unit. It has been reported earlier by Lakhotia and Mukherjee (1970) and by Mulder et al. (1968) that after the initiation of all the replication units is completed,i.e., when the chromosomes and the chromocenter are densely and continuously labelled, different groups of bands and interbands begin to show differentiation in the density of label. This differential unlabelling continues until all the chromosomes finally become unlabelled, and does not seem to be a function of the amount of DNA in the groups of bands, as shown by cytophotometric measurement of the ^{3}H-TdR labelled autoradiograms (Mulder et al., 1968). Whether this is due to the existence of a termination signal for the individual replicons has not been proved; but, as it appears from the study of Hand (1975) in L-929 cells, such a termination signal possibly does not exist (see also Callan, 1974). It is probable, therefore, that this type of asynchrony involves a regulation at the level of super-replicon organization that is at the level of the cluster of replicons (Hand, 1975; Blumenthal et al., 1974; Taylor et al., 1974).

It has been observed that the groups of bands on the polytene chromosomes which are the last to finish replication (e.g. the 3C, 11A, 15EF, on the X chromosome; 56EF, 59CD on the second chromosome; and 84A1-B2 on the third chromosome) are some of the late replicating sites (Lakhotia and Mukherjee, 1970; Arcos-Terán, 1972; Mukherjee et al., 1978). Late replicating sites may correspond to intercalary heterochromatin; these sites may contain repetitive DNA sequences (Evgenev et al., 1977). Therefore, it seems reasonable that heterochromatinization of DNA sequences, i.e. organization of the base sequences in some repetitive order, might have some regulatory function in the determination of the specific order of termination of the super-replicon complex, as suggested by Blumenthal et al. (1974).

Mattern and Painter (1977) showed that in CHO cells regions close to the origin and those far from it both had certain repetitive sequences. They suggested that while a class of repetitive sequences may be involved in the regulation of initiation of individual replicons, by keeping replicon clusters in a required conformation, certain others may involve regulation of whole replication.

There is a third level of asynchrony in the replication of DNA in *Drosophila*. This involves the termination or chain growth of the DNA during replication of different non-homologous chromosomes. The first of its kind was shown between the *X* chromosome and autosomes in male *Drosophila* by Berendes (1966). This was confirmed and elaborated by Lakhotia and Mukherjee (1970) and Mukherjee and his collaborators (Chatterjee and Mukherjee (1973; Mukherjee and Chatterjee, 1976, Chatterjee and Mukherjee, 1977a). This is considered to be the counterpart of the hyperactivity of the *X* chromosomes of the male *Drosophila* and therefore of dosage compensation (Mukherjee and Beermann, 1965). In different species of *Drosophila*, it has been shown by Mukherjee and his collaborators that the *X* chromosome in the male invariably has fewer labelled sites for a given pattern of the autosome (Figure 8), implying that it completes its replication earlier than the autosomes and by corollary, than the *X* chromosomes of the female. Upon the examination of the patterns of labelling during the early, middle and terminal parts of replication phase in *XX-XO* mosaic glands, and their intra-nuclear relationship as shown in Table 2, Chatterjee and Mukherjee (1977a) claimed that this early completion of replication of the DNA in the *X* chromosome of the male is associated with faster chain growth, and is not due to early initiation. Several alternatives have been discussed by them. Two of them will be discussed here. There may be either an initiation of DNA replication in the male *X* replicons earlier than other chromosomes or an induction of a larger number of initiation sites. The possibility of an early initiation can be ruled out on the basis of the finding that no nucleus from a male has been observed in which

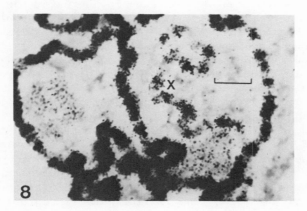

Figure 8: A ^{3}H-TdR labelled nucleus (*XO*) from (*XX-XO*) mosaic salivary gland of a larva of *D. melanogaster*. The X chromosome shows distinctly fewer labelled sites than the autosomes.

Table 2. ^{3}H-TdR labelling patterns of the X chromosomes in XX2A and XO2A nuclei for specific labelling of the autosome 2R showing intranuclear relationship (summarized from Chatterjee and Mukherjee, 1977a). Note that during DD-2C all 115 nuclei of XX2A and all 13 nuclei XO2A type have all nuclei with the labelling on the X same as that on the autosomes.

Labelling pattern of autosome	Number of nuclei (♀ + ♂)	Labelling pattern of X	
		XX2A	XO2A
DD to 2C	115 + 13	DD-2C (115)	DD-2C (13)
3C to 3C-3D	47 + 8	3C to 3C-3D (47)	3C-3D (1)
			3D (3)
			2D (4)
3D to 1D	195 + 19	3D to 1D (195)	2D (2)
			1D (12)
			unlabelled (5)
CL	8 + 0	CL (8)	0

the X is only labelled while the autosomes are unlabelled even under a pulse condition as low as 3 minutes. The possibility of induction of a larger number of initiation sites of replication is highly improbable in this case, as this would require the existence of an inducer specific only for the X chromosome, and that only in the XO cells in the mosaic glands. Existence of such inducers for the whole cells, such as the embryonic cells (Callan, 1972, 1974) is quite understandable; and as shown by Rao and Jahnsen (1970), Graves (1972), and others, such inducers of initiation are synthesized in G_1, are gradually accumulated through G_1 and reach a peak at G_1-S transition (Rao et al., 1977). In no case, however, has an inducer specific for a particular chromosome been reported. Therefore, it seems highly unlikely that the early completion of replication of the DNA replication in the male X is due to the higher number of initiation sites induced. A likely conclusion, therefore, is that the early completion of the X chromosomal replication in the male Drosophila is accomplished during the events of chain growth and probably by a faster rate of chain growth. This aspect can, however, be checked further in two ways: (1) by examining the size of the replicons specifically of the male X DNA in fiber autoradiograms and (2) by examining the number of sites labelled on the male X during the initial part of the labelling pattern. As the labelling is highly dispersed during the very early part of initiation, the latter experiment is extremely difficult from the point of view of resolution under the optical microscope. A third way of examining the possibility would involve the induction of new initiation sites under the condition of puromycin pretreatment (Chatterjee and Mukherjee, 1978). Puromycin treatment prior to pulsing induces new initiation on the chromosomes which are at the terminal part of replication. New initiation is seldom observed under normal conditions. If there is a possibility of the existence of a greater number of initiation sites in the male X, the puromycin-treated male X chromosome should show label over a larger number of sites distinct from the late replicating ones. This study is in progress.

The second category of asynchrony between non-homologous chromosomes is the inter-autosomal asynchrony reported by Chatterjee et al. (1976) in *D. pseudoobscura*. This category of asynchrony is revealed by out-of-phase termination of the sites of two whole chromosomes. Such asynchrony also involves the chain growth of the super-replicon complex. Whether such interchromosomal asynchrony exists in other species has to be examined.

These results on asynchronous termination implicate the existence of autonomous regulation at the level of chain growth and termination in a given replicon series or super-replicon complex (discussed below), and suggests that the control of initiation and termination may be two separate events.

The salivary gland nucleus of *Drosophila* larva contains a giant nucleolus. Under the light microscope it reveals a characteristic matrix consisting of RNA and protein and Feulgen-positive and acridine-orange fluorescing threads and beads scattered over the matrix (Figures 9a, b). The Feulgen-positive chromatin threads are presumably derived from the so-called nucleolar organizer of rDNA cistrons. The DNA very likely loops out from the nucleolar organizer, maintains the continuity with the chromatin DNA, and manifests itself as the thread-like structure (Figure 9a) (see also Choudhry and Godward, 1979). Perry et al. (1970) and Spear and Gall (1973) have shown that the nucleolar DNA in the giant cells is under-replicated. How this under-replication of the rDNA maintains a continuity in the organization of the chromatin DNA is yet to be solved. It has, however, been shown that the replication of chromatin DNA and that of rDNA is highly correlated with respect to their timing of initiation and termination (Rodman, 1968). Lakhotia and Roy's work (1979) also supports this idea. Our data show that the conformation of the nucleolar thread can be related to the sequential order of the chromatin replication (Figure 10).

Using α-amanitin, Chatterjee and Mukherjee (1977b) could show that the drug inhibits the initiation of both chromatin DNA replication and nucleolar DNA replication. It therefore appears likely that the mechanism of initiation of DNA replication of both DNA's is guided by similar factors.

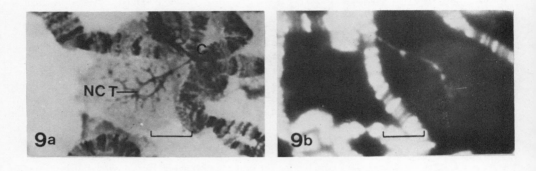

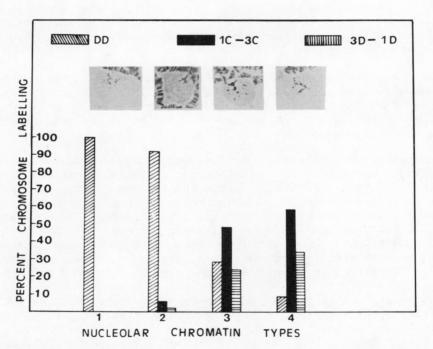

Figure 9: Salivary gland nucleus showing the nucleolar chromatin threads (NCT) emerging from the chromocenter (C). (a) A Feulgen-stained preparation, (b) an acridine orange stained preparation. The former has been photographed with Zeiss Photomicroscope III; the latter with NU-2 fluorescence microscope using the exciter filter No. B 223 G (2) and the barrier filter No. OG1.

Figure 10: Histogram showing the frequencies of different types of labelling patterns in relation to the NCT types.

REGULATION OF DNA REPLICATION

The specific components of the S phase, namely initiation, chain growth and termination can be characterized by examining the replication cytology through antibiotic screening. Specific inhibitors like α-amanitin, puromycin, cycloheximide, actinomycin D, mitomycin C, caffeine, etc. have been used for this purpose in this laboratory and elsehwere. Certain important results relevant to the regulation of DNA replication are discussed below.

Inhibition of DNA synthesis by α-amanitin has been shown in the mammalian system by Montecuccoli et al. (1973). They have suggested that the toxin inhibits the initial and middle parts of the S phase probably through the inhibition of the primer RNA. Chatterjee and Mukherjee (1977b) have confirmed the inhibition of DNA replication by α-amanitin in *Drosophila*. They showed that the initial (DD) and the middle (1C-3C) patterns are considerably inhibited by the toxin (Figure 11). Since both DD and 1C-3C labelling types are likely to have the initiation fragments of the newly synthesized DNA, it is very likely that the inhibition of DNA replication by α-amanitin may be mediated through the inhibition of the RNA primer synthesis. Egyhazi et al. (1972) have shown that α-amanitin inhibits the synthesis of small molecular weight RNA.

Reports on the effects of puromycin and cycloheximide on DNA synthesis are controversial (Painter, 1976). It has been shown in many

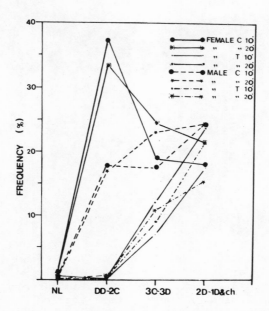

Figure 11: Frequencies of different labelling patterns in control and α-amanitin treated salivary glands of *Drosophila* larva.

eukaryotic systems that inhibition of protein synthesis before the
S phase delays or blocks the onset of DNA synthesis. Work on *Physarum*, mammalian cells and *Drosophila* suggest that both cycloheximide
and puromycin affect the chain growth or fork displacement, but not
the initiation (Mueller et al., 1962; Painter, 1970; Schneiderman
et al., 1971; Muldoon et al., 1971; Hyodo et al., 1971; Fujiwara,
1972; Gautschi, 1974; Gautschi and Kern, 1973; Chatterjee et al.,
1978). However, an opposite result has been reported by Hand and
Tamm (1973) and Hori and Lark (1973) on the effect of puromycin.

Results of Chatterjee et al. (1978) have shown that the frequency
of nuclei with DD type (initial) labelling rises immediately following the treatment with puromycin (100 μg/ml). This increase in
DD pattern is accompanied by a severe decrease in the terminal pattern (Figure 12). Our preliminary results on fiber autoradiography,
using a single pulse of ^{3}H-TdR following puromycin treatment, show
that the major modal class of labelled segments in the fiber autoradiograms lies in the range of 0.5 to 2.0 μm, which probably represents the initial fragments of the newly synthesized DNA (Figure
13).

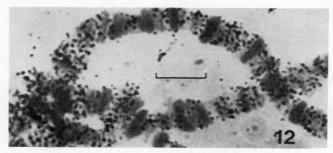

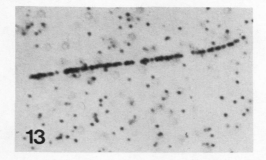

Figure 12: An autoradiogram of the salivary gland chromosomes of
Drosophila hydei, preincubated in puromycin (100μg/ml for 20 min) and
pulsed for 20 min in ^{3}H-TdR. The labelling pattern is typically DD,
with only the interband and puff sites labelled.

Figure 13: A fiber autoradiogram of DNA from puromycin-treated
Drosophila polytene chromosomes, showing only short stretches of
labelled segments.

The results from the puromycin experiments suggest that, by some mechanism, the drug results in a critical level of certain proteins which may be required to activate the potential initiation sites not activated normally. It has been observed that puromycin pretreatment drastically increases the incorporation of ^{3}H-uridine into the chromosomes (Chatterjee, 1975). It is therefore possible that the increase in the initiation by puromycin might occur via an increase in the activation of the potential initiation regions.

The effects of another drug, caffeine, have been examined. Caffeine is known to be an inhibitor of repair-synthesis in eukaryotic systems (Norman, 1971). It has been claimed that caffeine also induces unwinding of DNA template (Chetasanga et al., 1976). Treatment of salivary glands of *Drosophila* male larvae with caffeine (50 μg/ml) prior to pulsing the salivary glands with ^{3}H-TdR, induces initiation of DNA synthesis with a higher than normal frequency (Table 3), but it does not increase the frequency of initiation in the female. Although the investigation is not yet complete, the data suggest that the state of organization of the DNA in the chromatin along with the critical level of certain factors might be important to regulate the initiation.

Table 3. Effect of caffeine (50 μg/ml) on the initiation of DNA replication in *Drosophila melanogaster* males. Values given are mean percent of all nuclei.

Experimental Condition	Sex	Early DD-1C	Mid 2C-3C-3D	Terminal 3D-CL	Unlabelled
Control	Male	14.4	6.9	37.2	41.3
Control	Female	29.6	7.0	49.2	13.9
Treated	Male	37.0	3.3	24.2	35.6
Treated	Female	26.9	6.4	39.8	26.8

FIBER AUTORADIOGRAPHY OF *DROSOPHILA* DNA AND THE CONCEPT OF REPLICON AND SUPER-REPLICON IN EUKARYOTES

Analysis of the data from fiber autoradiography, velocity gradient sedimentation, and electron microscopy from different eukaryotic systems reveals certain common features distinct from prokaryotic systems. First, for all eukaryotes examined, the data provide substantial evidence for multiple initiation sites. Second, the number of replicons may vary from 10^3 to 10^5 (or larger) per cell. Third, the number of initiation sites varies in different tissues. Fourth, the size of the replicons may vary from 2 to 600 µm. Finally, the rate of fork displacement never exceeds 2 µm/min, as compared to a fast rate of about 14 µm/min for prokaryotes (Lewin, 1974). However, the replication in both pro- and eukaryotes is bidirectional (Hubermann and Riggs, 1968; Weintraub, 1972; Hubermann and Tsai, 1973; Kriegstein and Hogness, 1974). From this pattern of results, Blumenthal et al. (1974) and Painter (1976) proposed an operational definition specifically for the eukaryotic replicon. According to them "a eukaryotic replicon is a segment of DNA containing exactly one site, called the origin, at which initiation of DNA replication begins and from which both parental strands are duplicated". Earlier, Hubermann and Riggs (1968) proposed the existence of an origin and two termini for a eukaryotic replicon in a manner similar to that in prokaryotes (Cairns, 1963). But Blumenthal et al. (1974), Callan (1974), and Painter (1976) explained the difficulty that a chromosome would have if the idea of fixed termini was accepted. Hand (1975), using a double labelling technique with radioactive precursors of high and low specific activities, showed that the presence of the abrupt end of labelled segments, expected if fixed termini existed, was rather low. The frequency of such abrupt ends of labelled segments was in the same order as expected on the basis of some unidirectional chain growth. Hand claimed that this evidence rules out the existence of fixed termini, and suggested that any control that might be associated with the rate of chain growth or fork displacement must involve some other level of action.

Data obtained from various studies indicate that there is a great deal of variation in the number and size of replicons in eukaryotes. One constant feature perhaps is the rate of chain growth. According to Blumenthal et al.(1974), in *Drosophila* the rate of chain growth is 2.6 kb/min/fork. Although there is some variation, the rate is also more or less uniform for the mammalian cells (Painter, 1976). A comparison of the number and size of replicons and of the rate of chain growth in different eukaryotes in relation to *Drosophila* is presented in Table 4. It is seen from the table that the fork rates are not too widely departed from each other. The rate is somewhat low for most of the amphibians studied including the frog.

As discussed earlier, Mukherjee and his co-workers (Chatterjee and Mukherjee, 1975; Chatterjee and Mukherjee, 1977a) observed that in *Drosophila* polytene chromosomes a number of bands form a composite unit with respect to initiation and termination of replication. Such units are considered to constitute a super-replicon organization which is referred to as a replication series or unit. Blumenthal et al. (1974) described such a super-replicon organization as a "cluster of replicons". The existence of clusters of replicons with synchronous initiation has also been shown for L929 cells by Hand (1975).

Table 4. Comparative account of the number and size of replicons and rate of fork development in different animals

Parameters	Drosophila melanogaster	HeLa cells	L5178Y	CHO	Triturus	Chick Embryo Cells
Number of replicons	10^5	10^3-10^4	10^5	$(1.5-2.0) \times 10^5$	–	–
Size of replicons in μm	2.5 (Cleavage) 9.5 (Somatic)	15-30	200	15-30	600 (Somatic)	15-17
Fork rate μm/min	1.0	0.5 0.5-1.8 (Painter)	0.6-0.9	0.5-2.5 0.5-1.8 (Painter)	0.33	0.5

Our autoradiographic investigations reveal that such a cluster may be constituted of 1 to 8 bands (and their interbands). Thus, the segments 1A to 11A of D. melanogaster have been divided into 39 replication units, and the autosome segments of 2R (56F to 60F) into 20 units (Chatterjee and Mukherjee, 1977a). Similar division has also been made in other species of Drosophila (namely, pseudoobscura and hydei) (Chatterjee and Mukherjee, 1973 and 1975). In all these cases, both the initiation and the termination of the respective units are synchronized. Thus, we define a super-replicon complex (previously termed a replication unit) as a cluster of replicons in which the adjacent origins are activated at approximately the same time, and in which the termination events are set synchronously. Work in progress in this laboratory suggests that the organization may be further resolved into smaller complexes.

GENETIC DISSECTION OF SUPER-REPLICON ORGANIZATION AND LOCALIZATION OF REGULATION OF LATE REPLICATION

It has been mentioned earlier that different sites on the replicating chromatin in the giant cells of Drosophila, as also in other eukaryotes, complete their replication at different times of the S period. Those which finish the replication in the later stages are designated as late replicating sites, and probably contain repetitive DNA, while those which finish replication within the first half are known as early replicating sites. Since it has been claimed that the rate of chain growth under normal optimum conditions is more or less fixed (Blumenthal et al., 1974) and, as shown by Hand (1975), the existence of fixed termini in eukaryotes is doubtful, it is presumed that secondary regulation exists at the level of replicon cluster or super-replicon organization. We examined this possibility by genetically dissecting the replicon clusters. The idea of genetic dissection originated from the work of Arcos-Terán and Beermann (1968), who showed that certain factors which might be determinants of the late replication of the white region in *Drosophila melanogaster* are possibly located at 3C3-5. Our data, using deficiencies, translocations, inversions, and duplications, for regions in the X chromosome as well as the autosomes, show that such factors are inherently present in association with each cluster and occupy a specific sequence of the DNA segment. The factors act in a cis-dominant manner (Mukherjee et al., 1978; D. Majumdar, R.N. Chatterjee, C. Chatterjee, A.K. Duttagupta, and A.S. Mukherjee, in preparation). The results are briefly summarized below.

We analyzed the replication of the X chromosomes and chromosome 2 of *D. melanogaster*, in which the replication units complexes 1A1-6, 3C1-7, and certain others have been split at one or more regions by deficiencies, inversions, or duplications. As the results in general lead to the same inference in all cases, only three sets of them are presented here for the sake of brevity. They are the deficiency heterozygote for 1A1-3 $[R(1)2/+]$, the deficiency heterozygote for 3C1-4 $[Df(1)w^{vco}/+]$, the inversion homozygote for 3C1-7 $[In(1)w^{m51b19}]$ and the duplication $Dp(1;1)Co$.

Each of these sites forms one replication unit complex (Lakhotia and Mukherjee, 1970; D. Majumdar, R.N. Chatterjee, C. Chatterjee, A. K. Duttagupta, and A.S. Mukherjee, unpublished). In $Df(1)w^{vco}$ the complex 3C1-7 has been split at 3C4-5; in $In(1)w^{m51b19}$, the 3C1-7 has been split at 3C1-2; in $R(1)2$ the 1A has been split at 1A3-4 (1A1-3 missing); and in the duplication $Dp(1;1)Co$ the 3C1-7, 3D5-E2 have been split at 3C4-5 and 3D6-E1, respectively. The 3C5-3D6 segment has been duplicated in the last one. In all these altered sequences the terminal patterns of labelling of the altered homologues have been changed (Figures 14-17) without much effect on initiation. Results show that the alteration in the time of termination of the super-replicon complex, as a consequence of splitting up the sequence, depends largely upon the position with respect to the neighboring sequence. If the altered part of the replication unit complex in the new sequence is next to an early replicating unit complex, the new sequence (including the part from the originally late replicating segment) becomes early replicating (Figures 14-17). Analysis of the data on frequencies and intensities of grains on these sites in the autoradiograms suggests that the effect is, in most cases, on the termination. Results further show that the late replicating property of the 3C1-7 is regulated by a factor probably located at 3C3-4 band-interband. This finding strongly suggests that there exists a regulation, at the level of determination, of the timing of termination of the super-replicon and implies that such regulatory signals behave semi-autonomously and in a cis-dominant manner. The nature of such signals remains to be determined (Blumenthal et al., 1974).

REGULATION OF DNA REPLICATION 77

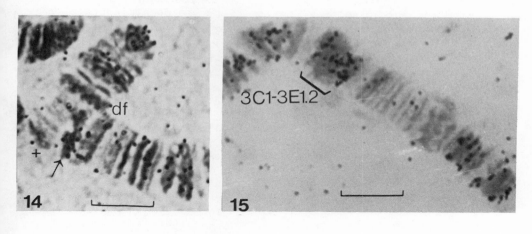

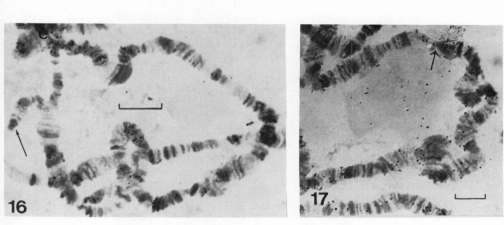

Figures 14-17: ^{3}H-TdR autoradiograms of altered genetic sequences in *D. melanogaster*, showing terminal patterns. Figure 14 - *Df(1) w^{vco}/+* heterozygous X chromosomes showing the intense labelling (arrow) on the normal homologue (+) at 3C1-7 and absence of labelling on the deficiency bearing homologue (df), 3C5-7. Figure 15 - ^{3}H-TdR labelling on the X chromosomes of homozygous *Dp(1;1)Co*, showing labelling on the segment 3C1-3E2 (bracketed). Figure 16 - *R(1)2/+*, showing the whole X chromosome. The tip part of the normal homologue is free and seen on the left (arrow). Note the label on the corresponding homologous tip of the ring-X at the point where it joins with the chromocenter (C). Figure 17 - Homozygous *In(1) w^{m51b19}* showing absence of labelling on the 3C1 (arrow) and poor labelling on the inverted 20D heterochromatin close to it.

CONCLUDING REMARKS

From the foregoing presentation it is clear that although a good deal of progress has been made during the last decade towards understanding the regulation of DNA replication in the eukaryotic genome, large lacunae still remain. However, it is apparent that regulation of DNA replication in these systems exists at different levels. For the sake of discussion four levels will be considered.

1. The first level involves factors that are required to control the switching on of the whole genome into S. Such factors would presumably determine the activation of new initiation. Blumenthal et al. (1974) suggest that the heterochromatic parts of the genome might act as such a determinant. The existence of mutants like "giant" *(gt)* in almost all the chromosomal arms of *Drosophila melanogaster* suggest that the nature of the determinants could not be anything more than intercalary heterochromatin. This series of mutants individually induces an extra round of DNA replication in the larvae.

2. There is a second level, whereby a regulated activation of limited but multiple primary initiation sites takes place. The number and location of these primary sites are probably fixed in the polytene chromosomes of *Drosophila*. Very likely these sites are present in the sequences of specific interband DNA. The regulation is probably mediated through a protein factor as suggested by Rao et al. (1977). The initiation of the remaining replicons would follow soon after the initiation of the primary ones. After these two sequential steps, the chain elongation would then proceed almost automatically. No signal for termination of the individual replicon would seem mandatory. Whenever and wherever two adjacent growing forks would meet, the termination of both the replicons would be achieved (Painter, 1976; Hand, 1975).

3. Yet a third level of regulation appears to be present in many eukaryotes and specially so for *Drosophila* polytene chromosomes. This regulation involves the control of a set of replicons forming the cluster, here termed <u>super-replicon complex</u>. Such a control should have the responsibility to determine the synchronous termination of the cluster of replicons. This would be possible by controlling the rate of fork development at the end without altering the average rate of chain growth of 2.6 kb/min/fork (Blumenthal et al., 1974). Our results on the genetic dissection of the super-replicon complex and the consequent pattern of change in the time of the replication of the complex seem to have indicated the existence of such factors.

4. Finally, there seems to exist yet another level of control that prevents the reinitiation of any potential site before completion of the replication of the whole genome. The possibility of disruption of such controlling machinery by such agents as puromycin or actinomycin D reiterates the existence of this level of control.

No doubt, we still have much to understand about the mechanism and nature of the interactions among these factors. The problem is intimately related to the knowledge of the organization of the eukaryotic genome. Parallel investigations on the two aspects would hopefully bring about the conclusion of the drama.

ACKNOWLEDGEMENTS

The work reported here has been supported by grants from various agencies to A.S.M. We gratefully acknowledge the financial support from Department of Science and Technology (Grant No. 11(13)/75-SERC), University Grants Commission (Grant No. F.23-201/75/SR II), Department of Atomic Energy (Grant No. BRNS/B & M/124/78), Council of Scientific and Industrial Research, Special Assistance from University Grants Commission. We also thankfully acknowledge the generous help of Prof. L.E. Feinendegen on various aspects. D.M. and M.G. were supported by Junior Research Fellowships from the C.S.I.R. Part of this work has also been supported by a grant (No. F. 23-457/76-SR II) from the University Grants Commission to A.K.D.G.

REFERENCES

Arcos-Terán, L., 1972, DNS-Replikation und die Natur der Spät replizierenden orte in X-chromosome von Drosophila melanogaster, Chromosoma, 37:233.

Arcos-Terán, L., and Beermann, W., 1968, Changes of DNA replication behavior associated with intragenetic changes of the white region in Drosophila melanogaster, Chromosoma, 25:377.

Berendes, H.D., 1966, Differential replication of male and female X-chromosome in Drosophila, Chromosoma, 20:32.

Blumenthal, A.B., Kriegstein, H.J., and Hogness, D.S., 1974, The units of DNA replication in Drosophila melanogaster chromosomes, Cold Spring Harb. Symp. Quant. Biol., 138:205.

Cairns, J., 1963, The bacterial chromosome and its manner of replication as seen by autoradiography, J. Mol. Biol., 6:208.

Callan, H.G., 1972, Replication of DNA in the chromosomes of eukaryotes, Proc. Roy. Soc. (London), B 181:19.

Callan, H.G., 1974, DNA replication in the chromosomes of eukaryotes, Cold Spring Harb. Symp. Quant. Biol., 38:195.

Chatterjee, R.N., and Mukherjee, A.S., 1977a, Chromosomal basis of dosage compensation in Drosophila. IX. Cellular autonomy of the faster replication of the X chromosome in haplo X cells of Drosophila melanogaster and asynchronous initiation, J. Cell Biol., 74:168.

Chatterjee, R.N., and Mukherjee, A.S., 1977b, Inhibition of initiation of DNA replication in Drosophila by α-amanitin and its possible significance, Ind. J. Exp. Biol., 15:973.

Chatterjee, S.N., 1975, Gene physiological studies on polytene chromosome and dosage compensation in Drosophila, Ph.D. Thesis, University of Calcutta.

Chatterjee, S.N., and Mukherjee, A.S., 1973, Chromosomal basis of dosage compensation in *Drosophila*. VII. DNA replication patterns chromosomes, Cell Differentiation, 2:1.

Chatterjee, S.N., and Mukherjee, A.S., 1975, DNA replication in polytene chromosomes of *Drosophila pseudoobscura*: New facts and their implications, Ind. J. Expt. Biol., 13:452.

Chatterjee, S.N., Mondal, S.N., and Mukherjee, A.S., 1976, Interchromosomal asynchrony of DNA replication in polytene chromosomes of *Drosophila pseudoobscura*, Chromosoma, 54:117.

Chatterjee, S.N., Chatterjee, C., and Mukherjee, A.S., 1978, Effect of puromycin on DNA synthesis in *Drosophila* polytene chromosomes: A probe into the control of replication, Ind. J. Exp.Biol., 16: 1027.

Chetasanga, C.J., Rushlow, K., and Boyd, V., 1976, Caffeine enhancement of digestion of DNA by Nuclease S_1, Mut. Res., 34:11.

Choudhry, A., and Godward, M.B.E., 1979, The nucleolus in telophase, interphase and prophase, The Nucleus, 21: (in press).

Egyhazi, E., Monte, B.D., and Edstrom, J.E., 1972, Effects of α-amanitin on *in vitro* labelling of RNA from defined nuclear components in salivary gland cells from *Chironomus tentans*, J. Cell Biol., 53:523.

Evgenev, M., Andrei, L., and Gubenko, I., 1977, Are late replicating regions in polytene chromosomes of *Drosophila* enriched by repeated nucleotide sequences? Nature, 268:766.

Fujiwara, Y., 1972, Effect of cycloheximide on regulatory protein for initiating mammalian DNA replication at the nuclear membrane, Cancer Res., 32:2089.

Gautschi, J.R., 1974, Effects of puromycin on DNA chain elongation in mammalian cells, J. Mol. Biol., 84:223.

Gautschi, J.R., and Kern, R.M., 1973, DNA replication in mammalian cells in the presence of cycloheximide, Expt. Cell Res., 80:15

Graves, J.A.M., 1972, Cell cycles and chromosome replication patterns in interspecific somatic hybrids, Expt. Cell Res., 73:81.

Hand, R., 1975, Regulation of DNA replication on subchromosomal units of mammalian cells, J. Cell Biol., 64:89.

Hand, R., and Tamm, I., 1973, DNA replication: Direction and rate of chain growth in mammalian cells, J. Cell Biol., 58:410.

Henning, W., and Meer, B., 1971, Reduced polytene of ribosomal RNA cistrons in giant chromosomes of *Drosophila hydei*, Nature New Biol., 233:70.

Hori, T., and Lark, K.G., 1973, Effect of puromycin on DNA replication in Chinese hamster cells, J. Mol. Biol., 77:391.

Hubermann, J.A., and Riggs, A.D., 1968, On the mechanism of DNA replication in mammalian chromosomes, J. Mol. Biol., 32:327.

Hubermann, J.A., and Tsai, A., 1973, Direction of DNA replication in mammalian cells, J. Mol. Biol., 75:5.

Hyodo, M., Koyama, H., and Ono, T., 1971, Intermediate fragments of newly replicated DNA in mammalian cells. II. Effect of cycloheximide on DNA chain elongation, Expt. Cell Res., 67:461.

Kriegstein, H.J., and Hogness, D.S., 1974, Mechanism of DNA replication forks and evidence for bidirectionality, Proc. Natl. Acad. Sci. U.S.A., 71:135.
Lakhotia, S.C., and Mukherjee, A.S., 1970, Chromosomal basis of dosage compensation in Drosophila. III. Early completion of replication by the polytene X chromosome in male: further evidence and its implications, J. Cell Biol., 47:18.
Lakhotia, S.C., and Jacob, J., 1974, EM autoradiographic studies on polytene nuclei of Drosophila melanogaster. II. Organization and transcriptive activity of the chromocentre, Expt. Cell Res., 86:253.
Lakhotia, S.C., and Roy, S., 1979, Replication in Drosophila chromosomes. I. Replication of intranucleolar DNA in polytene cells of D. nasuta, J. Cell Sci., 36:185.
Lewin, B., 1974, "Gene Expression - 2 Eucaryotic Chromosomes", John Wiley & Sons, London.
Majumdar, D., and Mukherjee, A.S., 1980, Existence of potential initiator puffs and their role in replication, Proc. IV All Ind. Cell Biol. Conf., Calcutta, (In press).
Mattern, M.R., and Painter, R.B., 1977, The organization of repeated nucleotide sequences in the replicons of mammalian DNA, Biophys. Journal, 19:117.
Montecuccoli, G., Novello, F., and Stirpe, F., 1973, Effect of α-amanitin poisoning on the synthesis of deoxyribonucleic acid and of protein in regenerating rat liver, Biochim. Biophys. Acta, 319:199.
Mueller, G.C., and Kajiwara, K., 1966, Early and late replicating DNA complexes in HeLa nuclei, Biochim. Biophys. Acta, 114:108.
Mueller, G.C., Kajiwara, K., Stubblefield, E., and Rueckert, R.R., 1962, Molecular events in the reproduction of animal cells. I. The effect of puromycin on the duplication of DNA, Cancer Res., 22:1084.
Mukherjee, A.S., and Beermann, W., 1965, Synthesis of ribonucleic acid by the X-chromosome of Drosophila melanogaster and the problem of dosage compensation, Nature, 207:785.
Mukherjee, A.S., and Chatterjee, S.N., 1975, Chromosomal basis of dosage compensation in Drosophila. VIII. Faster replication and hyperactivity of both arms of the X-chromosome in males of Drosophila pseudoobscura and their possible significance, Chromosoma, 53:91.
Mukherjee, A.S., and Chatterjee, S.N., 1976, Hyperactivity and faster replicative property of the two arms of the male X of Drosophila pseudoobscura, J. Microscopy, 106:199.
Mukherjee, A.S., Chatterjee, R.N., Chatterjee, C., and Majumdar, D., 1978, Is termination of replication of a chromosomal replication unit guided by its nearest neighbor?, Proc. XIV Intern. Congr. Genet., Moscow, Section C8, p. 194.
Muldoon, J.J., Evans, T.E., Nygaard, O.F., and Evans, H.H., 1971, Control of DNA replication by protein synthesis at defined times

during the S period in *Physarum polycephalum*, Biochim. Biophys. Acta, 247:310.

Mulder, M.P., van Duijn, P., and Gloor, H.J., 1968, The replication organization of DNA in polytene chromosomes of *Drosophila hydei*, Genetica, 39:385.

Norman, A., 1971, DNA repair in lymphocytes and some other human cells, (DNA Repair Mechanism), in: "Symposium Schloss Reinhartshausen", F.K. Schattaner, ed., Springer-Verlag, Rhein, Stuttgart pp. 9-16.

Painter, R.B., 1970, The molecular basis of changes in rate of mammalian DNA synthesis, J. Cell Biol., 47:153a.

Painter, R.B., 1976, Organization and size of replicons, in: "Handbook of Genetics", Vol. 5, R.C. King, ed., Plenum Press, New York, p. 169.

Pelling. C., 1966, A replicative and synthetic chromosomal unit - the modern concept of the chromomere, Proc. Roy.Soc. (London)., 164:279.

Perry, R.P., Greenberg, J.R., and Tartof, K.D., 1970, Transcription of ribosomal, heterogeneous nuclear and messenger RNA in eukaryotes, Cold Spring Harb. Symp. Quant. Biol., 35:577.

Plaut, W., Nash, D., and Fanning, T., 1966, Ordered replication of DNA in polytene chromosomes of *Drosophila melanogaster*, J. Mol. Biol., 16:85.

Rao, P.N., and Jahnsen, R.T., 1970, Mammalian cell fusion: studies on the regulation of DNA synthesis and mitosis, Nature, 225:159.

Rao, P.N., Sunkora, P.S., and Wilson, B.A., 1977, Regulation of DNA synthesis: Age dependent cooperation among G1 cells upon fusion, Proc. Natl. Acad. Sci. U.S.A., 74:2869.

Rodman, T.C., 1968, Relationship of developmental stages to initiation of replication in polytene nuclei, Chromosoma, 23:271.

Roy, S., and Lakhotia, S.C., 1979, Replication in *Drosophila* chromosomes: Part II - Unusual replicative behaviour of two puff sites in polytene nuclei of *Drosophila kikkawai*, Ind. J. Exp. Biol., 17:231.

Rudkin, G.T., 1972, Replication in polytene chromosomes, in: "Developmental Studies on Giant Chromosomes", W. Beermann, ed., Springer-Verlag, New York, pp. 59-85.

Schneiderman, M.H., Dewey, W.C., and Highfield, D.P., 1971, Inhibition of DNA synthesis in synchronized Chinese hamster cells treated in G1 with cycloheximide, Expt. Cell Res., 67:147.

Seale, R.L., 1978, Nucleosome associated with newly replicated DNA have an altered conformation, Proc. Natl. Acad. Sci. U.S.A., 75:2717.

Spear, B.B., and Gall, J.G., 1973, Independent control of ribosomal gene replication in polytene chromosomes of *Drosophila melanogaster*,, Proc. Natl. Acad. Sci. U.S.A., 70:1359.

Stubblefield, E., 1974, The kinetics of DNA replication in chromosomes, in: "The Cell Nucleus", Vol. II, H. Busch, ed., Academic Press, New York, pp. 149-162.

Taylor, J.H., Wu, M., and Erickson, L.C., 1974, Functional subunits of chromosomal DNA from higher eukaryotes, Cold Spring Harb. Symp. Quant. Biol., 38:225.

Weintraub, H., 1972, Bi-directional initiation of DNA synthesis in developing chick erythroblasts, Nature New Biol., 236:195.

A COMBINED GENETIC AND MOSAIC APPROACH TO THE STUDY OF OOGENESIS IN

DROSOPHILA

Eric Wieschaus

European Molecular Biology Laboratory
Heidelberg
Federal Republic of Germany

INTRODUCTION

An embryologist usually becomes interested in oogenesis with the realization that many of the decisions made by the cells in the early embryos reflect a distribution of determinants present in some form already in the egg prior to fertilization. This allows us to reduce the question of why different regions of the blastoderm have different embryonic fates into two subquestions: (1) How do the different regions of the egg cytoplasm become different from each other? (2) How are these initial cytoplasmic differences interpreted by the individual blastoderm cells in choosing their particular fates?

In this paper, I will deal mainly with the first question and in particular will describe a set of genetic experiments designed to study the establishment of dorsal-ventral polarity in the oocyte. One of the advantages of using *Drosophila* for this work was obviously the ease with which mutations affecting oogenesis can be obtained. By identifying specific functions necessary for oogenesis, such mutations allow us to reconstruct its course in a manner similar to a prokaryote geneticist's reconstruction of a biosynthetic pathway. Oogenesis, like most processes in higher organisms, involves a number of different cell types, each with its own defined role, roles which may depend on and interact with others in a defined temporal sequence. This points up a second advantage of *Drosophila*, namely the variety of techniques available for construction of genetic mosaics. By studying the egg production of mosaic females in which only specific cell types are mutant, we can assign oogenetic functions to specific parts of the ovary. Moreover, as will be described in the last section of this paper, such mosaics allow us to test whether mutations which normally have major effects on the viability of whole flies also play a role in oogenesis.

fs(1)K10 and the Establishment of Dorsal-Ventral Polarity in the Oocyte

In *Drosophila*, the dorsal side of the unfertilized egg can be most easily identified by the pattern of the outermost egg covering, the chorion. In addition to the more elongated appearance of the follicle cell imprints in this region of the chorion, the dorsal side of the egg also possesses two respiratory appendages which jut out diagonally to either side of the egg at about 30% egg length from the anterior end. Normally, the dorsal side of the embryo develops in the region of the egg cytoplasm underlying the dorsal side of the chorion, just as the anterior end of the embryo develops at the anterior end of the egg as defined by other chorion landmarks. Since it is possible to remove the chorion without affecting the course of embryonic development, it is unlikely that the chorion actively directs the development of the underlying embryo. Instead the correspondence between the chorion pattern and embryonic development probably indicates that both the secretion of the egg shells and the laying down of embryonic determinants are controlled by the same organizing principle during oogenesis: most likely, that they both result from a dorso-ventral polarity in the developing oocyte. A mutation which alters this initial organizing polarity would be expected to produce a consistent alteration in both the chorion pattern and the subsequent embryonic development. In the past ten years an impressive number of mutations have been isolated which cause female sterility and thus seem to play a major role in oogenesis. A small number of these affect the embryonic pattern and only one, *fs(1)K10*, has been reported to affect both embryonic development and chorion pattern.[1]

fs(1)K10 (= *K10*) is a recessive maternal effect mutation located at the tip of the X-chromosome. Homozygous females lay eggs which lack the two simple appendages normally found at the anterior end of the wild type egg (Figure 1). Instead, the dorsal appendage material is secreted in a ring around the egg as though the entire circumference has become dorsalized. Although the most obvious effect of the mutation is observed in the appendage material at the anterior end of the egg, it is also possible to detect alterations in other regions of the chorion by comparing the pattern of follicle cell imprints in wild type and *K10* eggs. Very little mutant effect is noted in the posterior regions of *K10* eggs but here the difference between dorsal and ventral sides of wild type eggs is very slight and a dorsalization would be difficult to detect. It is worth emphasizing that *K10* does not result in a total abolition of polarity since the dorsal side of a *K10* egg can always be recognized by the interruption in the appendage material and the somewhat more elongated appearance of the follicle cell imprints in that region. Instead, the mutation results in a reduction in the range of dorsal-ventral values such that the entire circumference of a *K10* egg represents values found only in the upper sixth of a wild type egg.

GENETIC AND MOSAIC STUDIES OF OOGENESIS

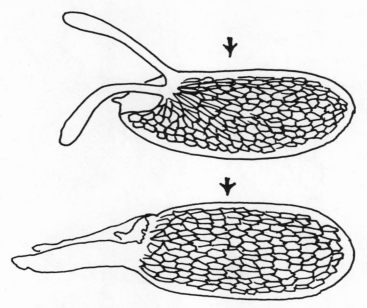

Figure 1: The chorion pattern of wild-type eggs (upper drawing) and *K10* eggs (lower drawing). The arrows indicate the dorsal side of each egg.

The excess of appendage material prevents most *K10* eggs from being fertilized. Those which are fertilized, however, show the same restricted dorsalized range seen in the chorion. The first sign of a dorso-ventral distinction in a normal embryo occurs at the onset of gastrulation when cells along the ventral midline fold in to give rise to mesoderm. *K10* embryos do not make this invagination and, in the anterior region of the embryo, the ventral and lateral regions of the egg are thrown into folds characteristic of the dorsal side. Although gastrulation is thus abnormal, the *K10* embryo does not die but continues to develop until it reaches an abnormal larval stage (Figure 2). The anterior ends of such animals contain no internal organs but consist instead of a hollow tube of skin with the hair pattern[2] characteristic of the dorsal side. Posteriorly the pattern of the larvae becomes more ventralized and typically the last few segments of a *K10* larva show the bands of denticles normally found only in the ventral hypoderm.

At present we have no explanation why the extent of dorsalization in the embryo varies along the anterior posterior axis. Given the lack of dorsal-ventral landmarks in the posterior chorion, it is possible that the same anterior-posterior variability applies to the chorion of *K10* eggs. In any case the general correspondence between

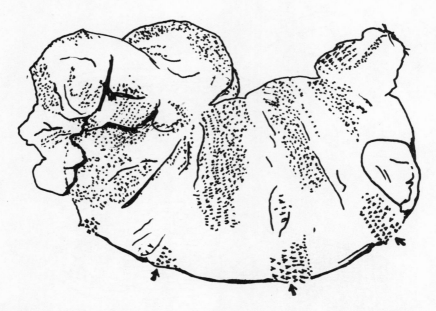

Figure 2: An abnormal larva which developed from a fertilized *K10* egg. The anterior end of the larva is at the upper right, the posterior end at the lower left. The ventral denticle belts are indicated with arrows.

the chorion and embryonic pattern indicates that *K10* may affect some aspects of the primary polarity in the developing oocyte and that the mutation may thus be useful in determining the cell type in the ovary responsible for that polariy.

The first obvious morphological effect of the mutation is observed in ovaries late during oogenesis at about stage 10 of King.[3] At this point each developing follicle consists of three cell types: the oocyte itself, the 15 interconnected nurse cells and the overlying follicle cell epithelium.[3,4] The oocyte and nurse cells are sister cells, still interconnected by a system of ring canals formed during the mitosis which gave rise to the cluster. The follicle cells are somatically derived and had initially covered the entire nurse cell oocyte complex with a single cell-layered epithelium. By stage 10 most of the follicle cells have migrated away from the nurse cells and surround exclusively the oocyte, except at the oocyte's anterior end where the ring canal maintains a connection with the nurse cells. The follicle cells will eventually secrete the egg coverings at the end of oogenesis.

The follicles of normal ovaries have a clear dorsal-ventral polarity at this stage (Figure 3). The oocyte nucleus is acentric

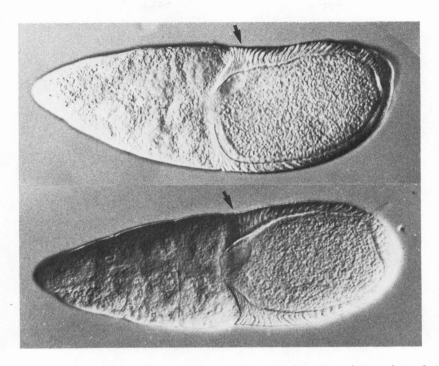

Figure 3: Stage 10 oocytes from wild type females (upper) and *K10* females (lower). The arrows indicate the dorsal side of the wild type oocyte and the presumed dorsal side of a *K10* oocyte judging from the position of the oocyte nucleus.

and lies near the surface at the dorsal side. The overlying follicle cells are much thicker in this region than those on the ventral side of the oocyte. This difference is most apparent in the anterior region of the egg where the dorsal appendage will eventually be secreted. In follicles from *K10* oocytes, most of the indications of dorsal-ventral polarity are absent. Although the oocyte nucleus is still acentric, the follicle cells over the entire circumference of the egg show the thickened appearance normally found only on the dorsal side. The shape of the oocyte is also different, probably due to the absence of the thinner ventral follicle cells which would separate the oocyte from the nurse cells.

The most obvious alteration in *K10* ovaries is the position of the follicle cells surrounding the oocyte. It would be false however to conclude that the primary effect of the mutation is on these cells, and indeed judging from the results of mosaic studies[1], the contrary seems to be the case. These mosaics were constructed by transplanting the germ cell precursors ("pole cells") between mutant and wild type embryos. Since the pole cells eventually give rise to the oocyte and nurse cells, but not to the follicle cells, it thus becomes possible to construct mosaic ovaries in which either the follicle cells or the underlying germ cell derivatives are mutant in genotype. In both types of mosaics, the final pattern of the egg depended only on the genotype of the germ cell and was independent of that of the follicle cells. When the germ cell was mutant, the eggs were of the *K10* type; when it was wild type, the eggs were normal.

It is also possible to make mosaic ovaries using mitotic recombination in females heterozygous for *K10*. Here a chromosomal exchange produces a homozygous germ cell in an ovary whose somatic cells have remained heterozygous and thus wild type. The eggs which develop from homozygous germ cells show the K10 phenotype, even when the homozygosity arose in a germ cell during the last stem cell division prior to the onset of oogenesis.[5] This confirms the results of the pole cell transplantation and also indicates that the necessary activity at the *K10* locus is probably restricted temporally to oogenesis.

The results of the pole cell transplantation experiment indicate that the *K10* mutation blocks a function which normally must occur in the germ cell in order that the developing egg shows normal dorso-ventral polarity. From the ovary preparations described above (Figure 3), it was concluded that the first morphological change observed in *K10* ovaries concerns the migration pattern of the follicle cells over the surface of the oocyte. Given the germline dependence of *K10* it is tempting to speculate that this migration is dependent on positional cues provided by the underlying germ cell, perhaps embedded in the surface of the oocyte itself.

If the follicle cells are indeed following germline cues in the secretion of the chorion, one would expect two classes of mutations affecting chorion pattern. The first class would include germline dependent mutations like *K10* which presumably alter the positional cues provided by the oocyte. The second set would include somatically dependent mutations affecting ability of the follicle cells either to follow the positional cues and make a normal pattern or their ability to secrete the final chorion. The number of mutations altering chorion morphology is at present very small. Of the even smaller number affecting chorion pattern, none has yet been tested for its somatic dependence. One female sterile mutation on the X-chromosome, *f(1)384*, causes homozygous females to produce eggs in which the dorsal appendages are rudimentary and the entire surface of the chorion appears vacuolated and abnormal.[6] The eggs usually collapse immediately after being laid and none has been observed to develop even to the earliest embryonic stages. By combining *384* with *K10*, we have been able to use mitotic recombination to demonstrate its somatic dependence.[7] Heterozygotes for both mutations (*K10 384/+*) were X-irradiated as young larvae. Homozygous clones produced in the germline of these individuals could be identified when the animals had developed to adults by the production of *K10* eggs. Since the clones should be homozygous for both mutants, if *384* like *K10* were also germline dependent, we would expect these clonally derived *K10* eggs also to show the chorion defects associated with *384*. This however was not the case, and all showed normal chorion differentiation, although of course arranged in the typical *K10* pattern. The *384* defects must therefore depend on somatic rather than germline derived cell type, and the best candidate for this role would be the follicle cells which actually secrete the chorion. That this is probably the case is indicated by a second class of mosaics produced in these experiments. These females laid non-*K10* eggs which showed partially or wholly the *384* phenotype (Figure 4). In the partial *384* eggs the boundary between mutant and wild type morphology is what one would expect of clones induced in the larval precursor for the follicle cell epithelium overlying the oocyte.

It is likely that *384* blocks some step in the synthesis or final differentiation of the chorion, and therefore it is perhaps not surprising that the phenotype of the mutation might depend on the follicle cells which actually do the final secretion. The more promising aspect of the experiments is that they demonstrate the feasibility of making clones in other cell types involved in oogenesis, thus leading to a refined analysis of the process. Such analysis should allow a more clear designation of different roles during oogenesis and potentially allow a study of the interaction between the germ cell and the follicle cell epithelium.

The Use of Mosaics to Study the Effects on Oogenesis of Lethal Mutations.

Although the most dramatic alterations in early embryonic development are probably produced by maternal effect mutations, a number of loci are known which alter the embyronic pattern but are dependent on the genotype of the embryo itself. An important example of such zygotic effects is provided by the embryonic lethal *Krüppel*.[8] Embryos homozygous for *Krüppel* develop into larvae with abnormal segmentation and polarity. A normal *Drosophila* larva possesses an involuted head, three thoracic and eight abdominal segments. *Krüppel* embryos have normal heads, a normal 8th and 7th abdominal segment, a somewhat enlarged sixth segment and sometimes a fifth. This is followed by a plane of mirror image symmetry, a reversed sixth and sometimes a reversed 7th. Occasionally, the mirror image symmetry is so complete in the dorsal region of the embryo, that posterior spiracles are found in the region of the embryo immediately adjacent to the head. Studies on the development of *Krüppel* indicate that the failure to form thorax and anterior abdomen is not due to death of the respective primordia during early embryogenesis. Instead, the final pattern in *Krüppel* embryos results from an alteration in the developmental program of the blastoderm cells, such that the region of the embryo which normally would give rise to three thoracic

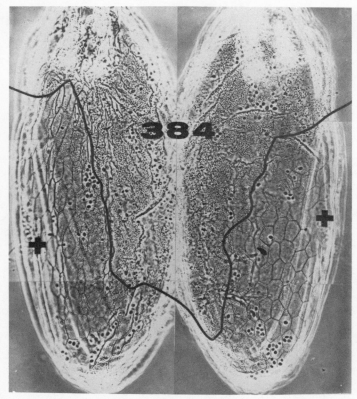

Figure 4: Mosaic chorion composed of regions of wild-type (+) and mutant (fs(1)384) morphology.

and eight abdominal segments, gives rise to a much reduced number of segments in reversed polarity.

The smaller number of segments and reversal of polarity associated with Krüppel is reminiscent of the effects of the maternal effect mutation, bicaudal[9,10]. It is therefore important to determine whether the effect of the mutation results strictly from the activity of the embryonic genome or whether the Krüppel locus also shows a maternal effect (i.e. whether the phenotype is dependent on the fact that homozygous Krüppel embryos are obtained from heterozygous mothers). The simplest approach to this problem involves a series of genetic crosses where duplications or translocations are used to obtain homozygous Krüppel embryos from duplication mothers which are essentially wild type with respect to Krüppel. The fact that the homozygous embryos from such mothers display exactly the same

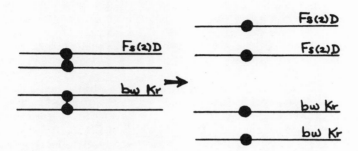

Figure 5: Schematic representation of mitotic recombination to produce germline clones homozygous for Krüppel (=Kr). The Kr chromosome has also been marked with brown (=bw) so that the clone will also be homozygous for that marker and all the clonally-derived progeny will have brown eyes.

phenotype as those from heterozygous mothers indicates that the phenotype depends on the genotype of the embryo and exclude the role of any dominant maternal effect in the final pattern.[11]

The cross outlined above does not exclude the possibility that Krüppel might play some role in oogenesis in addition to its effect during embryogenesis. Mosaics have been particularly useful in ruling out this possibility. Normally embryos homozygous for Krüppel die at the end of oogenesis. Heterozygotes, however, survive and it is possible to irradiate these individuals as larvae and produce clones in the germline homozygous for the mutation. The experiment is simplified by the introduction of a dominant female sterile mutation (Fs(2)D) on the homologous chromosome in trans with Krüppel (Figure 5). Under these conditions, the mitotic recombination which produces the Krüppel clone also results in the elimination of the female sterile effects. Only flies which have lost Fs(2)D from part of their germline will lay normal eggs; this provided a simple and practical means of identifying 8 mosaic females among the 200 flies in the irradiated population. Since all potentially fertile eggs from these mosaic females were derived from clones homozygous for Krüppel, the normal morphology and development of these eggs indicates that the wild type Krüppel genes play no necessary, irreplaceable function during oogenesis.

REFERENCES

1. E. Wieschaus, J.L. Marsh, and W.J. Gehring, *fs(1)K10:* A germ line-dependent female sterile mutation causing abnormal chorion morphology in *Drosophila melanogaster*, Wilhelm Roux Arch. Entwicklungsmech. Organismen 184:75 (1978).
2. M. Lohs-Schardin, C. Cremer, and C. Nüsslein-Volhard, A fate map for the larval epidermis of *Drosophila melanogaster*: localized cuticle defects following irradiation of the blastoderm with an ultraviolet laser microbeam, Devel. Biol. 73:239 (1979).
3. R.C. King, "Ovarian Development in *Drosophila melanogaster*", Academic Press, New York (1970).
4. K.S. Gill, Developmental genetic studies on oogenesis in *Drosophila melanogaster*, J. Exp. Zool. 152:251 (1963).
5. E. Wieschaus, and J. Szabad, The development and function of the female germ line in *Drosophila melanogaster*: A cell lineage study, Devel. Biol. 68:29 (1979).
6. M. Gans, C. Audit, and M. Masson, Isolation and characterization of sex-linked female sterile mutants in *Drosophila melanogaster*, Genetics 81:683 (1975).
7. E. Wieschaus, C. Audit, and M. Masson, in preparation.
8. Gloor, Schädigungsmuster eines Letalfaktors (Kr) von *Drosophila melanogaster*, Arch. Julius Klaus Stift. Vererbungsforsth. Sozialanthropol Rassenhyg., 25:38 (1950).
9. A. Bull, *Bicaudal,* a genetic factor which affects the polarity of the embryo in *Drosophila melanogaster*, J. Exp. Zool. 161:221 (1966).
10. C. Nüsslein-Volhard, Genetic analysis of pattern formation in the *Drosophila melanogaster* embryo. Characterization of the maternal effect mutant bicaudal. Wilhem Roux Arch. Entwicklungsmech. Organismen 183:249 (1977)
11. C. Nüsslein-Volhard, and E. Wieschaus, in preparation.

DEVELOPMENTAL ANALYSIS OF *fs(1)1867*, AN EGG RESORPTION MUTATION OF

DROSOPHILA MELANOGASTER

János Szabad and János Szidonya

Institute of Genetics
Biological Research Center
Hungarian Academy of Sciences
Szeged, Hungary

INTRODUCTION

The genetic dissection of reproduction started several years ago with the characterization of mutations which interefere with fertility (Gans et al., 1975; Postlethwait et al., 1976; Khipple and King, 1976; Landers et al., 1976; Rizzo and King, 1977; Bownes and Hames, 1978; King et al., 1978; Postlethwait and Handler, 1978; for reviews see King, 1970; King and Mohler, 1975). Among these the female sterile mutations identify genes which are responsible for the synthesis of molecules produced and used in oogenesis and embryogenesis. Special attention is paid to the maternal effect lethal (*mel*) mutations which cause failure of development in eggs laid by homozygous females (Rice and Garen, 1975; MacMorris Swanson and Poodry, 1976; Mohler, 1977; Fausto-Sterling et al., 1977; for review see King and Mohler, 1975). These eggs are defective in some compound which becomes part of the ooplasm during oogenesis and reveal genes which are determinative for the development of the embryo.

A number of mutations have been characterized which interfere with the development and function of those organs which play a role in the egg production of *Drosophila* females (King, 1970; King and Mohler, 1975; Postlethwait and Handler, 1978). Mutations in genes specific to oogenesis have probably no influence on viability and are without apparent defect in adult structure. Mutations in genes involved in functions other than oogenesis can also result in female sterility. Some female sterile mutations certainly have pleiotropic effects on the morphology of adult structures (King, 1970).

The genes essential to oogenesis can be expressed in different organs and tissues of the body. There are known female sterile mutations which affect genes active in the hormone-producing corpus allatum (*ap4*, Postlethwait and Weiser, 1973; Postlethwait et al., 1976; Postlethwait and Handler, 1978) and in the vitellogenin synthesizing cells of the fat body (Bownes and Hames, 1978; Postlethwait and Handler, 1978). Many genes are expressed in the ovary itself (King, 1970; King and Mohler, 1975; Postlethwait and Handler, 1978). Mutations in these genes might bring about female sterility by causing a defect in the differentiation of ovarioles (Bakken, 1973), in cystocyte divisions, nurse cell function (King, 1970), follicle cell development (Rizzo and King, 1977) or uptake of vitellogenin (Bownes and Hames, 1978), etc. For reviews see King (1970) and King and Mohler (1975).

The organ in which the mutated gene will be expressed (inside or outside the ovary) can be determined by a reciprocal transplantation of ovaries between the sterile and wild type flies (King and Bodenstein, 1965; Postlethwait and Handler, 1978). Whether the mutant phenotype depends on the genotype of the germ line or the follicular cells can be established by the analysis of the development of mosaic egg chambers where mutant germ line cells are surrounded by wild-type follicular cells (Illmensee, 1973; Marsh and Wieschaus, 1977; Marsh et al., 1977; Wieschaus et al., 1978; Regenass and Bernhard, 1978; Wieschaus and Szabad, 1979).

Among the EMS-induced X-linked recessive mutations which showed delayed development we found *fs(1)1867* which has proved to be female sterile. The *fs(1)1867* homozygous females do not lay eggs. Egg chamber development proceeds through stage 10 and thereafter the chambers become resorbed. *fs(1)1867* has pleiotropic effects on development as well as on bristle morphology. This report describes the characterization of the *fs(1)1867* mutation:
  -its effects on development and on bristle morphology
  -identification of the tissue in which the *1867+* gene is expressed

RESULTS

fs(1)1867 (abbreviated *fs1867*) is an EMS-induced X-linked recessive mutation. It maps to 62.9 (0.4 to the right of *car*). *fs1867* affects development, bristle morphology and causes female sterility. These phenotypes could be the result of two independent mutations which have not been separated to date. There is no sign of a deletion in the salivary gland chromosome. The expressivity and penetrance are 100%. The viability of the *fs1867* individuals is good.

Development and Bristle Morphology

The body size of the mutant flies is almost normal. The coloration is lighter than in wild type. Bristles are Minute-like: they are shorter and finer than wild type and are often missing. The bristle phenotype is best expressed on the scutellum.

The development of the mutant individuals and the autonomy of the bristle phenotype was tested by crossing y w fs(1)1867/Binsn females with In(1)w^{vC}/Y.y^+ males (for description of the mutations and chromosomes see Lindsley and Grell, 1968). Flies bearing the Binsn balancer chromosome and the y w fs(1)1867/In(1)w^{vC} females eclosed 10 days after oviposition (Figure 1). The development of the y w fs(1)1867 males took about 14 days (25°C); i.e. there is a

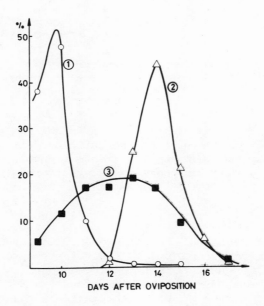

Figure 1: The time course of eclosion. Curves 1-3 show the 3 classes of flies eclosed from the cross: y w fs(1)1867/Binsn x In(1)w^{vC}/Y.y^+. (1) represents all Binsn-bearing flies plus y w fs(1)1867/In(1)w^{vC} females. (2) = y w fs(1)1867 males. (3) = y w fs(1)1867 gynandromorphs. (Based on 1939 flies).

4 day delay in the development of the fs1867 individuals. fs1867 gynandromorphs were generated among the y w fs(1)1867/In(1)w^{vC} zygotes by elimination of the unstable X chromosome In(1)w^{vC} for a review see Janning, 1979). The first gynandromorphs eclosed on the 10th day and the last ones 17 days after oviposition (Figure 1). In general the cuticle of the early gynandromorphs was mostly female, nonmutant as compared to the late ones whose cuticle was largely male (Figure 2). If we consider those gynandromorphs which

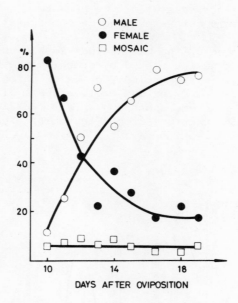

Figure 2: The contribution (in %) of mutant (male), non mutant (female) and mosaic structures to the cuticle of the y w fs(1)1867 gynandromorphs which eclosed on different days after oviposition. (Based on 121 gynandromorphs with 22 structures scored on each of them.)

eclosed on the same day, the different cuticular regions originating from different imaginal discs or histoblasts were mutant and nonmutant in equal frequency. This result suggests that there is no "focus" which would be responsible for the rate of development of the mutant individuals. Rather the ratio of the mutant and nonmutant tissues seems to be in correlation with the time of development. Whether competition occurs between mutant and nonmutant cells within compartments of a mosaic primordia (Ferrus and Garcia-Bellido 1977; Simpson, 1979) has not yet been decided.

Female Sterility

The $fs(1)1867$ homozygous females never lay eggs. The fertility of the males is unaffected. Both internal and external genitalia of the females are normal. The number of ovarioles does not differ significantly between $fs1867$ homo- and heterozygous individuals ($P>0.01$; 13.0 ± 1.7 versus 14.0 ± 2.3 ovarioles per ovary, n = 26). Egg chamber maturation proceeds normally up to stage 10 (according to King, 1970), when it is terminated. Egg chambers subsequently degenerate and become resorbed. Residual material is left at the terminals of the ovarioles.

Transplantation of imaginal ovaries. To check if the *fs1867* mutation affects the ovary or other internal tissues of the body, reciprocal transplantations of mutant and wildtype ovaries of young females (less than 2 hours after eclosion) were carried out (King and Bodenstein, 1965; Postlethwait and Handler, 1978; Bownes and Hames, 1978). The ovaries of the freshly eclosed females contain only immature egg chambers. In 2 to 3 days the implanted ovaries contain matured follicles indicating that the host environment can support egg development in the implanted ovary. The results of the imaginal ovary transplantations are summarized in Table 1. Donor ovaries were homozygous for the egg shape marker mutation *fs(1)K10* (abbreviated *K10*), which does not interfere with egg maturation (Wieschaus et al., 1978; Wieschaus and Szabad, 1979) and *white* (*w*), a marker of the eyes and the Malpighian tubules (Lindsley and Grell, 1968).

Table 1. Development of Egg Chambers in the Implanted Ovaries After Transplantation of Imaginal Ovaries.

Donor	HOST			
	+	n	*fs(1)1867*	n
+	Yes	9	Yes	5
fs(1)1867	No	53	No	5
	Yes[a]	13		

[a] After $\geq$ 5 days.

When *K10* ovaries were transplanted into the abdomen of young wild-type females, egg maturation was complete after 3 days. There were about 50 mature eggs observed in each of the implanted ovaries. The *fs1867* hosts could support development of the *K10* ovaries just as well as the wild-type milieu. However, the host *fs1867* ovaries did not develop; they showed the egg resorption phenotype. When *fs1867* ovaries were transplanted into wild-type flies, no egg maturation could be observed within 4 days (28 cases). However, when the implanted *fs1867* ovaries were cultured in the wild-type abdomen for $\geq$ 5 days in 13 out of 38 ovaries a few (2.2 ±1.7 per ovary) egg chambers reached stages 12-14. The internal structuring of these egg chambers was abnormal. These results indicate that the block in the reproduction cycle of the *fs1867* females is in the ovary.

Transplantation of larval ovaries. We wished to exclude the possibility that the failure of egg production of the *fs1867* females is due to malfunction of the oviducts or other derivatives of the genital discs (Shüpbach et al.,1978a). For this reason reciprocal transplantations of ovaries of $fs1867$ mutant and nonmutant larvae were performed (Clancy and Beadle, 1937; Garen and Gehring, 1972). The $fs1867$ chromosomes were marked by $K10$ and w. The $fs1867^+$ individuals were homozygous for the mutation mal, a marker mutation which allows histochemical identification of the mal tissues (Janning, 1976). The results of the larval ovary transplantations are summarized in Table 2. There were 3 kinds of transplantations done. (1) When $K10$ w ovaries were transplanted into mal host larvae, 9 females (out of 13) laid eggs of two types: $K10$ abnormally shaped eggs originating from the implanted donor and normal eggs from the host ovary. (2) The mal females which received $K10$ w $fs1867$ ovaries laid only host (mal) eggs, even though the $K10$ w $fs1867$ ovaries were attached to the oviduct (Figure 3). The $fs1867$ ovaries invariably showed the egg resorption phenotype. (3) Ovaries of mal larvae were transplanted into $K10$ w $fs1867$ host larvae. Each of the 4 females (out of five) where the donor ovary was attached to the $fs1867$ oviduct laid many eggs originating from the donor ovary but none from the host ovary. These results indicate that the $fs1867$ oviduct is able to support egg deposition, and confirm the conclusions of the imaginal ovary transplantations i.e. that the reason for the malfunction of egg production is an autonomous feature of the $fs1867$ homozygous ovaries.

Table 2. Egg Production of Females after Transplantation of Larval Ovaries

Donor	Host	n^a	Origin of Eggs	
			Donor	Host
+	+	9	Yes	Yes
fs(1)1867	+	4	No	Yes
+	fs(1)1867	4	Yes	No

[a] Cases when the implanted ovary was attached to the host oviduct.

Is the egg resorption phenotype germ line or follicular cell dependent? This question can be answered by an analysis of mosaic egg chambers in which mutant germ line cells are surrounded by nonmutant follicular cells. If an egg can develop from such a

Figure 3: Host (*mal*, non-staining) and implanted *fs(1)1867* (staining) ovaries. Transplantation was performed at the larval stage. One host and the implanted ovary are attached to the oviduct.

mosaic egg chamber it means, most probably, that the genotype of the germ line cells is irrelevant with respect to egg development; and, consequently, the mutant phenotype is follicular cell dependent. If, on the other hand, there are no eggs originating from the mosaic egg chambers, this is an indication that the mutant phenotype is germ line dependent. Mosaic egg chambers can originate after the transplantation of pole cells (progenitors of the germ line) between mutant and nonmutant embryos (Illmensee, 1973). An alternative method has been published recently (Wieschaus and Szabad, 1979); by the induction of mitotic recombination, cells homozygous for the female sterile mutation can be generated in the heterozygous germ line. These homozygous germ line cells will be surrounded by heterozygous follicular cells during oogenesis (King, 1970) and thus a mosaic egg chamber develops.

For the production of mosaic egg chambers in which the *fs1867* homozygous oocyte and nurse cells are surrounded by nonmutant follicular cells, mitotic recombination was induced by X-rays in *fs(1)K10 fs(1)1867/mal* larvae. *K10* (1-0.5) and *mal* (1-64.8) were used as germ line marker mutations; eggs originating from *K10* homozygous cells have abnormal chorion morphology (Wieschaus et al., 1978; Wieschaus and Szabad, 1979). *mal* homozygous cells lack aldehyde oxidase activity which allows their histochemical identification (Janning, 1976; Illmensee, 1973; Schüpbach et al., 1978b; Wieschaus and Szabad, 1979). As a control, mitotic recombination was induced in *K10/mal* larvae (Wieschaus and Szabad, 1979) by a 1000r of X-rays 48 hours after oviposition.

Upon irradiation mitotic recombination may take place (Figure 4). If the site of the mitotic recombination is in the proximal

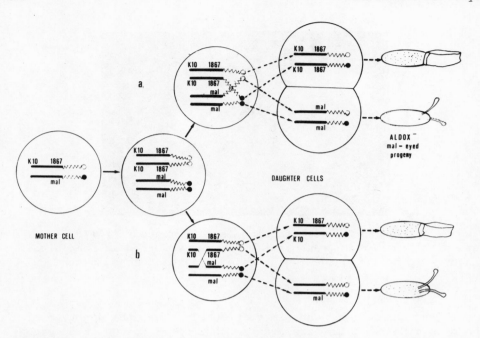

Figure 4: Schematic representation of induction of mosaicism in the germ line.
a) Mitotic recombination in the proximal heterochromatin results in fs(1)K10 fs(1)1867 and mal homozygous cells.
b) Mitotic recombination distal to mal results in fs(1)K10 homozyogous cells only. For details see text.

heterochromatin, a twin spot composed of K10 fs1867 and mal homozygous cells is generated (Figure 4a). If, on the other hand, the mitotic recombination takes place distal to mal, only K10 homozygous cells are produced (Figure 4b). The site of mitotic recombination can thus be determined by the presence or absence of the mal homozygous cysts in the ovarioles. The mosaicism of the germ line of those females which were irradiated as larvae was tested according to Wieschaus and Szabad (1979). As an indication of mitotic recombination in the germ line, the mosaic females laid K10 eggs. Ovarioles of these females were subsequently stained for aldehyde oxidase activity for the detection of mal cysts. The results of the mosaic analysis are shown in Table 3. K10 eggs producing mosaic females were recovered with the same frequency in the control as in the experimental series (23 mosaics were found among the 115 control and 15 among the 61 females which were heterozygous for the mutation fs1867). Six of the 10 K10 egg-producing control mosaic females had mal cysts. Out of 16 K10 egg-producing fs1867 heterozygous females, 11 had mal cysts in the ovarioles. These results clearly indicate that mitotic recombination took place in the proximal heterochroma-

Table 3. Germ Line Mosaicism in fs(1)K10 fs(1)1867/mal Heterozygous Females after Induction of Mitotic Recombination[a]

Genotype	K10-Mosaicism		mal-Mosaicism	
	Tested	Mosaic	Tested	Mosaic
fs(1)K10/mal[b]	115	23	10	6
fs(1)K10 fs(1)1867/mal	61	15	16	11

[a] 1000r; 50 kV, 1 mm Al filter, 330 r/min, 48 hours after oviposition.

[b] Data from Wieschaus and Szabad (1979).

tin in these 11 mosaic females, and thus the K10 eggs laid by these females had to be homozygous for fs1867. Consequently, egg chambers can develop if their germ line constituents are fs1867 homozygous and the follicular cells are normal in function, i.e. the fs1867 egg resorption phenotype is follicular cell dependent.

Mitotic recombination was also induced in young K10 fs1867/mal embryos. Of the 12 K10 producing mosaic females, 9 had mal homozygous cysts in their ovarioles. The frequency with which mal cysts bearing mosaic females were recovered (6/10; 11/16; 9/12) is as expected. It had been shown earlier that about 70% of the mitotic recombinations of the X chromosome take place in the proximal heterochromatin (Garcia-Bellido, 1972; Becker, 1977; Wieschaus and Szabad, 1979).

Histology

Egg chamber development proceeds normally in the fs1867 ovaries up to stage 10 (King, 1970). Typical oocytes and nurse cells develop. The deposition of yolk material into the cytoplasm of the oocyte can also be seen. Early follicles are covered by typical follicular cells. Follicular cells retract from the nurse cells and cover only the oocyte at the correct stage. The first signs of abnormalities are observed during the shape transition of the follicular cells which surround the oocyte; the formerly cuboidal follicular cells become columnar in shape in the case of the wild-type egg chambers. This step does not occur in the fs1867 egg chambers. Egg chamber disintegration is succeeded by the formation of agglomerates in the oocyte cytoplasm. This is followed by pycnotic processes which involve both nurse and follicular cells. As a result of egg resorption some residual material is left at the terminals of the ovarioles.

DISCUSSION

fs(1)1867 is a recessive mutation which influences developmental rate and bristle morphology and causes female sterility. The slow development and the bristle morphology phenotypes behave autonomously in gynandromorphs. There is no sign that the slow development is associated with any region ("focus") which would govern the rate of development. Rather the relative amount of the mutant tissues has determinative importance.

A focus could be found for the female sterility phenotype. It is located in the ovary, as shown by reciprocal transplantations of mutant and wild-type ovaries. That a few egg chambers of the mutant ovaries could develop in the wild-type abdomen might reflect some nonautonomy of the egg resorption phenotype; diffusible materials are produced in the wild-type host ovaries (in the oocyte nurse cells and/or in the follicular cells) which can rescue the mutant phenotype (Tokunaga, 1972; Garcia-Bellido and Merriam, 1973). In general, in the course of ovary transplantations, the mutant ovaries can develop in the wild-type milieu. There are two possible meanings of this result: 1) The defect is not in the mutant ovary (Postlethwait and Handler, 1978); 2) The defect is in the mutant ovary, but the phenotype is not cell autonomous. This latter possibility can be tested by the use of host females lacking either the germ line or the follicular cells or both components of the ovaries (Thierry-Mieg et al., 1972; Graziosi and Micali, 1974; Wieschaus and Szabad, 1979). An alternative solution is the transplantation of different wild-type organs into mutant females, expecting in this way to rescue the mutant phenotype.

Mosaic analysis of the egg resorption phenotype suggests that it is not germ line but rather follicular cell dependent; eggs did develop from those egg chambers in which the mutant oocyte and nurse cells were surrounded by nonmutant follicular cells. There are two reservations, however, which we must be aware of.

1. The *fs1867* phenotype might be germ line dependent but nonautonomous. In the mosaic analysis based on mitotic recombination, the majority of the follicles contain normal germ line cells. Diffusible materials from these cells might rescue the mosaic ones and consequently, eggs would develop from egg chambers where only the follicular cells were genotypically normal. The degree of rescue might depend on the site of the mutant and nonmutant tissues. In the case of ovary transplantations, the diffusible materials have to pass through the membranes covering both the host and the mutant ovarioles which may interfere with the rescue of the mutant phenotype. In the case of mosaic analysis some egg chambers contain mutant germ line cells; others, nonmutant ones. These line up within the ovarioles (Schüpbach et al., 1978b; Wieschaus and Szabad, 1979), a situation which may be favorable for the rescue of the mutant phenotype.

The role of follicle cells in the manifestation of the mutant phenotype could be directly established by the analysis of mosaic females whose ovaries contain only mutant follicular, and nonmutant germ line cells. Such mosaics can be produced by the transplantation of nonmutant germ line cells into mutant blastoderms. The role of germ line cells could be studied in mosaic females whose ovaries are made of nonmutant follicular and exclusively mutant germ line cells. Such mosaics could be generated by pole cell transplantations where mutant blastoderms serve as donors and blastoderms which lack pole cells (Thierry-Mieg et al., 1972; Wieschaus and Szabad, 1979) as recipients.

2. A perdurance of the 1867^+ gene might also rescue the follicles which contain mutant oocyte and nurse cells surrounded by nonmutant follicle cells. If the product of the 1867^+ gene is present in the cytoplasm of the $fs1867$ heterozygous mother cells at the time of induction of mitotic recombination, and does not dilute out from the $fs1867$ homozygous daughter cells by the time of follicle differentiation, expression of the mutant phenotype might not take place (Garcia-Bellido and Merriam, 1971). To allow enough time for cell proliferation, the $fs1867$ homozygous germ line cells were generated in young embryos, allowing at least 5 cell divisions before final differentiation (Wieschaus and Szabad, 1979). The fact that such cells were still able to support egg development indicates that either (1) the 1867^+ product has long perdurance, or (2) the $fs1867$ phenotype is follicle cell dependent. This latter possibility is also supported by the fact that the first visible sign of egg degeneration is seen in the follicular cells. It has to be mentioned, however, that the abnormal chorion morphology of the $K10$ mutation was originally expected to depend on the genoytpe of the follicle cells which produce the chorion. Despite expectations, $K-10$ is clearly germ line dependent (Wieschaus et al., 1978; Wieschaus and Szabad, 1979). There is most probably an extensive cooperation between the germ line and the follicular cells during oogenesis. Disharmony of this cooperation can lead to female sterility resulting in abnormal egg chamber development or death of the embryos.

SUMMARY

The X-linked 1867^+ gene seems to be a pleiotropic one. Mutation in this gene causes delay in development and abnormal bristle morphology. These phenotypes are expressed autonomously in genetic mosaics. There is no focus for the delay. The female sterility could be localized to the ovary (based on ovary transplantations). It seems that the 1867^+ gene is expressed in the follicular cells at one of the last steps of oogenesis. This is suggested by the results of mosaic analysis based on mitotic recombination. Possible drawbacks of the mitotic recombination type of analyses are also discussed.

REFERENCES

Bakken, A.H., 1973, A cytological and genetic study of oogenesis in *Drosophila melanogaster*, Develop. Biol., 33:100.

Becker, H.J., 1977, Mitotic recombination, in: "The Genetics and Biology of *Drosophila*", Vol. 1C, M. Ashburner and E. Novitski, eds., Academic Press, New York, pp.1019-1086.

Bownes, M., and Hames, B.D., 1978, Genetic analysis of vitellogenesis in *Drosophila melanogaster*. The identification of a temperature sensitive mutation affecting one of the yolk proteins, J. Embryol. exp. Morphol., 47:111.

Clancy, C.W., and Beadle, G.W., 1937, Ovary transplants in *Drosophila melanogaster*. Studies of the characters *singed, fused and female sterile*, Biol. Bull., 72:47.

Fausto-Sterling, A., Weiner, A.J., and Digan, M.E., 1977, Analysis of a newly-isolated temperature sensitive maternal effect mutation of *Drosophila melanogaster*, J. Exp. Zool., 200:199.

Ferrus, A., and Garcia-Bellido, A., 1977, Minute mosaics caused by early chromosome loss, Wilhelm Roux Archives, 183:337.

Gans, M., Audit, C., and Masson, M., 1975, Isolation and characterization of sex-linked female sterile mutants in *Drosophila melanogaster*, Genetics, 81:683.

Garcia-Bellido, A., 1972, Some parameters of mitotic recombination in *Drosophila melanogaster*, Molec. gen. Genet., 115:54.

Garcia-Bellido, A., and Merriam, J.R., 1971, Genetic analysis of cell heredity in imaginal discs of *Drosophila melanogaster*, Proc. Natl. Acad. Sci. USA, 68:2222.

Garen, A., and Gehring, W., 1972, Repair of the lethal developmental defect in *deep orange* embryos of *Drosophila* by injection of normal egg cytoplasm, Proc. Natl. Acad. Sci. USA, 69:2982.

Graziosi, G., and Micali, F., 1974, Differential responses to ultraviolet irradiation of the polar cytoplasm of *Drosophila* eggs. Wilhelm Roux Archives, 175:1.

Illmensee, K., 1973, The potentialities of transplanted early gastrula nuclei of *D. melanogaster*. Production of their imago descendants by germ line transplantations, Wilhelm Roux Archives, 171:331.

Janning, W., 1976, Entwicklungsgenetische Untersuchungen an Gynandern von *Drosophila melanogaster*. IV. Vergleich der morphogenetischen Anlagepläne larvaler und imaginaler Strukturen, Wilhelm Roux Archives, 179:349.

Janning, W., 1979, Gynandromorph fate maps in *Drosophila*, in: "Genetic Mosaics and Cell Differentiation", W. Gehring, ed., Springer-Verlag, New York, pp. 1-28.

Khipple, P., and King, R.C., 1976, Oogenesis in the female sterile /1/11304 mutant of *Drosophila melanogaster* Meigen, Int. J. Insect Morphol. Embryol., 5:127.

King, R.C., 1970, "Ovarian Development in *Drosophila melanogaster*", Academic Press, New York.

King, R.C., and Bodenstein, D., 1965, The transplantation of ovaries between genetically sterile and wild type *Drosophila melanogaster*, Z. Naturforschung.,20b:292.

King, R.C., and Mohler, J.D., 1975, The genetic analysis of oogenesis in *Drosophila melanogaster*, in: "Handbook of Genetics", Vol. 3, R.C. King, ed., Plenum Press, New York, pp. 757-791.

King, R.C., Bahns, M., Horowitz, R., and Larramendi, P., 1978, A mutation that affects female and male germ cells differentially in *Drosophila melanogaster* Meigen, Int. J. Insect Morphol. Embryol., 7:359.

Landers, M.H., Postlethwait, J.H., Handler, A., and White, J., 1976, Isolation and characterization of female sterile mutants in *Drosophila melanogaster*, Genetics, 83:s43.

Lindsley, D.L., and Grell, E.H., 1968, Genetic variations of *Drosophila melanogaster*, Carnegie Inst. Wash., Pub. No.627.

MacMorris Swanson, M., and Poodry, C.A., Ovary-autonomous inheritance at a developmental locus in *Drosophila*, Develop. Biol., 48:205.

Marsh, J.L., and Wieschaus, E., 1977, Germ-line dependence of the maroon-like maternal effect in *Drosophila*, Develop. Biol.,60:396.

Marsh, J.L., van Deusen, E.B., Wieschaus, E., and Gehring, W.J., 1977, Germ line dependence of the *deep orange* maternal effect in *Drosophila*, Develop. Biol., 56:195.

Mohler, J.D., 1977, Developmental genetics of the *Drosophila* egg. I. Identification of 59 sex-linked cistrons with maternal effects on embryonic development, Genetics, 85:259.

Postlethwait, J.H., and Weiser, K., 1973, Vitellogenesis induced by juvenile hormone in the female sterile mutant apterous-four in *Drosophila melanogaster*, Nature New Biol., 244:284.

Postlethwait, J.H., and Handler, A.M., 1978, Nonvitellogenic female sterile mutants and the regulation of vitellogenesis in *Drosophila melanogaster*, Develop. Biol., 67:202.

Postlethwait, J.H., Handler, A.M., and Gray, P.W., 1976, A genetic approach to the study of juvenile hormone control of vitellogenesis in *Drosophila melanogaster*, in: "The Juvenile Hormones", L. Gilbert, ed., Plenum Press, New York, pp. 449-469.

Regenass, U., and Bernhard, H.P., 1978, Analysis of the *Drosophila* maternal effect mutant *mat(3)1* by pole cell transplantation experiments, Molec. gen. Genet., 164:85.

Rice, T.B., and Garen, A., 1975, Localized defects of blastoderm formation in maternal effect mutants of *Drosophila*, Develop. Biol., 43: 227.

Ripoll, P., and Garcia-Bellido, A., 1973, Cell autonomous lethals in *Drosophila melanogaster*, Nature New Biol., 241:15

Rizzo, W.B., and King, R.C., 1978, Oogenesis in the *female sterile (1)42* mutant of *Drosophila melanogaster*, J. Morphol., 152:329.

Schüpbach, T., Wieschaus, E., and Nöthiger, R., 1978a, The primary organization of the genital disc of *Drosophila melanogaster*, Wilhelm Roux Archives, 185:249.

Schüpbach, T., Wieschaus, E., and Nöthiger, R., 1978b, A study of the females' germ line in mosaics of *Drosophila*, Wilhelm Roux Archives, 184:41.

Simpson, P., 1979, Parameters of cell competition in the compartments of the wing disc of *Drosophila*, Develop. Biol., 69:182.

Thierry-Mieg, D., Masson, M., and Gans, M., 1972, Mutant de stérilité a effet retardé de *Drosophila melanogaster*, C.R. Hebd. Seances Acad. Sci. Ser. D Sci. Nat., 275:2751.

Tokunaga, C., 1972, Autonomy or nonautonomy of gene effects in mosaics, Proc. Natl. Acad. Sci. USA, 69:3283.

Wieschaus, E., and Szabad, J., 1979, The development and function of the female germ line in *Drosophila melanogaster*. A cell lineage study, Develop. Biol., 68:29.

Wieschaus, E., Marsh, J.L., and Gehring, W., 1978, *fs(1)K10*, a germ-line dependent female sterile mutation causing abnormal chorion morphology in *Drosophila melanogaster*, Wilhelm Roux Archives, 184:75

DISCUSSION

A. Shearn: Is the mutation localized near any of the known Minute locations?

J. Szabad: There are nine Minute mutations known in this part of the chromosome, but none of these are associated with this mutation.

GENETICS OF PATTERN FORMATION

Peter J. Bryant and Jack R. Girton

Developmental Biology Center
University of California, Irvine
Irvine, California 92717 U.S.A.

INTRODUCTION

Over the last few decades, studies of pattern formation in *Drosophila* imaginal discs have been proceeding along two rather separate lines. One line of research includes the studies initiated by Vogt, Bodenstein and the Hadorn group on the experimental embryology of imaginal discs, where the ability of disc fragments to grow, to regenerate, and to duplicate has been documented (see Bryant, 1978 for review). This has led to a number of theories involving first, gradients of developmental capacity (Bryant, 1971) and second, a polar coordinate system in which any discontinuities are eliminated by intercalation (French, Bryant, and Bryant, 1976).

The other approach to pattern formation in *Drosophila* is one to which this organism is especially well suited. This involves studies of the effects of various mutations on patterns, either in entirely mutant animals or in a variety of genetic mosaic situations. This approach has led to the concept of a prepattern underlying pattern formation (Stern, 1968) and, more recently, to the idea of sequential gene activation coincident with progressive compartmentalization of the growing cell population (Garcia-Bellido, 1975).

The theories arising from these two approaches to pattern formation are quite different from one another, and it is difficult to see how they can be integrated. In this paper we do not propose a resolution to this problem, but we do describe recent work on some mutants which promises to bring the two approaches closer together.

SOME GENETIC ABNORMALITIES OF PATTERNS

We have chosen to concentrate our efforts on a selected group of mutations: those causing the types of abnormalities which might be expected according to some of the recent theories on pattern regulation in imaginal discs. The abnormalities are as follows:

(1) Pattern deficiencies. These are, of course, predicted by practically any theory of pattern formation, provided it is assumed that they are produced by degeneration and that this degeneration occurs so late in development that it cannot be compensated for by pattern regulation. Production of a partial pattern *ab initio*, without degeneration, is more difficult to explain according to most theories.

(2) Pattern deficiency/duplications. This refers to situations where part of an imaginal disc derivative is missing, as in the deficiencies, but where the remaining derivatives are duplicated with mirror-image symmetry. It is useful to call the part of the pattern with normal symmetry orthodromic, and the part with reversed symmetry antidromic (Figure 1a and b; Jurgens and Gateff, 1979). In some cases, mutations producing this phenotype can also produce simple pattern deficiencies, but this is not always the case. The deficiency/duplication phenotype is easy to understand for mutations which

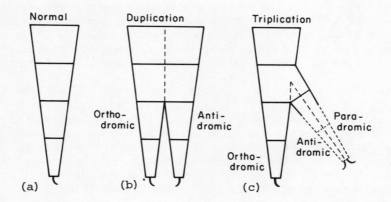

Figure 1: The morphology of abnormal legs in *Drosophila* mutants (Jurgens and Gateff, 1979). (a) Normal leg. (b) Deficiency/duplicaton. The part with normal asymmetry is termed orthodromic, whereas its partner, with reversed asymmetry, is termed antidromic. (c) Triplication. This is essentially a normal leg (the orthodromic part) carrying a symmetrical outgrowth (the anti- and paradromic parts).

remove part of the imaginal disc through degeneration, since surgical removal of parts of discs can also produce fragments which duplicate if given an opportunity for growth (see Bryant, 1978). Thus, any theory which can account for pattern duplication in imaginal disc fragments, such as the polar coordinate model (French et al., 1976) can also account for the pattern abnormalities caused by these mutations if it is assumed that extensive degeneration occurs so early in development that pattern regulation can occur in response to it.

(3) Pattern triplications. An abnormality of this type is equivalent to a completely normal imaginal disc derivative bearing an outgrowth which is itself mirror-symmetrical (Figure 1c). The part of the outgrowth which has reversed symmetry is called antidromic, whereas the part with normal symmetry is called paradromic (Jurgens and Gateff, 1979). Such defects have occasionally been produced surgically by cutting leg discs *in situ* (Bryant, 1971), but we have recently discovered that they can be more easily produced at a high frequency by applying pulses of high temperature to larvae carrying the temperature-sensitive cell-lethal mutation *l(1)ts-726* (J.R. Girton and P.J. Bryant, unpublished). In the latter experiment it is assumed, as with the deficiency/duplications, that the primary defect is local degeneration and that the symmetrical outgrowth is produced in response to the degeneration.

THEORY OF PATTERN FORMATION

The polar coordinate model for pattern formation (French et al., 1976) does not account directly for the symmetrical outgrowths produced by the temperature-sensitive cell lethal described above, nor does it account for the very similar symmetrical outgrowths produced from surgically-created symmetrical bases in cockroaches and amphibians (Bohn, 1965, 1972; S.V. Bryant, 1976; Bryant and Baca, 1978; French, 1978; Tank, 1978). This failure of the basic model led to the following revision of the part of it which deals with distal outgrowth (S.V. Bryant, P.J. Bryant and V. French, unpublished).

In an asymmetrical epimorphic field such as an amphibian or cockroach leg, or a lateral imaginal disc of *Drosophila*, when either distal or proximal parts of the pattern or presumptive pattern are removed, subsequent growth usually generates the parts of the pattern which are normally distal to the cut edge. When the piece from which this process occurs is a proximal piece, such as a peripheral ring from an imaginal disc, the result is regeneration of the missing distal parts. When the piece undergoing this process is a presumptive distal piece such as the central fragment of an imaginal disc, the result is usually duplication of the existing distal parts (see, for example, Bryant, 1975). In order to emphasize that this kind of pattern formation always generates more distal elements, it has been termed distal transformation (Rose, 1962). However, since the addition of distal parts appears to depend on addition of new cells rather than alteration of old ones, we call the phenomenon *distal outgrowth*.

We propose that the basic cell interaction leading to distal outgrowth is between cells having different circumferential positional values which come together as the amputation site heals (Figure 2a). This leads to circumferential intercalation according to the shortest intercalation rule (French et al., 1976), and if a complete circle of positional values was present at the amputation site, a new complete circle will be generated by this mechanism. An important point to recognize is that although a specific kind of wound-healing is used as an example in Figure 2a, the same outcome is predicted by practically any other set of circumferential confrontations that might occur. Any kind of wound closure must involve cell displacements that will lead to the generation of more complete circles.

We now propose that the new cells generated during circumferential intercalation at the growing tip of the appendage must adopt positional values which are more distal than the preexisting ones at the wound edge. Furthermore, we propose that this comes about as a result of strictly local interactions. If a cell which is produced during circumferential intercalation is assigned a circumferential positional value identical to that of an adjacent cell, we suggest that the new cell must adopt a radial positional value which is more distal. We will call this process *distalization*. Repeated rounds of circumferential intercalation with distalization, with some provision for stopping at the distal tip, will give an outgrowth which is both circumferentially and distally complete.

For symmetrical epimorphic fields such as amphibian tails, or in cases where growth occurs from a base which is made symmetrical by surgical means such as in double-half limbs in amphibians or cockroaches, the above model predicts that distalization will occur from the symmetrical partial circumferences, but that the extent of distalization will depend upon the orderliness and direction of wound-healing at the amputation site, on the number of different positional values present at the cut edge, and on the number of radial values to be replaced.

Figure 2b shows how the extent of distalization from symmetrical partial circumferences would depend upon the mode of wound-healing at the amputation site. With mode 1 healing there are no positional-value confrontations, and no growth or distalization is expected. The opposite extreme is mode 3, where the maximum degree of positional-value confrontation occurs. Circumferential intercalation with distalization gives a tapered symmetrical outgrowth which is distally incomplete in this case. We have arbitrarily adopted a model with twelve circumferential and five radial positional values, and models with different ratios of the two would predict different degrees of distalization. If, for example, there were only three radial positional values, then the mode 3 example in Figure 2b would be distally complete. An intermediate type of wound-healing is mode 2, which again produces tapered symmetrical outgrowths, but in this case, they are less distally complete than with mode 3.

Another variable which is expected to affect the extent of distalization from a symmetrical circumference is whether there is more or less than half of a complete circumference of positional values on either side of the line of symmetry. In cases where each side of the symmetrical circumference contains less than half of a complete circumference, as in the examples in Figure 2b, shortest intercalation between extreme values could never lead to generation of new positional values but only to duplication of existing ones. On the other hand, in cases where there is more than half of a complete set of positional values on each side of the line of symmetry, intercalation by the shortest route between extreme values would generate new positional values, as shown in Figure 2c. The two symmetrical complete circles which are thus produced are expected to lead to the generation of two complete distal outgrowths which are mirror-images of each other. At the point where the two complete circles are established, the appendage is expected to branch.

In general, the degree of distal outgrowth will depend upon the number of positional values present in each half of the double-half circumference, relative to the number of positional values in the proximal-distal sequence. The distalization rule makes the "complete circle rule" (French et al., 1976) dispensable, since the differences in degree of distalization under various circumstances now follow simply from the geometrical properties of complete circles compared with double partial ones.

THE DEFICIENCY/DUPLICATION PHENOTYPE

Many imaginal disc fragments produce mirror-image pattern duplications when they are given the opportunity for growth by culturing them in adult flies before transfer to larvae for metamorphosis (see Bryant, 1978). This duplication, as well as the regeneration which occurs in cultured complementary fragments, can be understood as a result of intercalation between different parts of the presumptive pattern which come together during wound healing and behave according to the shortest intercalation rule (French et al., 1976). In general, for duplication to occur the fragment must contain less than half of the circumferential positional values, even though, at least in the case of the leg disc, this need not correspond to less than half of the disc in physical terms (see Strub, 1977a).

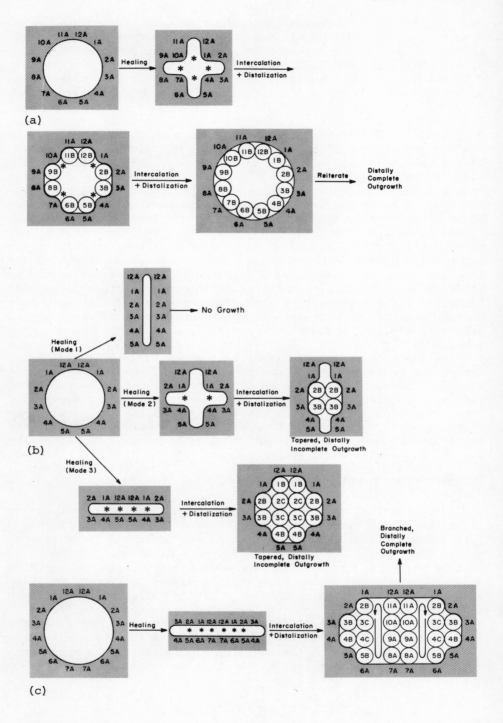

Figure 2: Model of distal outgrowth in asymmetrical and symmetrical epimorphic fields (S.V. Bryant, P.J. Bryant and V. French, unpublished). (a) The tissue remaining after removal of levels B, C, D and E of the pattern is shaded and the wound edge is outlined by the circle. This diagram could represent an imaginal disc with the center removed or the stump of an amphibian or cockroach leg after amputation of the terminal parts. It is proposed that during the process of wound healing, different parts of the circumference come into contact, and the second diagram shows one way in which this might occur. Circumferential intercalation (*) produces cells with circumferential positional values identical to those of pre-existing adjacent cells, so the new ones are forced to the next most distal level B (distalization). Subsequent intercalation completes the B level, and reiteration of the whole process generates the remaining distal levels. The process is essentially independent of variations in the directions of wound healing. (b) When the starting configuration is symmetrical, as in a double-half limb, the outcome depends on the kind of wound-healing which occurs. Mode 1 healing gives no positional-value confrontations and no distal outgrowth, whereas modes 2 and 3 give limited distalization corresponding to symmetrical and distally incomplete outgrowths. Mode 3 gives a more complete outgrowth than mode 2. (c) When the starting configuration consists of two symmetrical copies of more than half of the circumference, the shortest intercalation rule (French et al., 1976) predicts that certain kinds of wound healing (mode 3 of Figure 2b) will lead to the production of two symmetrical complete circles. This will give rise to a pair of diverging, branched, distally complete outgrowths.

The mutations *scalloped^{UCI}*, *vestigial*, *wingless* and *tetraltera* produce, in some animals, a deficiency/duplication phenotype in the wing disc derivatives. In most cases the wing itself is missing and the remaining notum or parts of it are duplicated with mirror-image symmetry (Figure 3b). The position of the line of symmetry relative to the pattern is variable in all of these mutants, and parts of the wing are sometimes included (A.A. James and P.J. Bryant, unpublished). In *scalloped^{UCI}* and *vestigial*, a range of different phenotypes is produced including a variety of simple deficiencies of the wing margin (Figure 3a), but only the deficiency/duplication phenotype is seen in *wingless* and *tetraltera*.

We have recently confirmed Fristrom's (1968) finding for *vestigial*, and shown for *scalloped^{UCI}*, that extensive degeneration occurs in the presumptive wing region of the wing disc during the third larval instar, thus accounting for the deficiency phenotype (Figure 4; A.A. James and P.J. Bryant, unpublished). We assume that more extensive degeneration accounts for the deficiency/duplication phenotype of the mutants. A study of the extent of the pattern deficiencies showed that a certain threshold level of deficiency had to be present in order for duplication to occur. This threshold level corresponds to about half of the circumference of the disc, which is predicted by the theory and which corresponds to the amount of the disc which has to be removed in the surgical experiments to give a fragment which will duplicate rather than regenerate. Presumably in the mutants when the deficiency is sub-threshold, any attempt to regenerate is counteracted by degeneration of that part of the presumptive pattern.

The deficiency/duplication phenotypes of *wingless* and *tetraltera* are also consistent with the known regulative behavior of the wing disc in that the part of the pattern which is present would be ex-

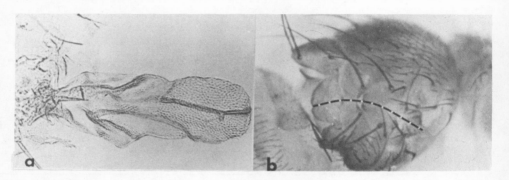

Figure 3: (From A.A. James and P.J. Bryant, unpublished).
(a) Deficiency of the wing margin produced by the *scalloped^{UCI}* mutation.
(b) Deficiency/duplication phenotype produced by the *scalloped^{UCI}* mutation. The wing and part of the notum are missing, and the remaining part of the notum is duplicated with mirror-image symmetry. Dashed line shows the position of the line of symmetry.

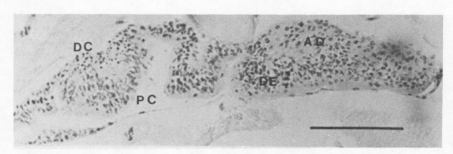

Figure 4: Localized cell death in a wing disc from *scalloped^{UCI}* (A.A. James and P.J. Bryant, unpublished). AD, adepithelial cells; DC, dead cells; PC, peripodial cavity. Bar represents 50 μm.

pected to be produced by a duplicating fragment. However, we have not detected excessive cell death during the third larval instar of these mutants and we have to conclude that if it occurs, it occurs only during early developmental stages. The fact that simple deficient phenotypes are not seen could be accounted for by assuming that the early degeneration would be followed either by complete duplication (giving the deficiency/duplication phenotype) or complete regeneration (giving a normal phenotype) depending on the size and location of the degeneration. However, a clonal analysis of *wingless* by Morata and Lawrence (1977) provided evidence against a degeneration mechanism for this mutant. More work needs to be done on *tetraltera* in this respect.

The deficiency/duplication phenotypes in leg-disc derivatives produced by the temperature-sensitive cell-lethal mutation *l(1)ts-726* have been studied in a similar way. Here it is known that the phenotype is a secondary response to degeneration in the leg disc of heat-treated larvae, since degeneration has been documented histologically (Clark and Russell, 1977) and in a mutant/non-mutant genetic mosaic, the antidromic part of the duplication can include non-mutant tissue (Russell, 1974). The deficiency/duplication phenotypes were shown to be consistent with the known regulative behavior of leg disc fragments, except that some duplications were found associated with surprisingly small deficiencies. This led Russell et al., (1977) to favor a radiating-gradient model for the leg disc, in which the high point of the gradient was located at the medial edge of the disc, in the deficient area associated with the largest duplications. A radical distortion of positional-value spacing would be necessary to bring these results into line with the predictions of the polar coordinate model.

DUPLICATION AND DISTAL OUTGROWTH IN IMAGINAL DISCS

Small fragments taken from the edge of the wing or leg imaginal disc generally duplicate during culture (see Bryant, 1978). Since such a fragment contains presumptive proximal parts of the pattern but is missing distal structures, it can be considered as potentially the base of a symmetrical distal outgrowth as in Figure 2, provided the appropriate wound-healing and cell interactions can occur. Conceivably, such distal outgrowth could occur before, during or after the circumferential duplication which occurs in these fragments. In the case of the wing disc, many presumptive proximal fragments (actually "edge pieces" obtained by a straight cut across the disc) underwent "conservative" duplication without showing distalization (Bryant, 1975). However, some of these fragments, though not others, regenerated a variable number of the more distal (and other proximal) structures during a more extended adult culture period (Duranceau, 1977; B.Kirby, personal communication; J. Karlsson, personal communication). The distal regeneration shown by these pieces may be analogous to the similarly variable distal regeneration shown by double half-limbs in amphibians and cockroaches. However, the problem has been studied directly by Schubiger (1971) and Strub (1977a,b) and Schubiger and Schubiger (1978) using the male foreleg disc. These authors studied the behavior of various sectors of the presumptive proximal part of the leg disc during adult culture and showed that even when these fragments underwent duplication (and by implication carried only a minority of the circumferential positional values), they nevertheless could regenerate distal structures at certain frequencies. Upper lateral or lower lateral ¼ proximal sectors showed no distal regeneration in Strub's (1977a,b) studies, but 16% of lower medial quarters regenerated claws (the most distal pattern element). When the starting fragment comprised half of the disc, claws were regenerated in 19% (lateral half) or 25% (lower half) of cases (Schubiger and Schubiger, 1978). With 3/4 of the circumference present, 87% of fragments produced claws (Schubiger and Schubiger, 1978). Since these fragments all duplicate their patterns, they are assumed to carry less than half of the circumferential positional values in both ortho- and antidromic halves, so the regenerates provide evidence for distal outgrowth from incomplete circles. Furthermore, the series of results taken together suggests that the probability and completeness of distal outgrowth is a function of the proportion of the circumference present in the presumptive proximal tissue. In addition, Schubiger and Schubiger (1978) found that both anterior and posterior compartments of the disc (Garcia-Bellido et al., 1973) must be present for distal outgrowth to occur, as also seems to be the case in the wing disc (J. Karlsson and M. Wilcox, personal communication). Distal outgrowth in wing-disc fragments can occur before circumferential duplication begins (J. Karlsson and M. Wilcox, personal communication), a result which is consistent with the view of distal outgrowth presented here.

Some further information on pattern regulation in the leg disc comes from surgical operations performed *in situ* on developing larvae (Bryant, 1971). Complete bisection of the leg disc gave one regenerating and one duplicating fragment as in the transplantation studies (Schubiger, 1971). However, these operations sometimes failed to completely sever the disc, and pattern triplications were obtained in several cases. Such a triplication consisted of a normal, asymmetrical and complete leg in normal orientation (the orthodromic part) together with a symmetrical outgrowth (the anti- and paradromic parts)(Figure 1c). These results can be interpreted according to the distalization model described earlier. An incision in the leg disc which is closed at both ends would have a wound surface which is symmetrical with respect to circumferential positional information, and which therefore could produce a symmetrical regenerate (Figure 5). Whether or not the regenerates were distally complete would depend on how much of the presumptive circumference was present at the wound site and on the speed and directions of wound healing before and during growth.

PATTERN FORMATION

Triplication of wing-disc derivatives with symmetry relationships similar to those in the leg have been produced by surgical operations or irradiation in *Drosophila* (Bryant, 1971; Postlethwait, 1975) and by surgical operations in *Ephestia* (Rahn, 1972).

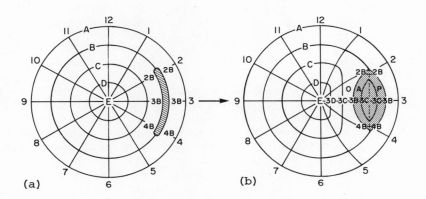

Figure 5: Diagram to show how a cut in an imaginal disc (a;shaded) could lead to a symmetrical, distally incomplete outgrowth (b; shaded) by the distalization process described in this paper. O, orthodromic; A, antidromic; P, paradromic sections of the pattern.

DUPLICATION AND DISTAL OUTGROWTH IN TEMPERATURE-SENSITIVE CELL-LETHAL MUTANTS

Several workers (Russell, 1974; Arking, 1975; Simpson and Schneiderman, 1975) have recently discovered that pattern duplications can be produced at high frequencies by applying pulses of high (restrictive) temperature to developing larvae of strains carrying mutations which cause cell death at the restrictive temperature. Two mutations of the *su(f)* locus have been studied in detail: *l(1)ts-726* (Russell, 1974; Russell et al., 1977; Postlethwait, 1978) and *l(1)mad^{ts}* (Jurgens and Gateff, 1979). Although direct evidence for temperature-induced cell death has been obtained only for *l(1)ts-726* (Russell, 1974; Clark and Russell, 1977), the results obtained with the two alleles are so similar that it seems likely that they cause similar primary defects. The most common result of an early restrictive-temperature pulse is leg duplication, where the abnormal leg comprises a bidorsally (Russell et al., 1977; Jurgens and Gateff, 1979) or bilaterally (Postlethwait, 1978) symmetrical partial leg with the ventral or medial structures, respectively, missing. That the orthodromic part of the pattern represents the surviving tissue after degeneration in the original leg disc, whereas the antidromic part has grown out from the orthodromic part by epimorphic duplication, seems highly likely from histological work (Russell, 1974; Clark and Russell, 1977) as well as genetic mosaic analysis (Girton and Russell, 1980). That being the case, the patterns found provide evidence about the circumferential requirement for distal outgrowth. In particular, Postlethwait (1978) showed that the antidromic part was distally incomplete in all cases where it was circumferentially incomplete (and in some cases where it was circumferentially complete) at the proximal level, thus supporting the idea of a complete circle requirement for complete distal outgrowth.

TRIPLICATIONS IN TEMPERATURE-SENSITIVE CELL-LETHAL MUTANTS

Another abnormality produced mainly by later restrictive-temperature pulses applied to temperature-sensitive cell-lethal mutants is triplication similar to that described above as resulting from surgery *in situ*. Such abnormalities are assumed to be complete legs (the orthodromic part) from which a symmetrical outgrowth (anti- and paradromic parts) has arisen(Figure 1c;Jurgens and Gateff,1979). We are presently testing this assumption by clonal analysis. The outgrowths produced after late heat pulses in *l(1)ts-726* are often converging; that is, the two symmetrical partial circumferences of the outgrowth become less complete (have fewer bristle rows) towards the distal end (Figure 6b; Postlethwait, 1978; Jurgens and Gateff, 1979; J.R. Girton and P.J. Bryant, unpublished). They are also sometimes diverging, indicated by the presence of more bristle rows in the two symmetrical partial circumferences towards the distal end (Figure 6a; J.R. Girton and P.J. Bryant, unpublished). Converging outgrowths usually terminate before the true distal tip is reached, and therefore do not produce a claw. Diverging outgrowths, on the other hand, usually diverge to the extent that they contain two complete circumferences (two full sets of bristle rows) and at that point they always branch (Figure 6a). The two branches almost always have normal morphology, are symmetrical with one another, and are distally complete as is indicated by the terminal claw. Occasionally, one of the branches bears an additional small converging outgrowth.

The diverging outgrowths, together with the normal leg from which they emanate, all provide a striking confirmation of "Bateson's rules", devised by William Bateson in 1894 to illustrate the symmetry relationships of supernumerary legs found in insects in nature. He found that triple extremities were one of the most common of the abnormalities and noted that the three distal patterns always followed certain rules of symmetry. The appendage next to the one in the normal position was always a mirror-image of it, and the one farther from the normal was always a mirror-image of the nearer. Bateson constructed a mechanical model to illustrate this principle, which also demonstrated that the symmetry relationships applied to longitudinal planes of symmetry at various directions in the appen-

PATTERN FORMATION 121

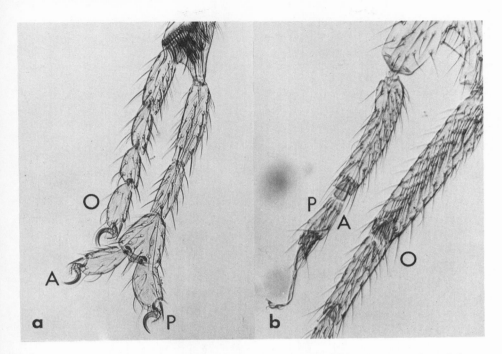

Figure 6: Outgrowth of parts of the *Drosophila* leg produced by a 48-hour pulse of 29°C applied to early third instar *l(1)ts-726* larvae (J.R.Girton and P.J. Bryant, unpublished). (a) Divergent, distally complete first leg outgrowth. (b) Convergent, distally incomplete third leg outgrowth. O,A,P, ortho-, anti-, and paradromic parts of the pattern.

dage. It is easy to see that these are exactly the symmetry relationships of the ortho-, anti- and paradromic elements of the triplications produced by surgical damage or temperature-sensitive cell-lethal mutations in *Drosophila* imaginal discs.

The similarities between the naturally occurring abnormalities described by Bateson and the responses to surgical damage in cockroaches (Bohn, 1972) and *Drosophila* (Bryant, 1971) suggest that many of the natural triplications might be explained as the results of accidental damage to the appendages earlier in the life of the animal. Most of the abnormalities produced in heat-pulsed *l(1)ts-726* animals can also be accounted for if it is assumed, as the histology indicates (Clark and Russell, 1977), that the primary effect of the mutation is to cause local degeneration in the imaginal disc. But whereas the deficiency/duplications produced by early heat pulses are probably the result of degeneration extending to the edge of the imaginal disc, it seems likely that the triplications arise from an internal zone of degeneration which does not reach the edge of the disc. We can then account for most of the outgrowth patterns if we assume that they represent the outcome of variable and symmetrical distal outgrowth from the margin of the degenerated zone as in Figure 5. If this is indeed the mechanism of outgrowth production, then several predictions can be made and tested:

(1) When the base of the outgrowth contains less than half of a circumference in each of its two symmetrical parts, the outgrowth will be either parallel (neither converging nor diverging) or converging and distally incomplete.

(2) When the base of the outgrowth contains more than half of a circumference in each of its two symmetrical parts, the outgrowth may converge and be distally incomplete, or it may diverge, branch and be distally and circumferentially complete in each of the two branches. Bateson's rules of symmetry should be followed.

(3) Outgrowths with distal bases should be more likely to diverge than those with more proximal bases. This is based on the assumption that the sizes of cell-death patches are approximately the same in presumptive proximal and distal segments. Since the presumptive segments are arranged concentrically in the leg disc, with the most distal presumptive segments towards the center, a cell-death patch of a given size would cover more of the circumference in distal presumptive segments than it would in more proximal presumptive segments.

The data from $l(1)ts-726$ are consistent with some of these predictions, although not all of them, and they show some additional interesting regularities. The data are summarized in Figure 7, where the arcs show the positions and sizes of the bases of (a) diverging and (b) converging outgrowths of the tarsus. It can be seen that diverging outgrowths always have relatively large bases but that a large base does not necessarily give a diverging

Figure 7: Arcs showing the positions and extents of outgrowth bases in legs of adults from heat-treated $l(1)ts-726$ larvae (from J.P. Girton and P.J. Bryant, unpublished). (a) Diverging outgrowths. (b) Converging outgrowths. The bases are plotted on a planar map of the tarsus, in which the more proximal segments are peripheral and the distal tip is central. Tarsal segments are numbered in the concentric circles, and bristle rows are numbered around the outside. The cross-hatching between rows 1 and 8 represents a characteristic trichome patch.

PATTERN FORMATION 123

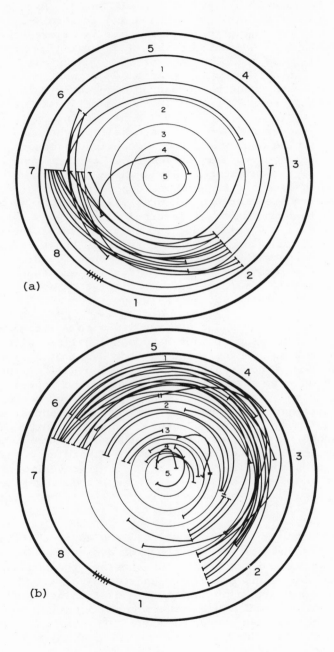

outgrowth. When the base is very small (¼ of the circumference or less), the outgrowth converges in all cases. Divergence is mostly limited to tarsal outgrowths and only rarely do those with more proximal bases diverge. All of this is as expected. However, the data seem to indicate that a diverging outgrowth can emanate from a base in which each of the symmetrical halves comprises less than half of a circumference of positional information. This can be clearly seen by comparing outgrowths in which the base contains bristle rows 7, 8 and 1, with those in which the base contains bristle rows 4, 5, 6 and 7 (row 7 being on the line of symmetry, at the extreme edge of each half of the base). These are nonoverlapping arcs which together make up less than a complete circumference, so that at least one of them must contain less than half of a circumference. Although such cases are not common, they are inconsistent with the model proposed earlier. They can **only be accommodated into the theoretical framework we have described** by assuming that the shortest intercalation rule is not always obeyed when the size of the positional information discontinuity is close to one half-circumference. This is actually known to be the case from strip-grafts with cockroach legs (French, 1978). A symmetrical base which comprises two halves of slightly less than a half-circumference could then sometimes undergo divergence and branching, but would be more likely to show convergence.

A surprising feature of our results with *1(1)ts-726* is that the bases of outgrowths are not randomly localized with respect to the circumference. On the one hand, the bases of all of the diverging outgrowths either include row 7 in each half or include row 7 on the line of symmetry. On the other hand the converging outgrowths never include row 7. Although this might reflect nonrandomness in the distribution of cell-death patches of different sizes or shapes, the fact that many converging outgrowths have a large base containing rows 2-6 suggests that large cell-death patches can occur in this part of the disc although they do not lead to diverging outgrowths. It therefore seems more likely that this result reveals inhomogeneity in the pattern regulation system that is responding to the cell death. We hope that a histological analysis of cell death in heat-pulsed *1(1)ts-726* larvae will resolve this question.

Another intriguing feature of our results is that in all diverging outgrowths and most of the converging ones, the bases extend into both anterior and posterior compartments, the boundaries of which are between rows 1 and 8 and between rows 3 and 4. This feature supports the conclusion of Schubiger and Schubiger (1978) that cells from both anterior and posterior compartments must be present in order for distal outgrowth to occur. If this is the correct intepretation, it must be the case that the contributions from the two compartments can be very unequal, since in some examples almost all of the outgrowth base is in one compartment.

CONCLUSIONS

The main advantages presented by the use of temperature-sensitive cell-lethal mutations in the study of pattern formation is the facility with which large numbers of defects can be produced, and the greater scoring accuracy which is possible with defects *in situ* compared to those in transplanted fragments. As we have shown in this paper, some aspects of the phenotypes produced are as expected from theory derived from surgical experiments. Others are yet to be fully explained and will undoubtedly contribute substantially to the development of future theories. Although with temperature-sensitive cell-lethal mutations it is possible to control the timing of events by varying the temperature treatments, the location and extent of the degeneration seems to be characteristic of the mutation itself. A detailed knowledge of the degeneration patterns in the mutants already studied, and comparable studies of the patterns produced by mutants with different degeneration patterns, will be necessary in order to determine which features of the results are unique to a given mutation, and which features will reveal important clues to the mechanism of pattern formation.

ACKNOWLEDGEMENTS

This work was supported by grants HD 06082 from the National Institutes of Health and PF-1599 from the American Cancer Society.

REFERENCES

Arking, R., 1975, Temperature-sensitive cell-lethal mutants of *Drosophila*: Isolation and characterization, Genetics, 80:519.

Bateson, W., 1894, "Materials for the Study of Variation", MacMillan, London.

Bohn, H., 1965, Analyse der Regenerationsfähigkeit der Insektenextremität durch Amputations- und Transplantationsversuch an Larven der afrikanischen Schabe (Leucophaea maderae Fabr.). II. Achsendetermination, Wilhelm Roux Arch., 156:449.

Bohn, H., 1972, The origin of the epidermis in the supernumerary regenerates of triple legs in cockroaches (Blattaria), J. Embryol. Exp. Morph., 28:184.

Bryant, P.J., 1971, Regeneration and duplication following operations *in situ* on the imaginal discs of *Drosophila melanogaster*, Develop. Biol., 26:637.

Bryant, P.J., 1975, Pattern formation in the imaginal wing disc of *Drosophila melanogaster*: Fate map, regeneration and duplication, J. Exp. Zool., 193:49.

Bryant, S.V., 1976, Regenerative failure of double half limbs in *Notophthalmus viridescens,* Nature, 263:676.

Bryant, S.V., and Baca, B., 1978, Regenerative ability of double half and half upper arms in the newt, *Notophthalmus viridescens,* J. Exp. Zool., 204:307.

Clark, W.C., and Russell, M.A., 1977, The correlation of lysosomal activity and adult phenotype in a cell-lethal mutant of *Drosophila*, Develop. Biol., 57:160.

Duranceau, C., 1977, Control of growth and pattern formation in the imaginal wing disc of *Drosophila melanogaster*, Ph.D. Thesis, University of California, Irvine.

French, V., 1978, Intercalary regeneration around the circumference of the cockroach leg, J. Embryol. Exp. Morph., 47:53.

French, V., Bryant, P.J., and Bryant, S.V., 1976, Pattern regulation in epimorphic fields, Science, 193:969.

Fristrom, D., 1968, Cellular degeneration in wing development of the mutant *vestigial* of *Drosophila melanogaster*, J. Cell Biol., 39:488.

Garcia-Bellido, A., 1975, Genetic control of wing disc development in *Drosophila*, Ciba Found. Symp., 29:161.

Garcia-Bellido, A., Ripoll, P., and Morata, G., 1973, Developmental compartmentalization of the wing disc of *Drosophila*, Nature New Biol., 245:251.

Girton, J.R., and Russell, M.A., 1980, A clonal analysis of pattern duplication in a temperature-sensitive cell-lethal mutant of *Drosophila melanogaster*, Develop. Biol., in press.

Jurgens, G., and Gateff, E., 1979, Pattern specification in imaginal discs of *Drosophila melanogaster* developmental analysis of a temperature-sensitive mutant producing duplicated legs, Wilhelm Roux Arch., 186:1.

Morata, G., and Lawrence, P.A., 1977, The development of *wingless*, a homoeotic mutation of *Drosophila*, Develop. Biol., 56:227.

Postlethwait, J.H., 1975, Pattern formation in the wing and haltere imaginal discs after irradiation of *Drosophila melanogaster* first instar larvae, Wilhelm Roux Arch., 178:29.

Postlethwait, J.H., 1978, Development of cuticular patterns in the legs of a cell lethal mutant of *Drosophila melanogaster*, Wilhelm Roux Arch., 185:37.

Rahn, P., 1972, Untersuchungen zur Entwicklung von Ganz-und Teilimplantaten der Flugelimaginalscheibe von *Ephestia kuhniella* Z., Wilhelm Roux Arch., 170:48.

Rose, S.M., 1962, Tissue-arc control of regeneration in the amphibian limb, Symp. Soc. Study Develop. Growth, 20:153.

Russell, M.A., 1974, Pattern formation in the imaginal discs of a temperature-sensitive cell-lethal mutant of *Drosophila melanogaster*, Develop. Biol., 40:24.

Russell, M.A., Girton, J.R., and Morgan, K., 1977, Pattern formation in a ts-cell-lethal mutant of *Drosophila*: the range of phenotypes induced by larval heat treatments, Wilhelm Roux Arch., 183:41.

Schubiger, G., 1971, Regeneration, duplication and transdetermination in fragments of the leg disc of *Drosophila melanogaster*, Develop. Biol., 26:277.

Schubiger, G. and Schubiger, M., 1978, Distal transformation in
 Drosophila leg imaginal disc fragments, Develop. Biol., 67:
 286.
Simpson, P., and Schneiderman, H.A., 1975, Isolation of temperature-
 sensitive mutations blocking clone development in Drosophila
 melanogaster, and the effects of a temperature sensitive cell
 lethal mutation on pattern formation in imaginal discs, Wilhelm
 Roux Arch., 178:247.
Stern, C., 1968, "Genetic Mosaics and Other Assays", Harvard University Press, Cambridge, Massachusetts.
Strub, S., 1977a, Pattern regulation and transdetermination in Drosophila imaginal leg disc reaggregates, Nature, 269:688.
Strub, S., 1977b, Developmental potentials of the cells of the male
 foreleg disc of Drosophila, Wilhelm Roux Arch., 182:75.
Tank, P.W., 1978, The failure of double-half forelimbs to undergo
 distal transformation following amputation in the axolotl,
 Ambystoma mexicanum. J. Exp. Zool., 204:325.

DISCUSSION

A. Shearn: Have you examined patterns of cell death?

P.J. Bryant: Clark and Russell have examined this for earlier pulses. The cell death is very patchy and the patches are in different places for different discs without much regularity. We have not looked at histology from later pulses yet.

A.S. Mukherjee: Does your model assume that the positions of the bristles on the tarsal segments are fixed at a very early stage?

P.J. Bryant: Not really. We are assuming that there is positional information on the disc at the time when it responds to a temperature pulse during the early third larval instar. In fact, there is other evidence that this is the case. When you transplant imaginal disc fragments from these stages, you produce specific structures.

D. Byers: Do you get triplications in the eye and the antenna?

P.J. Bryant: Eye and antennal duplications are produced by early pulses, but we have found triplications from later pulses.

THE CONTROL OF GROWTH IN THE IMAGINAL DISCS OF *DROSOPHILA*

Pat Simpson[1] and Ginès Morata[2]

[1] Céntré de Genetique Moléculaire
C.N.R.S., 91190
Gif-sur-Yvette, France
[2] Centro de Biologia Molecular,
C.S.I.C.
Universidad Autonóma
Madrid, 34, Spain

INTRODUCTION

Very little is known about the way in which growth is monitored in developing animals. The control of growth and size, although an important subject, is poorly amenable to experimental analysis. Different aspects of this topic such as: the control of different events in the cell cycle (Hartwell, 1974), changes in the cell surface throughout development and the possible intermediary role of the cytoskeleton between cell surface receptors and the nucleus (Edelman, 1976; Nicholson, 1976) have been investigated. However, since cell-cell interactions are of prime importance throughout growth and many events must occur at the level of the entire cell population (Wigglesworth, 1959), one rather neglected aspect of the problem has been the study of the behavior of cell populations.

The imaginal discs of *Drosophila* offer a good model system for studying growth from this point of view. Here we will briefly review evidence that growth is controlled intrinsically by the imaginal discs, and that compartments (Garcia-Bellido et al., 1973) may represent the units for size control. The phenomenon of cell competition (Morata and Ripoll, 1975) and its possible relation to growth control will be described and discussed.

CELL COMPETITION

Cell competition was first described by Morata and Ripoll (1975). These authors studied a group of mutations in *Drosophila*, the <u>Minutes</u>

and showed that their extended period of development was the result
of a cell-autonomous slower rate of division. Flies bearing Minute
mutations are of normal size, show moderate to good viability and
have no abnormalities other than smaller, thinner bristles (the term
"Minute" refers to the bristle phenotype). When clones of Minute
cells are produced in non-Minute animals however, they do not survive.
The elimination of these slowly-dividing non-lethal cells was termed
cell competition.

When Minute$^+$ clones are produced in a Minute animal, they grow
to be very large as a consequence of their advantageous growth rate.
During the growth of a large Minute$^+$ clone, the surrounding Minute
cells in the same compartment are progressively eliminated by cell
competition in the same way that Minute clones are in a background of
Minute$^+$ cells. Mosaic compartments are always normal in size and
shape. Cell competition appears to be necessary to obtain appendages
of normal size, since Minute$^+$ clones grow to be so large. The process
of cell competition is therefore closely correlated with the control
of size.

The Intensity of Competition is a Function of the Disparity Between Rates of Cell Division.

Recent unpublished observations have revealed that cell competition is more intense the greater the disparity in growth rate.
Clones of two different Minutes, $M(1)Bld$ and $M(2)c$, were induced together with control Minute$^+$ clones, in Minute$^+$ flies. $M(1)Bld$ is a
very extreme Minute with a developmental delay of 76 hours, whereas
$M(2)c$ is delayed by 49 hours (Ferrus, 1975). The results of this
experiment are shown in Figure 1, where it can be seen that clones
of the more extreme $M(1)Bld$ are eliminated sooner than those of the
less extreme $M(2)c$.

To test further the hypothesis that competition is a result of
the slower rate of growth of the Minute cells, we produced $M(2)c$
clones within $M(3)^{i55}$ animals. These two Minutes develop at approximately the same rate. Flies bearing both $M(3)^{i55}$ and $M(2)c$ develop
at the same rate as either $M(3)^{i55}$ or $M(2)c$ flies. As can be seen
in Figure 2, in this instance the $M(2)c$ cells are not outcompeted presumably because the $M(3)^{i55}$ cells around them are also dividing at a
reduced rate.

Competition is a Result of Local Cell Interactions

Preliminary results in our laboratories show that competition
occurs where there is contact between Minute and non-Minute cells.
We have obtained evidence for regional differences in growth within
large Minute$^+$ clones. Apparently more growth occurs at the edges of
such clones than in the middle. Cell competition occurs at those
edges where the cells are rapidly dividing. Minute cells further

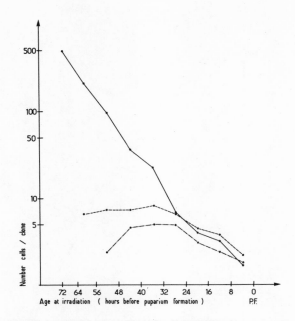

Figure 1: Semilogarithmic plot of the average sizes of clones resulting from X-irradiation (1000r) of females of the genotype $y\ f^{36a}/T(1;2)Bld/pwn\ ;\ mwh/+$, at different times throughout development of the wing discs. These females are both $M(1)Bld^+$ and $M(2)c^+$. The solid line represents mwh control Minute$^+$ clones that grow normally. The dashed-dotted line represents $pwn\ M(1)Bld$ clones and the dashed line $f^{36a}\ M(2)c$ clones; these clones are eliminated by cell competition. mwh, pwn and f^{36a} are cuticle cell markers.

away from this mosaic borderline are not eliminated. The process therefore relies upon local cell interactions.

At the mosaic borderline where competition is occurring, certain morphological changes can be seen. The outlines of Minute$^+$ clones growing in Minute flies are far more indented than those of Minute clones growing in Minute flies. In some cases small islands of Minute cells are seen trapped within the Minute$^+$ clone. In a similar fashion, Minute clones induced in Minute$^+$ animals were also observed to become dispersed (Simpson, 1979). Figure 3 illustrates this clone dispersion.

The Pattern of Overgrowth of Minute$^+$ Clones

Minute$^+$ clones growing in a Minute compartment grow first in a proximo-distal direction, i.e. parallel to the long axis of the wing, and then later fill out breadthways (unpublished observations). Such

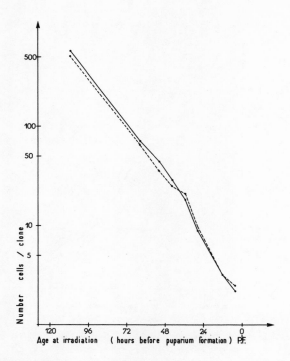

Figure 2: Semilogarithmic plot of the average sizes of clones resulting from X-irradiation (1000r) of females of the genotype $Dp(3;1)$ $mwh^+ ve^+ y\ f^{36a}/T(1;2)\ sc^{S2}/+\ ;\ mwh\ M(3)\ i55/mwh\ ve\ h$, at different times throughout the development of the wing discs. These females are $M(3)\ i55$ and $M(2)c^+$. The solid line represents mwh control recombinants that remain $M(3)\ i55$ like the rest of the animal. The dashed line shows $f^{36a}M(2)c$ clones that are also $M(3)\ i55$. It can be seen that there is no difference between the growth of the $f^{36a}M(2)c$ clones and the control clones. f^{36a} and mwh are cuticle cell markers.

a pattern of growth explains the distribution of remaining Minute cells in compartments mainly occupied by large Minute⁺ clones. The Minute patches are seen mostly at the edges of the compartments: costal region, alar lobe and along the anterior-posterior compartment boundary. It was observed earlier that Minute clones in Minute⁺ flies survive longest at the edges of compartments (Simpson, 1979).

We visualize that cell competition takes place in the following way: the Minute cells, which are dividing more slowly than the Minute⁺ cells around them, dissociate from one another upon contact with the Minute⁺ cells and are dispersed and then later eliminated. This process of clone dispersion may result from a difference in cell affinities. It is possible that because of their slow growth the Mi-

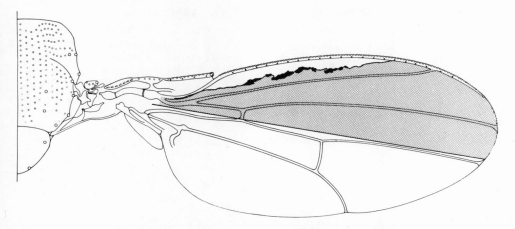

Figure 3: Clones produced on the wing of a female of genotype $y\ f^{36a}/M(1)n\ ;\ mwh/+$. The shaded area shows a large $y\ f^{36a}\ M(1)n^+$ clone induced by X-irradiation (1000r) at 48 hours after egg laying. The black patches represent a single $mwh\ M(1)n$ clone induced by a second irradiation at 120 hours after egg laying. It can be seen that the cells of Minute constitution become dispersed where they are in contact with Minute$^+$ cells.

nute cells become less cohesive, perhaps because they are developmentally younger relative to the Minute$^+$ cells.

There is No Competition Across Compartment Boundaries

The imaginal discs of *Drosophila* are progressively subdivided into compartments during development (Garcia-Bellido et al., 1973 and 1976). Compartments are believed to be independent units of development and to be under the control of a discrete set of selector genes (Garcia-Bellido, 1975; Morata and Lawrence, 1977). Although cell competition relies upon cellular interactions, there is no competition across compartment boundaries. Cells of different compartments are known to be touching one another, so physical proximity, although necessary for cell competition, seems not to be the only requisite. It is possible that compartments may be the units for size control in *Drosophila* (see next section). If this is the case, then competition across compartment boundaries would not be expected. This supports the idea of a relationship between cell competition and growth control.

Compartments as the Units of Size Control

It was suggested by Crick and Lawrence (1975), that compartments may be the units of size control. Some evidence for this is available from observations of gynandromorphs in which the female tissue is normal but the male tissue hemizygous for a temperature-sensitive mutation, *l(1)ts 1126*, which reduces mitotic rate (Simpson and Schneiderman, 1976). In gynandromorphs that pupariate at the normal time when grown at the restrictive temperature, the size of the anterior or posterior wing compartments that are composed entirely of *l(1)ts 1126* tissue is drastically reduced relative to those of wild type tissue (see Figure 4). However, when pupariation is delayed, both compartments, mutant and wild type, are of normal size. The anterior and posterior compartments must therefore grow independently from one another. No evidence of a similar kind is available for other, late-occurring, compartments. However, since these are clonally restricted in the same way as the anterior and posterior (Garcia-Bellido et al., 1976), it is probable that they also grow independently.

Growth is Controlled Autonomously by the Imaginal Discs

It is important to know to what extent growth of the imaginal discs is governed by the hormonal environment in the larva. The role of hormones during growth in *Drosophila* and particularly the necessity for ecdysone, has been extensively reviewed elsewhere (Wigglesworth, 1959; Oberlander, 1972; Doane, 1973; Slama et al., 1974). Our only purpose here is to examine those aspects pertaining to the final size of the imaginal discs. The discs stop dividing at the onset of pupariation. We have recently shown, however, that the intervention of pupariation is itself dependent upon growth of the discs. In the following discussion we will argue that the cessation of growth is independent of hormonal control.

Use was made of the fact that experimentally-induced lesions in the imaginal discs *in situ* lead to extra growth during the ensuing regeneration, which in turn retards pupariation (Russell, 1974; Simpson and Schneiderman, 1975). Studies on gynandromorphs bearing sex-linked recessive temperature-sensitive cell lethal mutations have revealed that it is sufficient for only two of the discs of an animal to be damaged in order for larval development to be prolonged. The complete absence of one or more discs in a larva however, is without any effect upon the time of pupariation (Simpson et al., 1980). Therefore, it appears that discs that have not completed a certain amount of growth are able to inhibit pupariation.

Pupariation can only be delayed by such experimental intervention at or before a certain critical stage in development, around the beginning of the third larval instar (Simpson and Schneiderman, 1975 and 1976), when the imaginal disc cells are still actively dividing. This critical period coincides with the release of a small peak of ecdysone, and experimentally delayed pupariation is accompanied by a similarly retarded secretion of this ecdysone peak (Berreur et al., 1980). Therefore, the direct relationship between growth of the imaginal discs and the timing of pupariation may involve the prevention of an elevated ecdysone titer by immature discs.

GROWTH CONTROL IN IMAGINAL DISCS 135

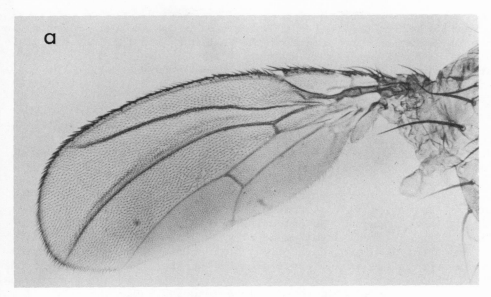

Figure 4a: Photograph of the wing of a $R(1)2w^{vC}/l(1)ts\ 1126\ f^{36a}$ gynandromorph grown at 22°C for three days and then shifted up to 29°C. The anterior compartment is female and is composed of wild type tissue. The posterior compartment is male, composed of $l(1)ts\ 1126\ f^{36a}$ tissue and is considerably smaller.

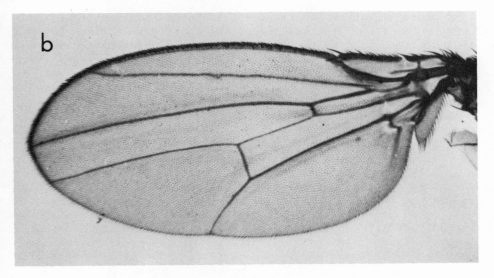

Figure 4b: Wing of wild type Drosophila.

It is noteworthy that the duration of the extended period of larval life is correlated with the amount of extra growth that occurs during the regeneration. The nature of the feedback from the imaginal discs to the hormonal centers is unknown in the Diptera. A similar situation prevails in cockroaches where there is evidence that leg autotomy may result in a nervous stimulation of the brain affecting the subsequent release of brain hormone (Kunkel, 1977).

A large number of mutants have been described which display abnormal discs. In these mutant pupariation, may, or may not, be delayed or deficient. It is worth pointing out that the phenotype of many of these is consistent with the above scheme. In mutants in which there is a hypertrophy of the imaginal discs, pupariation is very delayed or does not take place (Bryant and Schubiger, 1971; Stewart et al., 1972). Although some may have pleiotropic effects on both the discs and the endocrine organs, we would predict that the prolonged larval development of many such mutants is a result of the extra disc growth, and that it is not a continued hormonal stimulus that causes the extra growth. On the other hand, mutations leading to a complete absence or a total degeneration of the imaginal discs are generally able to pupariate (Hadorn, 1961; Shearn et al., 1971; Stewart et al., 1972). In one case it was shown that the larvae of a "disc-less" mutant were capable of supporting development of implanted wild-type discs (Shearn and Garen, 1974). This phenotype is also consistent with the idea that growing imaginal discs prevent pupariation.

In a fly in which pupariation is retarded because of lesions in only some of the imaginal discs, the other intact discs have an extended period of time for growth before the onset of metamorphosis. They do not, however, give organs or appendages that are any larger than the usual size (Simpson et al., 1980). The hormonal conditions necessary for growth are fulfilled nonetheless, since growth and regeneration proceed in the damaged discs. Experimental evidence for the production of giant adult insects by prolonged feeding due to juvenile hormone treatment is not convincing. In all cases growth of the larval body decreases and eventually stops with each supernumerary larval instar. Less than 1% of the large larvae metamorphose and when they do, the resulting adults are abnormal (Slama et al., 1974). The one reported instance of an additional moult in the Diptera was not followed by adult differentiation (Zdarek and Slama, 1973). The so-called giant mutant in *Drosophila* owes its slightly enlarged body size to an increased cell size and not to an additional number of cells (unpublished observations). In cases in which enlarged imaginal discs have been described, these discs either differentiate abnormal or duplicated patterns (Bryant and Schubiger, 1971; Martin et al., 1977) or else prove incapable of differentiation (Gateff and Schneiderman, 1969; Stewart et al., 1972; Garen et al., 1977).

Although there is no doubt that the correct hormonal stimulus is indispensable for growth to take place, it appears that the signal for the cessation of growth is not dependent upon humoral factors. Therefore, we conclude that the control of growth is intrinsic to the imaginal discs.

CONCLUSION

It can be seen from the foregoing observations that pattern development is independent of the number of mitoses accomplished by any particular cell, since Minute/Minute+ mosaic compartments are of normal size and pattern. So far it has not been possible to dissociate pattern and normal size in order to obtain experimentally a larger pattern size; extra growth inevitably results in pattern abnormalities. The feedback interactions between growth and pattern size in *Drosophila* imaginal discs are independent of humoral controls; they may result from the same type of local interactions that initially regulate the pattern itself.

ACKNOWLEDGEMENTS

We thank Drs. P. Berreur and M. Gans for useful discussions.

REFERENCES

Berreur, P., Porcheron, P., Berreur-Bonnenfant, J., and Simpson, P., 1980, Ecdysone levels and pupariation in a temperature-sensitive mutation of *Drosophila melanogaster*, J. Exp. Zool., in press.

Bryant, P.J., and Schubiger, G., 1971, Giant and duplicated imaginal discs in a new lethal mutant of *Drosophila melanogaster*, Devel. Biol., 24:233.

Crick, F.H.C., and Lawrence, P.A., 1975, Compartments and polyclones in insect development, Science, 189:340.

Doane, W.W., 1973, Role of hormones in insect development, in: "Developmental Systems: Insects, Vol. 2", J.C. Counce and C.H. Waddington, eds., Academic Press, p. 291.

Edelman, G., 1976, Surface modulation in cell recognition and cell growth, Science, 192:218.

Ferrus, A., 1975, Parameters of mitotic recombination in *Minute* mutants of *Drosophila melanogaster*, Genetics, 79:589.

Garcia-Bellido, A., 1975, Genetic control of wing disc development in *Drosophila*, in: "Cell Patterning", Ciba Foundation Symposium 29, Associated Scientific Publishers, p. 161.

Garcia-Bellido, A., Ripoll,P., and Morata,G., 1973, Developmental compartmentalization of the wing disc of *Drosophila*, Nature New Biology, 245:251.

Garcia-Bellido, A., Ripoll, P., and Morata, G., 1976, Developmental segregations in the dorsal mesothoracic disc of *Drosophila*, Devel. Biol., 48:132.

Garen, A., Kauvar, L., and Lepesant, J.A., 1977, Roles of ecdysone in *Drosophila* development, Proc. Nat. Acad. Sci. U.S.A., 74:5099.

Gateff, E., and Schneiderman, H.A., 1969, Neoplasms in mutant and cultured wild-type tissues of *Drosophila*, Nat. Cancer I. Monogr., 31:365.

Hadorn, E., 1961, Developmental Genetics and Lethal Factors, John Wiley, New York.

Hartwell, L.H., 1974, Genetic control of the cell division cycle in yeast, Science, 183:46.

Kunkel, J.G., 1977, Cockroach molting. II. The nature of regeneration-induced delay of molting hormone secretion, Biol. Bull., 153:145.

Martin, P., Martin, A., and Shearn, A., 1977, Studies of $l(3)C43^{hs1}$, a polyphasic, temperature-sensitive mutant of *Drosophila* with a variety of imaginal disc defects, Devel. Biol., 55:213.

Morata, G., and Lawrence, P.A., 1977, Homeotic genes, compartments and cell determination in *Drosophila*, Nature, 265:221.

Morata, G., and Ripoll, P., 1977, Minutes: Mutants of *Drosophila* autonomously affecting cell division rate, Devel. Biol., 427:211.

Nicholson, G., 1976, Transmembrane control of the receptors on normal and tumor cells. I. Cytoplasmic influence over cell surface components, Biochim. Biophys. Acta, 457:57.

Oberlander, H., 1972, The hormonal control of development in imaginal discs, in: "Results and Problems in Cell Differentiation", H. Ursprung and R. Nothiger, eds., Springer Verlag, New York, p. 155.

Russell, M., 1974, Pattern formation in imaginal discs of a temperature-sensitive cell lethal mutant of *Drosophila melanogaster*, Devel. Biol., 40:24.

Shearn, A., Rice, T., Garen, A., and Gehring, W., 1971, Imaginal disc abnormalities in lethal mutants of *Drosophila*, Proc. Nat. Acad. Sci. U.S.A., 68:2594.

Shearn, A., and Garen, A., 1974, Genetic control of imaginal disc development in *Drosophila*, Proc. Nat. Acad. Sci. U.S.A., 71:1393.

Slama, K., Romȃnuk, S., and Sorm, R., 1974, "Insect ormones and Bio-analogues", Springer Verlag, New York.

Simpson, P., 1979, Parameters of cell competition in the compartments of the wing disc of *Drosophila*, Devel. Biol., 69:182.

Simpson, P., and Schneiderman, H.A., 1975, Isolation of temperature-sensitive mutations blocking clone development in *Drosophila melanogaster*, and the effects of a temperature-sensitive cell lethal mutation on pattern formation in imaginal discs, Wilhelm Roux Arch., 178:247.

Simpson, P., and Schneiderman, H.A., 1976, A temperature sensitive mutation that reduces mitotic rate in *Drosophila melanogaster*, Wilhelm Roux Arch., 179:215.

Simpson, P., Berreur, P., and Berreur-Bonnenfant, J., 1980, The initiation of pupariation in *Drosophila*: dependence on growth of the imaginal discs. Submitted for publication.

Simpson, P., 1976, Analysis of the compartments of the wing of *Drosophila melanogaster* mosaic for a temperature-sensitive mutation that reduces mitotic rate, Devel. Biol., 54:100.
Stewart, M., Murphy, C., and Fristrom, J.W., 1972, The recovery and preliminary characterization of X chromosome mutants affecting imaginal discs of *Drosophila melanogaster*, Devel. Biol., 27:71.
Wigglesworth, V.B., 1959, "The Control of Growth and Form: A Study of the Epidermal Cell in an Insect", Cornell University Press, New York.
Žďárek, J., and Slama, K., 1972, Supernumerary larval instars in cyclorhaphous *Diptera,* Biol. Bull., 142:350.

DISCUSSION

P.J. Bryant: I would like to comment on your last point. We have looked at two mutations which have overgrowing imaginal discs during the extended larval period. To determine whether this is a hormonal defect or an intrinsic disc defect we have transplanted discs into adult hosts. In both cases, the growth control defects were autonomous to the discs. Therefore, we believe that the extended larval period results from a disc defect.

Y. Hotta: I would like to comment regarding this disc defect autonomy. Such a defect in a disc may not necessarily mean a primary disc defect. The possibility is not excluded that the defect occurs very early in disc formation. Here the effect would show autonomous behavior although it lies outside the disc. There are autonomous disc defects which map, by mosaics, to a place away from the disc.

A. Shearn: What happens in control experiments where you do not have the Minute mutation and are just looking at multiple wing hair clones? Is there any difference in the size and shape of clones that are found in different parts of the wing?

P. Simpson: I don't think there are significant differences but I have not really looked at a large enough number.

AN ANALYSIS OF THE EXPRESSIVITY OF SOME BITHORAX TRANSFORMATIONS

Gines Morata and Stephen Kerridge

Centro de Biologia Molecular
Universidad Autónoma de Madrid
Madrid-34
Spain

INTRODUCTION

Of the homeotic mutants known in *Drosophila*, those of the bithorax system are by far the best known genetically, thanks to the extensive analyses by E.B. Lewis (1963, 1964, 1967, 1978). The interest of their study lies not only in their effect on the development of certain segments (see Morata and Lawrence, 1977) but also in the genetic structure of the system; it has been shown that the bithorax genes are organized in a gene complex which includes several adjacent loci: bithorax (*bx*), Contrabithorax (*Cbx*), Ultrabithorax (*Ubx*), bithoraxoid (*bxd*), postbithorax (*pbx*), Hyperabdominal (*Hab*), Ultraabdominal (*Uab*), and probably others. These genes seem to be under the same control elements. (See Lewis, 1978 for a recent review.) This gene complex is one of the few examples studied and characterized in eukaryotes. In general (but not always) each of these loci is associated with one particular type of homeotic transformation; a specific segment or half-segment is transformed into another.

Mutations in the locus *bx*, for example, produce a transformation of the anterior metathoracic into the anterior mesothoracic segment (Figure 1). The extent of the transformation depends on the allele and genetic combination used and ranges from very weak showing only few mesothoracic elements, to very strong producing almost a replica of the anterior mesothorax. A similar transformation is also produced by the alleles of the gene *Ubx*. In this paper we describe in detail the mesothoracic transformation produced by different alleles of the two loci and by combinations between them.

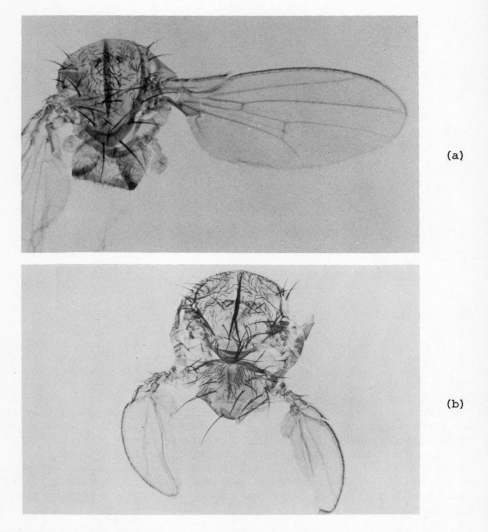

Figure 1: (a) The dorsal part of the adult mesothoracic and metathoracic cuticle. The dorsal mesothorax consists of a proximal region called notum (N) and the wing blade. The wing contains bristles in the margin. Those in the most proximal region, Costa (Co) bristles, are different from those of the medial and distal part of the margin that form the triple row (TR). The dorsal metathorax consists of a proximal part, the metanotum (MT), a featureless piece of cuticle devoid of bristles and trichomes, and the distal appendage or haltere (H). (b) Example of maximal transformation of anterior metathorax into anterior mesothorax produced by bx^3/Ubx^{130}. The metanotum becomes almost identical to the notum and the anterior haltere is transformed into anterior wing. In the fly shown in the picture the wings of the normal mesothorax have been removed.

The aim of this analysis was to characterize the mutant alleles both genetically and developmentally.

MATERIALS AND METHODS

The degree to which each genetic combination shows the mutant phenotype ("expressivity") has been evaluated by estimating the amount of anterior mesothoracic structures that appears in the mutant metathorax. We have considered three clearly distinguishable mesothoracic patterns that extend along the proximo-distal axis: the notum, the costa and the triple row (Figure 1). Individual elements of these three patterns can be readily identified in isolation. To quantify the amount of mesothoracic tissue we have counted the number of notum, costa and medial triple row bristles. We are aware of the problem that the number of bristles may not give an accurate estimation of the extent of mesothoracic transformation, especially for the distal transformation into wing. This is because the margin of the haltere (where the bristles appear) is transformed more often than the more internal cuticular regions which, if transformed, only produce wing trichomes (Morata and Garcia-Bellido, 1976). However, the transformation cannot be estimated in terms of number of wing trichomes because very often these are intermediate between haltere and wing (see Morata, 1975). The number of mesothoracic bristles is given as a percentage of the average number of bristles in the same structures in the normal mesothorax. Except when otherwise stated, all the crosses were performed at 25°C.

We have used three recessive alleles of the bithorax locus; bx^1, bx^3 and bx^{34e} are all considered to be point mutants and are not associated with gross chromosomal aberrations (Lindsley and Grell, 1968). For the Ultrabithorax locus we have studied four mutant alleles. Two of these, Ubx^1 and Ubx^{9-22} (recently isolated in the Madrid laboratory) are apparently point mutants with normal polytene chromosomes and the other two, Ubx^{130} and Ubx^{6-26} (also newly induced) are chromosomal rearrangements with breakpoints in (or very close to) the region 89E1-4 where the bithorax genes are located.

All the Ubx mutations are recessive lethals but show the bithorax transformation when in a "trans" configuration with the bx alleles. They also show the respective mutant phenotype over the recessive alleles of the mutants postbithorax and bithoraxoid (Lewis, 1963 and 1964). Thus, the Ubx alleles behave as recessive markers with respect to bx, pbx and bxd alleles so that they can be used for complementation analysis. They are considered as dominant only because when heterozygous they produce a slight enlargment of the haltere that makes the mutation detectable in one dose.

We have used the chromosome $Df(3R)bxd^{100}$ (which is deficient for the bithorax and Ultrabithorax loci) to examine the chosen bx pseudoalleles in hemizygous condition.

RESULTS

Genetic Interactions

The expressivity of the transformations produced by the three mutants has been studied in homozygous and hemizygous conditions. (The latter was in "trans" over the physical deletion of the wild-type gene, in this case $Df(3R)bxd^{100}$). The results are shown in Table 1. In general bx^1 is the weakest mutant; it shows mainly notum and little costa and medial triple row. The amount of notum and costa in bx^{34e} is similar to bx^1 but the percentage of medial triple row is greater. The bx^3 allele is the strongest for the three structures considered.

In combination with the $Df(3R)bxd^{100}$, both bx^{34e} and bx^3 show transformations stronger than in the respective homozygotes, although in the case of bx^3 there is little difference between the two phenotypes. These results indicate that the two mutants are hypomorphic, bx^3 being closer to the amorphic condition. However, when bx^1 is over the deficiency, it shows a phenotype weaker than

Table 1. Expressivity of the *bx* alleles

Genotype	n	N	Co	MTR
Wild-type Mesothorax	36	100(114± 9)	100(59±4)	100(87± 6)
bx^1/bx^1	70	41(46±28)	15(9±9)	13(11±18)
bx^{34e}/bx^{34e}	36	26(30± 5)	13(8±1)	29(25± 5)
bx^3/bx^3	18	56(63± 4)	39(23±8)	72(63± 4)
$bx^1/Df\ bxd^{100}$	59	3(4±11)	0.8(0.5±3)	8(7± 7)
$bx^{34e}/Df\ bxd^{100}$	24	50(56± 5)	27(16±3)	58(51± 6)
$bx^3/Df\ bxd^{100}$	25	73(83± 5)	49(29±3)	87(76± 3)

n = number of hemithoraces studied; N = notum; Co = costa; MTR = medial triple row.
The average number of bristles of each type are expressed as a percentage of the number found on wild-type mesothorax. The numbers in parentheses give the average number of bristles ± the standard deviation.

when homozygous. Clearly bx^1 is a different kind of mutant that cannot be considered hypomorphic despite its weak phenotype.

Close examination of the mutant structures reveals some clear differences between bx^1 on the one hand and bx^{34e} and bx^3 on the other. The first difference is that while bx^{34e} and bx^3 show a very consistent transformation with little variability (note the low standard deviation values in Table 1), the mutant bx^1 shows a highly variable and erratic phenotype so that even in the same fly one side may be almost wild-type and the other may show a large mesothoracic transformaton. (Note also the high standard deviation values of bx^1 in Table 1.) Another difference between the two sets of mutants is the expression of the transformation in the trichomes; the trichomes of the normal wing are longer and more widely spaced than those of the haltere. In the case of bx^3 the trichomes produced in the anterior haltere are all equal in size and indistinguishable from those of the wing. Those of bx^{34e} are all equal and of intermediate size between haltere and wing. However, those of bx^1 are not intermediate; although most of them are normal haltere trichomes, when they are transformed , they usually form perfect wing trichomes. These wing trichomes are often invaginated inside the haltere territory (Figure 2) probably as a result of differential cell affinities (Morata and Garcia-Bellido, 1976; Garcia-Bellido and Lewis, 1976). It is also common to find bx^1 halteres where the only mesothoracic structures that can be seen are the bristles. This differential behavior of bristles with respect to trichomes has been described for Contrabithorax (Morata, 1975).

Each mutant also produces different regions in the homeotic mesonotum. In the case of bx^{34e} there is preferential production of scutellum and also the part of the notum close to the wing hinge. In bx^1 the different notum structures appear with the same probability.

The combinations of the different bx alleles produce only anterior transformations. The expressivity of the transformation varies with the different genotypes (see Table 2). In general the results obtained are in accordance with the expectations from the expressivity of the individual mutants as shown in Table 1. The only exception is the combination bx^1/bx^{34e} which presents a phenotype weaker than those of either mutant in homozygous condition. This result suggests that there is partial complementation.

As a rule, all the trans combinations of bx and Ubx alleles show mesothoracic transformation. The expressivity of the different genotypes is shown in Table 3. There are two main points that should be emphasized from these data: the first is that the transformations produced by the two point Ubx alleles are clearly weaker than those produced by the Ubx rearrangements. The second point is that the interactions of the Ubx alleles with bx^1 are quite erratic. Three

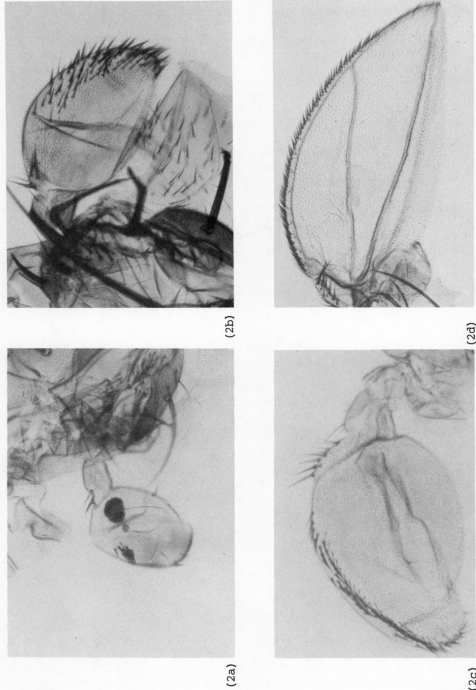

(2a)
(2b)
(2c)
(2d)

Table 2: Trans combinations of bx alleles

Genotype	n	N	Co	MTR
bx^1/bx^{34e}	56	9 (11±15)	0.3 (92±0.5)	0.5 (0.5±0.8)
bx^1/bx^3	54	37 (42±28)	4 (3±2)	7 (6 ±5)
bx^{34e}/bx^3	20	49 (56± 7)	15 (9±2)	59 (51 ±5)

See footnote to Table 1 for explanation of column headings.

combinations (bx^1/Ubx^1, bx^1/Ubx^{9-22} and bx^1/Ubx^{130}) are weaker than bx^1 homozygotes, whereas bx^1/Ubx^{6-26} is stronger. In the case of bx^1/Ubx^1 and bx^1/Ubx^{9-22} the phenotype is close to that of the Ubx heterozygote.

The analysis of the type of mesothoracic structures produced by the different combinations reveals that the two rearrangements Ubx^{130} and Ubx^{6-26} give rise to strong transformations over bx^3 and bx^{34e}. The amount of notum, costa and medial triple row is increased when compared with the homozygotes. The percentages of transformation are similar to those found for the combinations of bx^3 and bx^{34e} with the $Df(3R)bxd^{100}$ (compare with Table 1). However, over the same two mutants Ubx^1 and Ubx^{9-22} show a different pattern of regional transformation. While the amount of medial triple row that they produce is not such less than Ubx^{6-26} or Ubx^{130} (Table 3) and equal or greater than the corresponding homozygotes (bx^{34e} or bx^3), the amount of notum tissue is much reduced. bx^{34e}/bx^{34e}, for example, produces 26% notum and 29% medial triple row (Table 1), but bx^{34e}/Ubx^{9-22} produces 2% notum and 53% medial triple row (Table 3). Therefore, the transformation produced by Ubx^{9-22} is weaker for the notum and stronger for the medial triple row.

Figure 2: Different degrees of perfection in the formation of the TR pattern. a) bx^1 homozygous haltere showing a group of unpatterned MTR bristles. b) An imperfect pattern of TR bristles in a bx^{34e} homozygous haltere. Although the relative position of each type of bristle is like the normal wing, the bristles are not following straight lines. c) Bristle alignment is better in bx^3/bx^{34e} halteres and virtually perfect in bx^3/Ubx^{100} (d).

Table 3. Trans combinations of *bx* and *Ubx* alleles.

Genotype	n	N	Co	MTR
Ubx^1/bx^1	64	0.0	0.0	10 (9± 6)
Ubx^1/bx^{34e}	26	0.6 (0.7± 1)	19 (11± 2)	54 (48± 4)
Ubx^1/bx^3	40	21 (24 ±11)	26 (16± 3)	66 (57± 4)
Ubx^{9-22}/bx^1	40	0.0	0.0	0.5 (0.4±1)
Ubx^{9-22}/bx^{34e}	40	2 (3 ± 2)	20 (12± 2)	53 (46± 5)
Ubx^{9-22}/bx^3	37	26 (30 ±11)	29 (17± 3)	63 (55± 7)
Ubx^{130}/bx^1	49	4 (5 ±11)	0.0	8 (7± 4)
Ubx^{130}/bx^{34e}	24	59 (67 ± 4)	37 (22± 2)	76 (67± 4)
Ubx^{130}/bx^3	16	72 (82 ± 3)	60 (36± 4)	86 (75± 3)
Ubx^{6-26}/bx^1	32	61 (69 ±25)	30 (18±10)	27 (23±14)
Ubx^{6-26}/bx^{34e}	34	69 (79 ±11)	41 (24± 3)	68 (59± 6)
Ubx^{6-26}/bx^3	37	84 (96 ±13)	68 (40± 3)	86 (75± 6)

See footnote to Table 1 for explanation of column headings.

Characteristics of the Mesothoracic Patterns Produced in the Mutant Metathorax

In the case of extreme combinations like bx^3/Ubx^{130} or $bx^3/Df(3R)bxd^{100}$, the mesothoracic structures developed in the mutant are virtually identical to the normal ones (Figure 1). However, in the case of the weak or intermediate mutations there is a reduced number of elements and consequently the mesothoracic pattern is incomplete. We have studied some of these defective patterns in detail on the presumption that we could learn something about how normal patterns are formed. In particular, we have analyzed the

Figure 3: Sorting out of wing and haltere cells. In bx^1 homozygous halteres it is frequent to find wing trichomes forming: a) invaginations that eventually separate completely from the haltere trichomes; b) forming isolated spheres of wing trichomes inside the appendage. In other cases there are haltere trichomes which are invaginated inside the appendage (notice the sensilla trichoidea, typical of the haltere) c) or evaginated outside d).

EXPRESSIVITY OF BITHORAX TRANSFORMATIONS 149

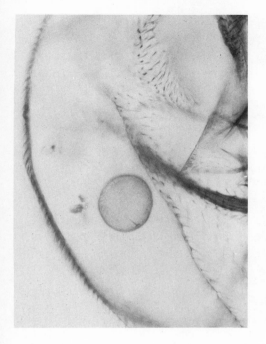

(3b)

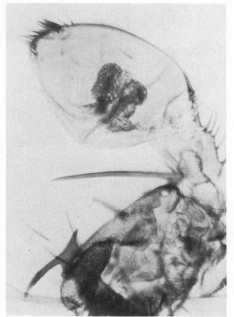

(3d)

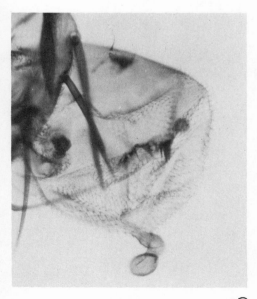

(3a)

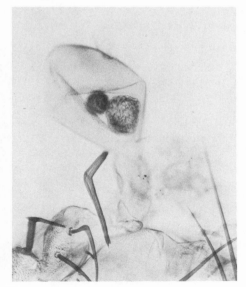

(3c)

type of triple row structures formed in the mutant halteres.

Normally the triple row (TR) pattern is formed by one row of bristles differentiated by the ventral wing compartment (ventral triple row or VTR) and two rows from the dorsal wing compartment (medial and dorsal triple rows, respectively MTR and DTR). It is a highly ordered and invariant pattern; the number and arrangement of the elements is constant. The three rows of bristles form perfect straight lines (Figure 3).

Flies mutant for bx^1 and bx^{34e} and other combinations of them show a variable number of elements of the TR in their halteres, and we can ask several questions about the type of patterns that they form. For example, do the individual elements of each type of differentiation appear in isolation? As can be seen in Figure 3, in some cases mutant halteres show only MTR bristles; thus individual elements can differentiate autonomously. This is not surprising since it is known that mutant bx^3 bristle cells can differentiate autonomously even when completely surrounded by haltere cells (Morata and Garcia-Bellido, 1976).

Do the individual elements form straight lines as in the normal TR? The answer to this question is shown also in Figure 3. Although the individual differentiation of the bristles is perfect, they do not form a straight line. The appearance of unpatterned groups of TR bristles is very common in the case of mutations producing weak wing transformations such as bx^1. This observation suggests that there are two separate processes during the formation of the TR: 1) the differentiation of the individual elements and 2) the spatial arrangement of them.

As the number of TR elements increases the pattern becomes more normal. The gradation towards the normal pattern is shown in Figure 3. In the case of bx^{34e}, where the MTR transformation is about 28% (Table 1), the TR pattern is still very imperfect although it can be seen that relative position of the three types of bristles is like in the normal TR; the medial bristles have the dorsal and ventral ones on each side (Figure 3). However, the bristles do not form straight lines; instead they are spread over an area. In bx^3/bx^{34e} or bx^{34e}/Ubx^1, where the amount of MTR is 59% and 54%, respectively, and the TR pattern is more perfect than in bx^1, the bristles still do not align properly. In the case of the most extreme combinations, the TR pattern is indistinguishable from the normal one (Figure 3).

Effect of Temperature

We have done experiments to study the effect of different temperatures on the expressivity of some of these combinations. In general although we have detected differences in the expression of some mutant combinations, the pattern of temperature effect is not

Table 4. Effect of the temperature on the expressivity of bx^1/bx^1

Temperature	n	N	Co	MTR
20	72	36 (41±27)	3 (2± 4)	1 (1± 4)
25	70	41 (46±28)	15 (9± 9)	13 (11±18)
30	70	34 (38±29)	32 (19±12)	33 (29±35)

See Footnote to Table 1 for explanation of column headings.

very consistent and therefore we have not followed this in detail. Only in the case of bx^1 homozygotes have we found a clear-cut and consistent effect of the temperature. As shown in Table 4, at 20°C bx^1 homozygotes show notum structures and practically no costa or MTR. However, at 25°C and especially at 30°C the amount of costa and MTR increases substantially while the notum remains constant. Thus, there is a specific effect of the temperature on the distal components of the transformations.

DISCUSSION

The expressivity of bx^3 and bx^{34e} (in homozygous condition, in combinations between them, and in interaction with Ubx alleles and over the deficiency) suggest they are typical hypomorphic mutants. The classical deficiency test suggests that bx^3 is closer to the amorphic condition. The following results obtained with these two mutants are consistent with this conclusion: 1) they are recessive; 2) the combinations over the deficiency are more extreme than when the alleles are homozygous; and 3) the expressivity of the trans heterozygote bx^3/bx^{34e} is intermediate between those of the respective homozygotes.

A different picture emerges from the study of bx^1. It has some of the features associated with hypomorphic mutants: it is completely recessive and its phenotype is weak and affected by the temperature. But its interactions with the other bx and Ubx alleles and the deficiency are different from those of bx^{34e} and bx^3. The phenotype over the deficiency is weaker than the homozygotes. It also shows very weak phenotype over most of the Ubx alleles and shows partial complementation with bx^{34e}. Thus, bx^1 is a different type of mutant and cannot be considered as hypomorphic although it certainly affects (at least in part) the same function impaired by bx^{34e} and bx^3.

All the Ubx alleles studied show bx phenotypes over the bx alleles. The two mutants associated with rearrangements, Ubx^{130}

and Ubx^{6-26}, show very strong transformations similar to those produced by the deficiency (compare Table 1 and Table 3). Thus, these mutants can be considered as genetic deficiencies for the bx gene. They behave like completely amorphic bx mutants. They probably also represent the amorphic condition for pbx and bxd as well, since they also fail to complement pbx and bxd alleles (Lewis, 1963 and 1964). The two point alleles, Ubx^1 and Ubx^{9-22}, show weaker phenotypes over the bx mutants (Table 3). However, when compared with the Ubx rearrangements, the reduction in expressivity affects mostly the notum structures. Consider, for example, the phenotypes over bx^{34e}; in bx^{34e}/Ubx^{130} the notum transformation is 59% and 76% for the MTR (Table 3). In the case of bx^{34e}/Ubx^1 the notum transformation is dramatically reduced to 0.6% whereas the MTR is 54%, not much less than in bx^{34e}/Ubx^{130}. The other point mutation Ubx^{9-22} gives similar values (2% and 53% respectively) to those of Ubx^1.

The comparison of the phenotypes of the point Ubx alleles over bx^{34e} with the bx^{34e} homozygotes (Tables 3 and 1) shows a different pattern of mesothoracic transformation; the two Ubx alleles over bx^{34e} show more MTR but less notum than bx^{34e} on its own.

In conclusion, the effect of the Ubx rearrangements is indistinguishable from that of the deficiency and affects equally all the mesothoracic structures considered. By constrast the two point Ubx alleles affect almost exclusively the production of the most distal mesothoracic structures: the costa or MTR. We do not know if this is a general effect of all the point Ubx alleles. It is worth pointing out that although the phenotype that the Ubx point mutants produce over bx alleles is weak, they also interact with the genes bxd and pbx (Lewis, 1963 and 1964). By contrast, a strong bx allele such as bx^3 has no effect on the other genes of the system.

The amount of mesothoracic tissue produced in the bithorax mutants is very variable; some combinations like bx^1/Ubx^1 or bx^1/bx^{34e} are almost wild-type; whereas in others like bx^3/Ubx^{130} or $bx^3/Df(3R)bxd^{100}$ the anterior metathorax virtually disappears and is replaced by a replica of the anterior mesothorax. Nevertheless, it is not possible to establish a gradation of transformation produced by the genotypes under study. The reason is that there is not a consistent pattern of mesothoracic transformation. Some genotypes such as bx^1/bx^{34e} or bx^1/bx^3 produce basically notum transformation and show very little or no transformation of the haltere into wing. On the contrary bx^{34e}/Ubx^1 or bx^{34e}/Ubx^{9-22} only show haltere transformation into wing, and there is practically no notum transformation. Thus, we cannot compare directly these genotypes since their effects are different according to position.

The fact that we cannot order these genotypes into a hierarchy of effects, immediately suggests that the bithorax function is of a complex nature. We can distinguish between an effect of notum transformation and an effect of wing (costa and MTR) transformation. While certain genotypes show the two effects, others predominantly only have one. The distinction of these two effects is also supported by the temperature effect on bx^1. The fact that the wing transformation is temperature-sensitive while the notum transformation is not, suggests a difference in the gene function in the two regions. One might speculate that the bx gene could be composed of two genes: one responsible for the proximal and other for the distal metathoracic development. It has been shown (Garcia-Bellido et al., 1973 and 1976) that there is a compartment boundary separating distal and proximal mesothoracic structures. The idea of genes controlling the development of specific regions is not unreasonable; two well-defined genes of the bithorax system, bx and pbx, are in charge of the development of two different regions within the metathoracic segment.

The perfection of the mesothoracic patterns in the mutant mesothorax increases with the number of mesothoracic structures. We can establish a distinction between differentiation of individual pattern components and the arrangement of them. Clearly, the differentiation of MTR bristles is a cell autonomous character which is not affected by the cellular environment (Morata and Garcia-Bellido, 1976), but the disposition of the elements seems to depend to some extent on the genotype of neighboring cells. This is suggested by the observation that while the TR pattern in the extreme bx genotypes is very similar to the TR pattern normally produced in the wing, clones of cells of the same genotypes are not able to reproduce the normal TR pattern, presumably because they are surrounded by metathoracic cells (Morata and Kerridge, unpublished).

In the clonal analysis of the TR in normal wings it has been suggested that movement of bristle cells is associated with the acquisition of the final pattern (Lawrence et al., 1979). Similar observations have been made for the sexcomb of the foreleg (Tokunaga, 1961) and the tarsus (Lawrence et al., 1979). It could be that the imperfect TR patterns found in the bx mutants results from some abnormality of migration of presumptive bristle cells.

REFERENCES

Garcia-Bellido, A., Ripoll, P., and Morata, G., 1973, Developmental compartmentalization of the wing disc of *Drosophila*, Nature New Biol., 245:251.
Garcia-Bellido, A., Ripoll, P., and Morata, G., 1976, Developmental compartmentalization in the dorsal mesothoracic disc of *Drosophila*, Devel. Biol., 48:132.
Garcia-Bellido, A., and Lewis, E.B., 1976, Autonomous cellular differentiation of homoeotic bithorax mutants of *Drosophila melanogaster*, Devel. Biol., 48:400.

Lawrence, P.A., Struhl, G., and Morata, G., 1979, Bristle patterns and compartment boundaries in the tarsi of *Drosophila*, J. Embryol. Exp. Morph., 51:195.

Lewis, E.B., 1963, Genes and developmental pathways, Amer. Zool., 3: 33.

Lewis, E.B., 1964, Genetic control and regulation of developmental pathways, in: "The Role of Chromosomes in Development", M. Locke, ed., Academic Press, New York, pp. 232-251.

Lewis, E.B., 1967, Genes and gene complexes, in: "Heritage from Mendel", A. Brink, ed., University of Wisconsin Press, Madison, pp. 17-41.

Lewis, E.B., 1978, A gene complex controlling segmentation in *Drosophila*, Nature, 276:565.

Lindsley, D.L. and Grell, E.H., 1968, Genetic variations of *Drosophila melanogaster*, Carnegie Inst. Washington, Publ. No. 627.

Morata, G., 1975, Analysis of gene expression during the development in the homoeotic mutant *Contrabithorax* of *Drosophila*, J. Embryol. Exp. Morph., 34:19.

Morata, G., and Garcia-Bellido, A., 1976, Developmental analysis of some mutants of the *bithorax* system of *Drosophila*, Wilhelm Roux Arch., 179:125.

Morata, G., and Lawrence, P.A., 1977, Homoeotic genes, compartments and cell determination in *Drosophila*, Nature, 265:211.

Tokunaga, C., 1961, The differentiation of a secondary sex comb under the influence of the gene engrailed in *Drosophila melanogaster*, Genetics, 46:157.

WHAT IS THE NORMAL FUNCTION OF GENES WHICH GIVE RISE TO HOMEOTIC MUTATIONS?

Allen Shearn

Department of Biology
The Johns Hopkins University
Baltimore, Maryland

WHAT IS A HOMEOTIC MUTATION?

A homeotic mutation may be defined as one which causes a transformation of the determined state of some or all of the cells of a particular imaginal disc or disc compartment of the determined state characteristic of some other disc or disc compartment. Mutations such as these have been intriguing geneticists and developmental biologists for more than fifty years because they seem to offer insight into how genes control developmental processes. A fundamental question raised by their very existence concerns the nature of the normal functions of those genes which can give rise to such homeotic mutations.

WHAT PROCESSES OCCUR DURING NORMAL IMAGINAL DISC DEVELOPMENT?

Table 1 lists those processes which we believe occur during normal imaginal disc development and indicates the class and time of action of genes whose function we believe is essential for each process.

During oogenesis, under control of the maternal genome, positional information is encoded in the cortex of the egg. Totipotent nuclei which migrate to the cortex incorporate this information when they become cellularized. This information must then be interpreted into a form which can be transmitted by cell heredity. The resulting cells are said to be determined. During the proliferative phase of imaginal disc development which occurs throughout the larval period, the determined state is transmitted from one cell generation to the next, but in addition, differently determined cells express different sets of gene functions. One obvious manifestation of this

Table 1. Processes which occur during normal imaginal disc development.

Process	Gene Class*	Time of Action	Expect Homeosis ?
Placement of positional information	Maternal (Inductor)		Yes
Interpretation of positional information	Zygotic (Activator)	Early embryonic	Yes
Transmission of determined state	Zygotic (Selector)	Late embryonic-puparium formation	Yes
Expression of determined state	Zygotic (Realizator)	Late embryonic-puparium formation	?
Terminal differentiation	Zygotic	After puparium formation	No

* The names in parentheses are those given by Garcia-Bellido (1977) for equivalent classes of genes.

differential gene expression is the distinctive size and shape of each pair of imaginal discs. During metamorphosis proliferation ceases and the process of terminal differentiation occurs. Each disc evaginates and secretes a specific pattern of bristles, hairs and sense organs.

DEFECTS IN SEVERAL OF THESE PROCESSES COULD LEAD TO HOMEOSIS

It is easy to imagine that genetically caused defects in the encoding or decoding of positional information could lead to homeotic transformations. Indeed it is quite possible that the specificity of such positional information can be inferred from the specificity of homeotic transformations caused by maternal-effect mutations. In fact, however, few mutations of this type have been identified. The maternal-effect mutations discussed in this volume by Eric Wieschaus may be examples of mutations affecting the encoding of positional information. If so, they seem to indicate that the positional information provided to the embryo is not extremely fine-grained.

It is not necessary that defects in the transmission or expression of the determined state should lead to homeosis. If, for example, each determined state were independent of each other, mutations in genes which act during the larval period would be expected

to cause developmental defects, but would not be expected to cause
switching to different determined states. However, nearly all of the
homeotic mutations that have been examined are in genes which normally act during the larval period (reviewed by Shearn, 1978). From
this we infer that each determined state is not independent. The
fact that defects in the transmission or expression of the determined
state can lead to alternate states implies that there exists underlying regulatory switches which relate each determined state to the
other (Kauffman, 1973).

TRANSDETERMINATION ALSO SUGGESTS THE EXISTENCE OF SUCH SWITCHES

Aside from homeosis there is another phenomenon called transdetermination which can lead to changes in the determined state of
imaginal disc cells (reviewed by Hadorn, 1978). This occurs when
mature discs are prevented from continuing their normal course of
development and instead their proliferative phase is extended by
culturing them either *in vivo* in the abdomen of adult females or
in vitro in a tissue culture medium we developed (Davis and Shearn,
1977). Transdetermination is detected by injecting such cultured
discs into metamorphosing larvae which allows terminal differentiation to occur.

As with homeosis, the mechanism of transdetermination is not
yet known. Nevertheless, the fact that it occurs also points to
the existence of mechanisms for switching between alternative states
of determination. The similarities in the directions of the observed transformations via homeosis or transdetermination suggest
that the same switches are involved in both cases (Ouweneel, 1976).
One argument that could be raised in opposition to that view is
that while the patterns are similar, they are not identical. I
believe that this is only because the complete range of transformations which could be caused by transdetermination (or homeosis) is
not yet known. For example, transdetermination from leg to antenna
or to wing has been known to occur for years; however, only recently
we observed that leg discs can also transdetermine (*in vivo* or *in
vitro*) to genital structures (Shearn et al., 1978a). So more transformations may yet be found.

HOMEOTIC MUTATIONS MAY AFFECT SWITCHES INDIRECTLY

Once we accept the idea of switches linking different determined
states, it is easy to imagine that genetically caused defects in such
switches could lead to homeotic transformations. I will call this
direct homeosis. An example of this type might be mutations at the
bithorax locus which are discussed in this volume by Dr. Morata. I
will call indirect homeosis those transformations which result from
genetically caused defects in the metabolism or behavior of imaginal
disc cells. Indirect homeosis may be considered to be *in situ* transdetermination. I will discuss one example of this type.

Several years ago we induced a set of lethal mutations on the third chromosome, screened them for ones which die after the second larval molt and examined these late lethals for imaginal disc abnormalities (Shearn et al., 1971). The largest class of mutations recovered caused a small-disc phenotype. All of the discs in such lethal larvae are smaller than those in normal larvae and cannot differentiate recognizable adult structures when injected into normal, metamorphosing larvae. Recently we isolated temperature-sensitive alleles of one of these small-disc mutants and found that exposure to the restrictive temperature for two days during the larval period caused all of the legs to be transformed to first legs (Shearn et al., 1978b). This is a clear example of what I call indirect homeosis.

I would like to summarize these general remarks with three points:

A. Both homeosis and transdetermination indicate the existence of regulatory switches which connect alternative states of determination.

B. Most homeotic mutations known are in genes which are essential continuously during the period of imaginal disc proliferation which suggests that these switches must be reset after each cell division. The phenomenon of transdetermination suggests that the resetting is not as accurate in cells which have divided much more than normal.

C. These switches may be revealed even by homeotic transformations which are caused by mutations in genes whose products may not be directly involved in the mechanism of such switches.

To understand how determined states are maintained and transmitted we must discover how many of these switches there are and which gene functions are directly involved in the switch mechanisms. But in addition to understanding how the switches work we must discover what kinds of perturbations in cell metabolism and/or cell behavior can interfere with their normal action.

STUDIES OF LETHAL HOMEOTIC MUTATIONS.

I would like now to describe some of the work we are doing with lethal homeotic mutations. The phenotype of these mutations has revealed several homeotic transformations not previously observed as well as many that are familiar. I will present evidence that these mutations are in zygotically acting genes which normally act continuously during the larval period, and that these mutations cause cell autonomous switches in developmental pathways. According to the scheme presented in Table 1 this means that these mutations are in genes that are involved either in the maintenance or the expression of the determined state. I will conclude by presenting our approach towards distinguishing whether these mutations act directly or indirectly on switches of the type that have been hypothesized.

The four mutations which I will discuss were from the same group of third chromosome late lethal mutations mentioned above (Shearn et al., 1971). Complementation tests indicate that these four mutations are actually two pair of alleles. One pair maps to a locus on the proximal left arm and the other to a locus on the distal right arm (Shearn, 1974). The imaginal disc phenotype caused by these mutations has been examined by injecting mutant discs into normal metamorphosing larvae and by examining the morphology of pharate adults for those alleles which survive to that stage. The entire range of transformation is summarized in Table 2. Every disc but the eye shows some kind of transformation. The wing and second leg do not show inter-disc transformations, but their posterior compartments are transformed to anterior as is found among *engrailed* mutants (Morata and Lawrence, 1975). All of the other discs tested do undergo inter-disc homeotic transformations. Some of these transformations (Table 2, row B) are also found in other mutants. Mutant antenna discs can give rise to tarsal structures as found in *Antennapedia* and *aristapedia* mutants (Le Calvez, 1948; Balkashina, 1929); mutant labial discs can give rise to antenna and leg structures as found in *proboscipedia* mutants (Bridges and Dobzhansky, 1933); mutant prothoracic discs can give rise to wing structures as found in *Hexaptera* mutants (Herskowitz, 1949); mutant haltere discs can give rise to wing structures as found in *bithorax* mutants (Lewis, 1963); and mutant third leg discs can give rise to second leg structures as found also in *bithorax* mutants. In addition to these, some of the transformations caused by mutations at the *III-10* and *703* loci have not been found to be caused by mutations at other loci (Table 2, row A). Mutant haltere discs can give rise to an antennal structure; mutant first leg discs can give rise to second leg structures and at a lower frequency, antennal structures; and mutant genital discs give rise to second leg and antennal structures. If it is true that homeosis and transdetermination are probes of the same regulatory mechanism, then the transformations from genital to leg and to antenna should have been expected since the genital disc transdetermines to leg and to antenna (Hadorn, 1963). These transformations indicate that the mutations we are studying do cause switches from one developmental pathway to another.

Clonal analysis has shown that these mutations cause cell autonomous transformations and are in genes which act continuously during the larval period. To perform this analysis we produced male flies hemizygous for y on the X chromosome and heterozygous for $y^+M(3)55$ and *mwh III-10* third chromosomes (Figure 1). Among the wild-type male progeny we recovered after γ irradiation we observed wing-like patches on halteres and first and third legs with apical bristles. These transformed patches were in areas of yellow bristles and/or *mwh* trichomes. Similar transformed clones were recovered on flies which had been irradiated at various larval stages.

Table 2. Homeotic transformations caused by *III-10* and other mutations.

	Eye	Ant	Lab	Pro	Wing	Hal	L1	L2	L3	Gen
A. *III-10* but not others						Ant	L2 Ant			L2 Ant
B. *III-10* and others		L(?)	Ant L(?)	Wing	P→A	Wing		P→A	L2	
C. others but not *III-10*	Wing	Gen			Hal	Abd		L1 L3	L1	
D. transdetermination, but not mutation					Ant Leg		Wing			

Based on these experiments we conclude that these mutations cause switches to alternate developmental pathways, are cell autonomous, and are in genes required throughout the proliferative stage. These characteristics, however, do not distinguish between direct or indirect homeosis. We are using genetic approaches in order to make such a distinction.

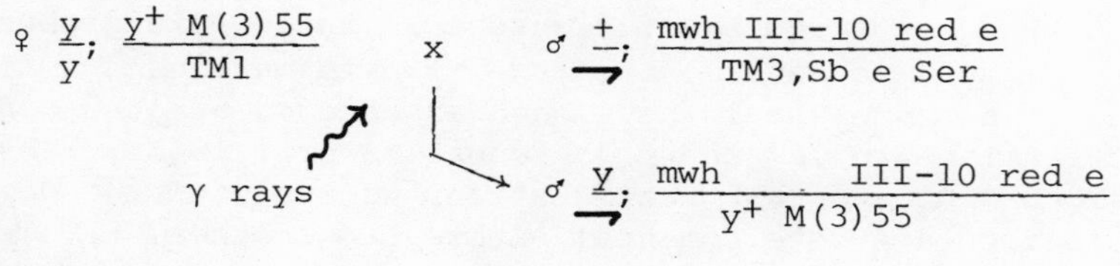

$$\female \; \frac{y}{y}; \; \frac{y^+ \; M(3)55}{TM1} \quad \times \quad \male \; \frac{+}{\rightarrow}; \; \frac{mwh \; III\text{-}10 \; red \; e}{TM3, Sb \; e \; Ser}$$

$$\gamma \text{ rays} \searrow \male \; \frac{y}{\rightarrow}; \; \frac{mwh \qquad III\text{-}10 \; red \; e}{y^+ \; M(3)55}$$

Figure 1: Genetic cross used to generate heterozgyous mutant progeny for testing cell autonomy

If mutations at both loci caused direct homeosis, then one might expect that homzoygosis for mutations at both loci would cause a phenotype similar to homzoygosis for either single mutation since their phenotypes are similar to each other. All possible double mutants have been examined and they all have more extreme phenotypes than any of the single mutants. In fact one of the double mutants expresses a small-disc phenotype. This probably means that mutations at one of the loci, at least, must cause indirect homeosis.

If mutations at either locus cause indirect homeosis, then non-leaky or amorphic alleles might be expected to express a more severe phenotype. To examine this point, we have begun to isolate additional alleles at both loci. So far we have five new alleles at the *III-10* locus and one new one at the *703* locus. Most of these alleles are not more extreme, but one of the *III-10* alleles does appear to be more extreme. *RF605* causes lethality in the first larval instar. Our work in this regard is still preliminary. But I believe that the analysis of multiple alleles should be adequate for distinguishing between direct and indrect homeosis.

To summarize, we have isolated mutations in two genes which reveal several new switches. If, as seems likely, these will be shown to cause indirect homeosis, then we will begin to examine the nature of the primary cellular defect which these mutations cause to probe the perturbations to which these switches are sensitive.

REFERENCES

Balkaschina, E.I., 1929, Ein fall der erbohomöosis (die genovariation "aristapedia") bei *Drosophila melanogaster*, Roux Archiv. Entwicklungsmech. Org., 115:448.

Bridges, C.B., and Dobzhansky, T., 1933, The mutant *proboscipedia* in *Drosophila melanogaster* -- a case of hereditary homeosis, Roux Archiv. Entwicklungsmech. Org., 127:575.

Davis, K.T., and Shearn, A., 1977, *In vitro* growth of imaginal disks from *Drosophila melanogaster*, Science, 196:438.

Garcia-Bellido, A., 1977, Homeotic and atavic mutations in insects, Amer. Zool., 17:613.

Hadorn, E., 1963, Differenzirungsleistungen wiederholt fragmentierter teilstücke männlicher genitalscheiben von *Drosophila melanogaster* nach kultur *in vivo*, Devel. Biol., 7:617.

Hadorn, E., 1978, Imaginal Discs: Transdetermination, in:"The Genetics and Biology of *Drosophila*, Vol. 2c", M. Ashburner and T.R.F. Wright, eds., Academic Press, London.

Herskowitz, I., 1949, *Hexaptera*, a homeotic mutant in *Drosophila melanogaster*, Genetics, 34:10.

Kauffmann, S., 1973, Control circuits for determination and transdetermination, Science, 181:310.

LeCalvez, J., 1948, $In(3R)ss^{Ar}$: Mutation aristapedia héterozygote dominant, homozygote léthale chez *Drosophila melanogaster*, Bull. Biol. France Belg., 82:97.

Lewis, E.B., 1964, Genetic control and regulation of pathways, in: "The Role of Chromosomes in Development", M. Locke, ed., Academic Press, New York.

Morata, G., and Lawrence, P.A., 1975, Control of compartment development by the *engrailed* gene in *Drosophila*, Nature, 255:614.

Ouweneel, W.J., 1976, Developmental genetics of homeosis, Adv. Genet., 18:179.

Shearn, A., 1974, Complementation analysis of late lethal mutants of *Drosophila melanogaster*, Genetics, 77:115.

Shearn, A., 1978, Mutational dissection of imaginal disc development, in: "Genetics and Biology of *Drosophila*, Vol.2C", M. Ashburner and T.R.F. Wright, eds., Academic Press, London.

Shearn, A., Davis, K.T., and Hersperger, E., 1978a, Transdetermination of *Drosophila* imaginal discs cultured *in vitro*, Devel. Biol., 65:536.

Shearn, A., Hesperger, G., and Hersperger, E., 1978b, Genetic analysis of two allelic temperature-sensitive mutants of *Drosophila melanogaster* both of which are zygotic and maternal effect lethals, Genetics, 89:341.

Shearn, A., Rice, T., Garen, A. and Gehring, W., 1971, Imaginal disc abnormalities in lethal mutants of *Drosophila*, Proc. Natl. Acad. Sci. U.S.A., 68:2594.

DISCUSSION

P.A. Lawrence: What is the operational distinction between direct and indirect homeosis?

A. Shearn: Among genes in which mutations can cause cell-autonomous homeosis, null alleles of those which give rise to direct homeosis should only express extreme homeosis, whereas null alleles of those which give rise to indirect homeosis should express other phenotypes such as discless or small discs. Thus to make such a distinction it is necessary to isolate a number of alleles, to identify those which are nulls and to examine their phenotype.

I. Kiss: Is it possible that some aspects of imaginal disc determination depends upon non-imaginal, larval tissue?

A. Shearn: Certainly it is possible. Although there are no examples known, if such a non-autonomous influence caused a homeotic phenotype it would be a clear example of indirect homeosis.

GENETIC AND DEVELOPMENTAL ANALYSIS OF MUTANTS IN AN EARLY ECDYSONE-INDUCIBLE PUFFING REGION IN *DROSOPHILA MELANOGASTER*

István Kiss[1], János Szabad[1], Elena S. Belyaeva[2], Igor F. Zhimulev[2], and Jenő Major[1]

[1] Institute of Genetics, Biological Research Center
of the Hungarian Academy Sciences,
Szeged, Hungary

[2] Institute of Cytology and Genetics
Academy Sciences of U.S.S.R.
Siberian Branch, Novosibirsk, U.S.S.R.

INTRODUCTION

Metamorphosis is a dramatic phase of the holometabolous insect life cycle, during which a reconstruction of the whole body morphology takes place. The transformation process involves degeneration and autolysis of most of the larval tissues as well as growth and differentiation of the adult organ rudiments. In the Diptera, which show the most extreme example of metamorphosis, both the initiation and the later processes are basically regulated by the steroid hormone ecdysone (Zdarek and Fraenkel, 1972). Metamorphosis probably involves stage-specific gene activation and repression, as exemplified by the ecdysone-induced puff-series on the salivary gland giant chromosomes of Dipteran larvae (Ashburner and Richards, 1976).

The studies reported on here deal with the initial regulation of metamorphosis, and were made by the Szeged and Novosibirsk laboratories in cooperation. Although the initial interests of the two groups were different, their experimental approaches eventually converged in an analysis of the genetic and developmental functions of the 2B5-6 early ecdysone-specific puff and the surrounding region. The present paper summarizes the preliminary results and tries to follow the "historical path" at the same time.

MATERIALS AND METHODS

Stocks were maintained on a standard yeast-cornmeal-agar medium at 25°C.

Mutations were induced with ethyl-methanesulfonate (EMS), according to the method of Lewis and Bacher (1969). Mutants were recovered on the X-chromosome using standard selection schemes as previously described (for the *npr* mutants: Kiss et al., 1976a; for the 2B1-10 region: Belyaeva et al., 1978). Mutant chromosomes were marked with y or y and w and maintained over the *Binsn* or *FM6* l^{69j} balancers (for the genetic symbols, see Lindsley and Grell, 1968).

Cytological localization was made by crossing y $1/FM6$ l^{69j} females with males carrying different deletions and/or duplications (see Figure 2).

For the complementation analysis, y $1^a/FM6$ l^{69j} females were crossed with y $1^b/Dp$ y^2 $Y67g$ males. In the case of complementing mutants, y $1^a/y$ 1^b females remained alive in the F_1 generation.

Genetic mapping was done using a y cv v f car standard chromosome.

Gynander mosaics were produced using the unstable ring-X chromosome $R(1)w^{vC}$ in the cross y w $1/Binsn$ ♀♀ X $R(1)w^{vC}/y^+Y$ ♂♂. F_1 progeny were screened for adult and (in the case of *npr* mutants) larval-puparial mosaics. The "expected number" of y w $1/R(1)w^{vC}$ adult mosaics was calculated by using the ratio betweeen non-lethal y $w/R(1)w^{vC}$ and $Binsn/R(1)w^{vC}$ gynanders which was found to be 1.7 (Kiss et al., 1976a). Adult mosaics were scored with respect to 30 landmarks on the body surface (Table 5), and the mosaic frequencies were used to delineate the mutant foci (Janning, 1978).

RESULTS AND DISCUSSION

The Non-Pupariating (*npr*) Phenotype

If we accept that metamorphosis involves stage-specific action of genes, we can suppose that the initiation of the process depends on the action of such genes. The function of these genes could be connected with the regulation of the high ecdysone titer immediately before puparium formation (Hodgetts et al., 1977) or with the early steps of the hormone action - provided that the ecdysone induces a cascade-type series of reactions. Mutations in such genes would either inhibit, or seriously delay, the onset of metamorphosis. Expression of the mutant phenotype would also be stage-specific; such mutants should have a normal embryonic and larval development but should be blocked in puparium formation. With this in mind, we looked for mutants on the X-chromosome which have a normal larval development but are inhibited or delayed in pupariation.

After screening more than 1500 independent lethal isolates, we found 3 *npr* mutants (Kiss et al., 1978). Their basic characteristics are shown in Table 1. They all have a good larval viability (the frequency of *y w 1/Y* mutant larvae on the 4th day after hatching was near to the 25% maximum). *npr* larvae survive 10-15 days after pupariation of their normal heterozygous sibs. The larval anatomy is by and large normal on the 4th day of larval life except for the imaginal discs which show various abnormalities (Table 1). Discs of *npr-1* and *nrp-2* seemed to be normal on the 4th day. By the 6th-7th day they became bloated with a short, distorted structure inside which was probably the result of a mutant evagination stopped at a very early phase (Figure 1A). A thin, structureless cuticle was deposited inside these discs after the 10th day of larval survival (Figure 1B). It probably represented the pupal cuticle. No cuticle was deposited on the abdomen. Apart from the imaginal discs, no signs of metamorphosis could be observed in other tissues of these larvae.

Autonomy of the mutant phenotype was tested in gynander mosaics, and by implanting wild type ring glands (the main ecdysone-producing organ in Dipteran larvae) into mutant larvae. As Table 2 shows, not one adult gynander was found in all the mutants although a significant number was expected. This probably means that the focus of the adult lethality is very large or multiple, and that viable mosaic combinations are therefore very rare. In the case of *npr-1* and *npr-2*, however, larval-puparial gynanders (LPGs) were recovered (Kiss et al., 1976b; Kiss et al., 1978). In such specimens, the cuticle remained soft and colorless on the mutant parts while it became normally tanned and sclerotized on the wild type parts. The boundary between them was rather sharp. Coincidence of the mosaic boundaries in the cuticle and the larval epiderm was proved by using the *mal* marker mutation which lacks aldehyde oxidase activity (Kiss et al., 1976b).

In older LPGs, mosaicism of tissues other than the epiderm also became apparent because metamorphosis began in the wild type parts while mutant parts showed the *npr* mutant phenotype. Wild type imaginal discs evaginated and secreted the pupal (sometimes even the adult) cuticle. Mutant discs showed instead the characteristic mutant morphology. Pigments were deposited in the wild type eye discs. Apolysis and secretion of the pupal cuticle could be observed on wild type parts of the abdomen. Mosaicism of the histolysis of fat body and salivary glands was also apparent. These observations suggest that: (i) the *npr-1* mutant function is autonomously expressed in different larval and imaginal tissues inhibiting or delaying the tissue-specific metamorphosis in a pleiotropic way; (ii) the mutant phenotype is not caused by the lack of the necessary ecdysone concentration, as there was clearly enough ecdysone in LPGs to induce pupariation and metamorphosis of the wild type parts.

Table 1. General Characteristics of the *npr* Mutants.

Mutant	Developmental Time[a]	Frequency of Pupariation	Imaginal Discs (size, structure)	Capabilities at Metamorphosis			Map position
				pre-pupal molt	pupal molt	pupal histolysis	
l(1)npr-1	4	0	normal[b] large, abnormal[c]	0	0	0	0.2
l(1)npr-2	4	0	normal[b] large, abnormal[c]	0	0	0	0.4
l(1)npr-3	4	0	very small, structureless[b] very small, structureless[c]	0	0	0	65.8

[a] The time needed to reach, under optimal conditions, the size of a mature Oregon-R larva.
[b] On day 4 of larval life.
[c] Data refer to the 8th-10th day of larval life.

Table 2. Autonomy of npr Mutant Phenotypes

Mutant	Gynander mosaicism			Implantation of wild type ring gland		
	Mosaics found		Mosaics expected[a]	No. of Larvae Surviving Injection		% Pupariation (within 5 days after transplant)
	Adult	Larval-Puparial		Experimental	Control	Experimental Control
1(1)npr-1	0	71	165	16	7	0 0
1(1)npr-2	0	8	10	33	9	0 0
1(1)npr-3	0	0	41	31	11	26 0

[a] No. of $y\ w\ 1/R(1)w^{vC}$ mosaics expected = 1.7 × (No. of $Binsn/R(1)w^{vC}$ mosaics found) (See Kiss et al., 1976a.)

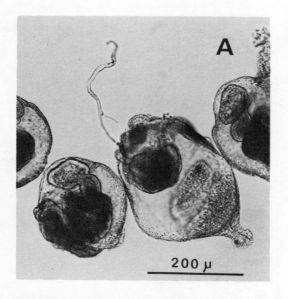

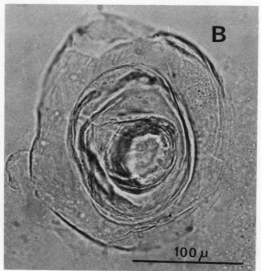

Figure 1: Imaginal discs of old *l(1)npr-1* larvae.
A: Leg and wing discs of a 7 day-old larva.
B: Pupal cuticle secreted inside a leg disc.

AN EARLY ECDYSONE-INDUCIBLE PUFFING REGION

The latter conclusion was further supported by the implantation of wild type ring glands. While this treatment induced puparium formation and histolysis in *npr-3* larvae, it remained ineffective in *npr-1* and *npr-2* (Table 2).

In another experiment, *npr-1* and *npr-2* mutant ring glands were implanted into *npr-3* larvae. Similarly to the wild type ring glands, they were able to induce pupariation, i.e. to secrete ecdysone. This fact, taken together with the deposition of pupal cuticle in the *npr-1* imaginal discs *in situ*, suggests that there is an internal ecdysone production in *npr-1* larvae, and the lack of metamorphosis is caused either by a relative insensitivity of the tissues to the hormone or by a general block of its early mechanism of action.

Genetic Structure of the 2B1-10 Region

To study the early action of ecdysone, the Novosibirsk group made a more direct approach; they chose one of the early ecdysone-specific puffs (2B5-6) on the X-chromosome and "saturated" its region with lethal as well as morphological mutations.

In the map of Bridges (see Lindsley and Grell, 1968), there are 10 bands in this region. Using the alcohol:acetic acid (3:1) fixative, only 6 bands could be detected (Figure 2): 4 thick (2B1-2, 3-4, 7-8, 9-10) and 2 thin ones (2B5 and 6).

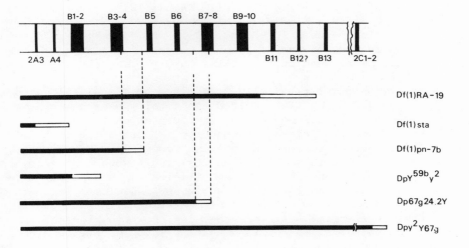

Figure 2: Cytology of the 2B1-10 region and right break points of the rearrangements used. Unfilled bars represent the uncertainty of determination.

The latter two can be seen only at puparium formation (puffing stage PS10-11), when the region is maximally condensed. Arrangement and morphology of the 6 bands show some kind of mirror-image symmetry, suggesting the presence of a "reverse repeat" (Bridges, 1935).

At 96-100 hours after egg-laying, when the giant chromosomes show the "larval" puffing pattern (PS1), one small puff can be found in the region; it apparently includes the decondensed 2B5 and 2B6 bands (Ashburner, 1972). At PS2, when the early ecdysone-specific puffs appear, decondensation of the region reaches its maximum and the neighboring thick bands also become loosened up. Following incorporation of ^{3}H-uridine, the label is uniformly distributed through the whole 2B1-10 zone. At PS4 a new puff starts to develop near the region of the 2C1-2 band. By PS9-10 the puff in 2B1-10 completely disappears and the other one in 2C1-2 decreases. A third puff becomes evident at this time in 2B11. The latter two puffs were described as one (2B13-17) previously (Ashburner, 1972). At the moment of pupariation (PS11), all 3 puffs regress and incorporation of ^{3}H-uridine ceases. A second wave of puffing activity appears in the region at PS18-20 (8-12 hours after puparium formation). A detailed description (including electron microscopic studies) of the structure and puffing activity of the region will be published elsewhere (E.S. Belyaeva, M.G. Aizenson, V.F. Semeshin, I. Kiss, I.F. Zhimulev, and K. Koczka, in preparation).

By screening 17,120 mutagenized X-chromosomes, 98 mutants were recovered within the territory of *Df(1)RA-19* (1F1-2 - 2B10, see Figure 2). *Df(1)sta* (1E1-2 - 2A3-4) uncovered 64 of them, i.e. they were located between 1F1-2 and 2A3-4, while the other 34 mutants were located outside of the deficiency between 2A3-4 and 2B10. The present study is focused on this latter group of mutants.

As previously known, alleles of *dor* and *br* are also located between 2A3-4 and 2B10. This raised the total number of mutants in our study to 42. According to the complementation test, they belong basically to 3 complementation units (Figure 3): 2 of them (*dor* and *swi* groups) seem to be "regular" complementation groups, while the third is a large complex containing 4 subgroups (*rbp, br, l(1)pp-1, l(1)pp-2*) and "overlapping" alleles which do not complement with alleles of more than one subgroup. Alleles belonging to different subgroups show a full complementation with each other. As for the "overlapping" alleles, they can show two kinds of complementation patterns; there are 5 alleles which do not complement with any other mutation in the complex ("long overlapping" alleles). *npr-1, npr-2* and *l(1)d.norm.-1^a* (the latter isolated in J.W. Fristrom's laboratory, see Stewart et al., 1972) belong to this class. Another 4 alleles overlap the *rbp* and *br* groups only ("short overlapping" mutants). The complementation pattern seems to be the same, irrespective of the temperature (18, 25 or 30°C) the crosses were

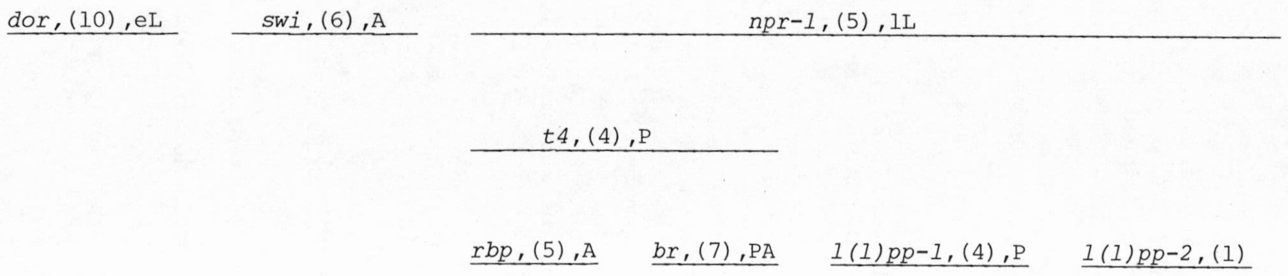

dor,(10),eL   swi,(6),A   npr-1,(5),1L
_____   _____   _____

                              t4,(4),P
                              _____

                       rbp,(5),A   br,(7),PA   1(1)pp-1,(4),P   1(1)pp-2,(1)
                       _____   _____   _____   _____

Figure 3: Mutant complementation groups in the 2B1-10 region. The numbers in brackets refer to the number of alleles in the group. Effective lethal phase is shown as A, pharate adult; eL, early 3rd instar larva; lL, late 3rd instar larva, P, pupa; PA, pupa-adult (see text).

made at, except that at 30°C a reduced viability was found in heterozygotes carrying t4 and l(1)pp-1 alleles. This suggests a partial overlap between the two groups.

While the majority of the mutants are lethal in homozygous or hemizygous condition, there are alleles which are viable and which show morphological abnormalities. (Some of the groups were named after these aberrations, e.g. swi: "singed wings", rbp: "reduced bristle number on palpus", etc.) Seven mutants are lethal or show a morphological phenotype in heterozygotes with Df(1)RA-19 but are viable and normal in homo- or hemizygous condition ("haplo-mutants"). The l(1)pp-2 complementation group is represented by a single "haplo-mutant" allele. Since the positions relative to each other of the complementation groups are not known, their sequence given in Figure 3 is arbitrary.

A more precise cytological localization of the complementation groups was made by using different duplications and deficiencies (Figure 2). All the groups were overlapped by $Dp67g24.2Y$ but none of them was complemented by $DpY59by2$. Heterozygotes carrying Df(1)pn-7b and mutations of the dor, swi, l(1)pp-1, or l(1)pp-2 groups are viable and normal. However, the "overlapping" mutants and the br and rbp alleles are lethal or have a mutant phenotype when heterozygous with Df(1)pn-7b. This interaction might indicate an overlap between the deficiency and rbp and br, but the existence of an independent br or rbp mutation on the Df(1)pn-7b chromosome can also not be excluded. According to the right breakpoints of the rearrangements used, the probable boundaries of the mutant region can be delineated. They can be 2B5 or the right half of 2B3-4 as well as 2B6 or the left half of 2B7-8. It is interesting that, by using a large deletion such as Df(1)RA-19, all the mutants seem to be located in the center of the region. Detailed results of the above-mentioned studies will be described elsewhere (E.S. Belyaeva, M.G. Aizenson, V.F. Semeshin, I. Kiss, I.F. Zhimulev and K. Koczka, in preparation).

Lethal Phenotypes in the Complementation Groups

Developmental characteristics of the representative alleles of each complementation group are summarized in Table 3. Homo- and hemizygous lethal alleles showing the most severe phenotype were mostly chosen. The l(1)pp-2 group is not included because its single allele is homozygous viable.

t81 (dor) hemizygous larvae grow slowly and die at a developmental age corresponding to the early 3rd istar. None of the larvae reach larval maturity and form puparium.

t200 (swi) develops normally up to adult eclosion. The pharate adult seems to be perfect but so weak that it is usually unable to

Table 3. Developmental Characteristics of Mutants of the 2B1-10 Region[1]

Mutant	Larval development Time (days)	Larval development Final Size	Pupari- ation	Prepupal Moult	Disc evagin- ation	Pupal Cuticle	Adult Cuticle	Histo- lysis	Testis (spermatid- elongation)
t324	4	n	0	0	0	+	0	0	$\pm$
t4	4	n	+	+	+	+	0	+	0
(rbp) t99	4	n	+	+	+	+	+	+	0
t35 (br)	4	n	$\pm$	$0(+)^2$	$\pm$	$\pm(+)^2$	$\pm(+)^2$	+	+
t10 (l(1)pp-1)	4	n	+	+	+	+	0	+	+
t81 (dor)	5	< n	0	0	0	0	0	0	0
t200 (swi)	5	n	+	+	+	+	+	+	+

[1] Abbreviations used: +, yes; $\pm$ not complete; 0, none; n, normal
[2] Data in brackets refer to the abdomen (see text)

leave the puparial case or dies soon thereafter.

As for the complex group, normal larval development is a common characteristic of all the mutants. The effective lethal phases are different, however.

The "long overlapping" alleles show a phenotype very similar to that of *npr-1* (see *t324* in Table 3). Similar symptoms were also described for *1(1)d.norm.-1^a* by Murphy et al., (1977).

The "short overlapping" alleles *(t4)* are early pupal lethals; their development is blocked right after the pupal moult. The testicular cysts remained completely larval. No sign of spermatid differentiation could be observed.

t99 (rbp) is a late pupal lethal; most of the animals reach an advanced stage of adult differentiation or the pharate adult stage.

t35 (br) shows a peculiar phenotype. After a normal larval development, puparial contraction is normal but tanning and sclerotization of the puparial case is rather imperfect; the cuticle remains yellowish and rather soft. Puparia dry out quickly. Imperfect puparium formation seems to be autonomously expressed in gynander mosaics. The imaginal discs seem to be normal at pupariation but 24 hours later become strongly bloated with a half-evaginated structure inside the peripodial membrane (Figure 4A). In 5-6 days, pupal and adult cuticles were laid down and tanned inside the discs (Figure 4B); eye discs became pigmented. Continuous pupal and adult cuticle was never found on the head and thorax. This is probably caused by the abortive morphogenesis of the imaginal discs which prevents the formation of a continuous adult epiderm. On the contrary a normal pupal and, sometimes, adult cuticle was found on the abdomen, forming a sack-like structure all around with a sharp boundary at the thorax. This suggests a relatively normal function/differentiation of the abdominal histoblasts. Histolysis of the internal larval tissues takes place normally but the adult organ rudiments apparently do not grow and differentiate.

t10 (1(1)pp-1) is an early pupal lethal. However, the testes became slightly coiled containing elongated sperm bundles.

Trans-heterozygotic combinations containing an "overlapping" and an "overlapped" lethal allele remain lethal and generally show the phenotype of the more advanced partner. There is one exception, however; *npr-1/t10* trans-heterozygotes develop into an apparently perfect adult except that the head is not everted but differentiates in its original position. Such specimens cannot eclose. Since *npr-1* is a larval and *t10* an early pupal lethal, the adult development in the heterozygote can be regarded as partial complementation.

AN EARLY ECDYSONE-INDUCIBLE PUFFING REGION   175

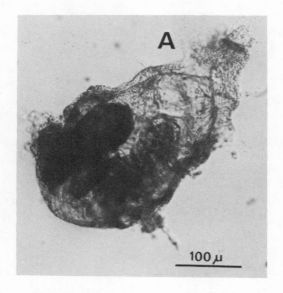

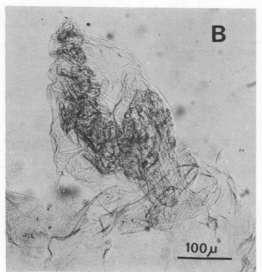

Figure 4: Imaginal discs of *t35* mutant pupae.
           A: Leg disc 4 days after pupariation. A half-evaginated
              structure is visible inside.
           B: Pupal and adult cuticle inside a leg disc 10 days
              after pupariation.

On the basis of the above-mentioned facts we can conclude that, at least in the complex group, the mutations are expressed in a stage-specific way and the developmental block occurs during pupariation or after the pupal molt, i.e. following the main active periods of the 2B5-6 puff (late larva and late prepupa).

Gynander Mosaic Analysis of the Mutant Groups

As contrasted with the "long overlapping" alleles which did not survive in adult gynanders (see *npr-1* and *npr-2* in Table 2), lethal alleles of all the other groups gave viable mosaic flies. This made a comparative mosaic analysis of the different mutant groups possible. Our aim was to see whether lethal foci characteristic for the subgroups of the complex could be determined, and if they showed different tissue specificities.

Table 4. % Viability of Mutant Gynanders

Cross: $\dfrac{y\ w\ sn^3\ 1}{Binsn}$ ♀ X $\dfrac{R(1)w^{vC}}{y+Y}$ ♂

Mutant	$\dfrac{y\ w\ sn^3\ 1}{R(1)w^{vC}}$ ♀♀		$\dfrac{Binsn}{R(1)w^{vC}}$
	Expected[1]	Found (%)	
Control ($y\ w\ sn^3\ 1$)	–	371	219
t4	119	14 (11.7)	70
t99	44	23 (52.3)	26
t336	73	15 (20.5)	43
t10	124	60 (48.4)	73

[1] The same as in Table 2

Viability of the mutant gynanders carrying representative alleles of the different groups was drastically decreased (Table 4). This suggests that only certain mosaic combinations can survive and that the lethal phenotypes have some kind of tissue specificity in their expression.

To examine the data from gynanders in more detail, 29 adult surface structures evenly distributed on the blastoderm fate map were scored for being mutant or mosaic. A representative example of these results (with *t10*) is given in Table 5. Two conclusions can be drawn if we compare the mutant data with those of the control: (i) the *t10* mutation has a general influence on all the structures scored, decreasing the viability when they are mutant, and (ii) this effect is more pronounced in the abdomen, reaching its maximum in the region

Table 5: % Frequency of Surface Structures Being Mutant or Mosaic in Control and *t10* Gynanders

Region	Landmark Structure	$y\ w\ sn^3\ 1^+$ control		$y\ t10\ w\ sn^3$	
		Mutant	Mosaic	Mutant	Mosaic
Head	postorbital bristle	58.7	–	20.9	–
	vibrissa	61.7	–	17.0	–
	antenna	58.0	2.6	17.1	3.4
	anterior orbital bristle	58.9	–	20.2	–
	outer vertical bristles	54.9	1.4	19.1	3.5
	postvertical bristle	55.9	–	23.4	–
	proboscis	56.3	6.7	16.4	2.3
Dorsal thorax	humeral bristles	64.7	0.8	18.4	0.0
	anterior scutellar bristle	68.9	–	22.6	–
	posterior supraalar bristle	67.9	–	21.1	–
	anterior notopleural bristle	69.8	–	23.9	–
	wing	67.0	1.0	18.6	1.2
Ventral thorax	sternopleural bristles	67.0	4.5	18.5	1.5
	leg 1	60.4	12.5	14.1	10.4
	2	61.7	5.7	15.7	6.5
	3	61.0	6.3	10.9	11.9
Dorsal abdomen	tergite 1	60.9	5.6	11.6	4.2
	2	56.9	11.8	11.5	13.1
	3	55.8	12.7	10.5	19.9
	4	51.9	15.5	11.1	20.0
	5	51.7	14.4	11.0	20.4
	6	49.3	15.0	7.9	14.3
	7	46.9	10.4	3.7	4.8
Ventral abdomen	sternite 2	60.2	6.9	8.6	3.2
	3	59.8	8.8	8.5	7.5
	4	57.1	9.1	7.5	7.5
	5	53.7	8.4	5.9	9.6
	6	50.8	7.2	4.3	5.9
	anal plate	51.5	7.5	2.3	2.9
	external genitalia	40.6	20.9	0.0	2.9

Table 6. Average Frequency (%) of Structures Being Mutant in Gynanders

Mutant	Number of gynanders scored	Head	Thorax		Abdomen		Anal plate + genitalia
			Dorsal	Ventral	Tergites	Sternites	
Control ($y\ w\ sn^3\ l^+$)	371	58.7	66.6	66.2	59.5	59.5	53.1
t4	120	6.0	9.5	7.7	20.0	15.2	29.4
t99	115	13.6	30.0	29.6	33.5	26.0	35.0
t336	95	8.0	7.3	6.3	17.7	13.3	21.0
t10	123	20.1	20.6	18.6	16.5	10.3	2.6

% Frequency of landmarks being mutant or mosaic[1]

[1] Before calculating the average %, half of the mosaic value was added to the mutant value for each structure. See Table 5 for the landmarks scored.

of the anal plate and genitalia, i.e. some kind of lethal focus can be confined to this region. Both effects are also shown by the other 3 mutants tested, except that the separation of the lethal "focus" is not so clear as in the case of *t10*. The general effect of the mutations is further supported by the observation that the size of the mutant spots was generally decreased in all the cases irrespective of their positions on the body.

Average frequencies of structures being mutant in gynanders are given in Table 6. As it can be seen, the lowest mutant percentage was found in different body parts of the different mutants. The sensitive area was located in the head and thorax for *t4* and *t336*, in the head for *t99* and near the end of the abdomen for *t10*. Although we do not know which tissues/organs are included in the "sensitive area", the mutants clearly form two groups in this respect: one includes *t4, t99* and *t336*; the other *t10*. The shows an interesting parallelism with the complementation map (Figure 3): *t99 (rbp)* and *t336 (br)* are "overlapped" by *t4*, while *t10 (l(1)pp-1)* can complement all these three and is clearly separated from them. Another case of parallelism is shown by the "long overlapping" alleles (e.g. *npr-1*) which do not complement with any other subgroup in the complex and do not yield any viable gynander mosaic combination.

CONCLUSIONS

Before drawing conclusions, we can summarize the facts as follows:

1. There is an early ecdysone-inducible puff in the region studied which apparently develops from the 2B5 and 2B6 bands and includes, during its maximal extension, the 2B1-10 region. The main active periods of this puff are a few hours before puparium formation and pupal moult, i.e. immediately before the major events of the early phase of metamorphosis.

2. Mutants isolated in the region belong basically to 3 complementation groups. One of them shows a complex complementation pattern. All the mutants could be confined to the region between 2B3-4 - 2B7-8.

3. Lethal phenotypes within the complex group are expressed in a stage-specific way and show two common characteristics: (i) larval development is always normal, (ii) effective lethal phases occur just before puparium formation or after the pupal moult, i.e. after the main active periods of the 2B5-6 puff. The lethal phenotype (inhibition of metamorphosis) is autonomously expressed in the different tissues of the "long overlapping" alleles suggesting the parallel expression of the mutant gene in the different tissues at the same time.

4. Gynander mosaic analysis revealed that all the mutations influenced the viability of the mutant surface structures in two ways: (i) a general decrease of viability was expressed, irrespective of the position of the mutant structure in the body; (ii) viability of certain parts was significantly lower than the average. The positions of these "sensitive areas" in the different mutants showed an interesting parallelism with the complementation map.

Two matters should now be discussed with respect to these results: the genetic organization of the complex locus as well as the possible relation between the mutants and the 2B5-6 puff.

As for the organization of the complex locus, two possibilities can be considered: either all the mutations belong to the same gene and the complementation pattern results from intracistronic complementation; or the different subgroups represent distinct cistrons, and the complementation observed is the result of intergenic interaction. In the latter case, however, the nature of the "overlapping" alleles has to be explained. They could represent either deletions or mutations with a polar effect. The deletion hypothesis is contradicted by two facts. First, no deletion could be detected cytologically in these alleles; second, $npr-1/t10$ trans-heterozygotes showed a partial complementation which would not be expected if $npr-1$ were a deletion. The complex locus might be an operon-like structure including several cistrons under a common control and the "overlapping" alleles would represent regulatory mutants. Phenotypic complementation and gynander mosaic data are equally compatible with both possibilities. To decide between them will require new genetic variants and new experimental approaches, e.g. rearrangements with breakpoints dividing the complex locus or determination of the distance between the two extreme points of the complex by fine structure mapping. Work is in progress in both directions.

Coincidence of the mutant groups and the puffing function may be significant. It is supported by the cytological position of the mutant groups and, in the case of the complex locus, by the phenotypic characteristics of the mutants. In either case the 2B5-6 puff remains inducible in $npr-1$ (unpublished observations, as well as personal communication by Dr. M. Ashburner).

We believe that elucidating the genetic organization and the developmental function(s) of this interesting region will give deeper insights into the mechanisms regulating the initiation of metamorphosis and the action of ecdysone.

ACKNOWLEDGEMENTS

The authors are indebted to H.J. Becker, J.W. Fristrom and M.M. Green for kindly providing *Drosophila* stocks.

REFERENCES

Ashburner, M., 1972, Puffing patterns in Drosophila melanogaster and related species, in: "Developmental Studies on Giant Chromosomes", W. Beermann, ed., Springer-Verlag, Berlin-Heidelberg-New York.

Ashburner, M., and Richards, G., 1976, The role of ecdysone in the control of gene activity in the polytene chromosomes of Drosophila, in: "Insect Development", P.A. Lawrence, ed., Blackwell Scientific Publications, London, pp. 203-225.

Belyaeva, E.S., Aizenson, M.G., Ilina, O.V., Zhimulev, I.F., 1978, Cytogenetic analysis of the puff in the 2B1-10 region of the X-chromosome of Drosophila melanogaster, Dokl. Akad. Nauk. U.S.S.R., 240:1219.

Bridges, C.B., 1935, Salivary chromosome maps, J. Heredity, 26:60.

Hodgetts, R.B., Sage, B., and O'Conner, J.D., 1977, Ecdysone titers during postembryonic development of Drosophila melanogaster, Develop. Biol., 60:310.

Janning, W., 1978, Gynandromorph fate maps in Drosophila, in: "Genetic Mosaics and Cell Differentiation", W.J. Gehring, ed., Springer-Verlag Berlin-Heidelberg-New York.

Kiss, I., Bencze, G., Fekete, E., Fodor, A., Gausz, J., Maroy, P., Szabad, J., and Szidonya, J., 1976a, Isolation and characterization of X-linked lethal mutants affecting differentiation of the imaginal discs in Drosophila melanogaster, Theoret. Appl. Genet., 48:217.

Kiss, I., Bencze, G., Fodor, A., Szabad, J., and Fristrom, J.W., 1976b, Prepupal-larval discs in Drosophila melanogaster, Nature, (Lond.),262:136.

Kiss, I., Major, J., and Szabad, J., 1978, Genetic and developmental analysis of puparium formation in Drosophila melanogaster, Molec. Gen. Genet., 164:77.

Lewis, E.B., and Bacher, F., 1969, Method of feeding ethyl methane sulfonate (EMS) to Drosophila males, Dros. Inf. Serv., 43:193.

Lindsley, D.L., and Grell, E.H., 1968, Genetic variations of Drosophila melanogaster, Carnegie Inst. Wash. Publ., 627.

Murphy, C., Fristrom, J.W., and Shearn, A., 1977, Aspects of development of imaginal discs in a nonpupariating lethal mutant in Drosophila melanogaster, Cell Diff., 6:319.

Stewart, M., Murphy, C., and Fristrom, J.W., 1972, The recovery and preliminary characterization of X-chromosome mutants affecting imaginal discs of Drosophila melanogaster, Develop. Biol., 27:71.

Zdarek, J., and Fraenkel, G., 1972, The mechanism of puparium formation in flies, J. exp. Zool., 179:315.

DEVELOPMENTAL AND GENETIC STUDIES OF THE INDIRECT FLIGHT MUSCLES OF
DROSOPHILA MELANOGASTER

I.I. Deak*, A. Rahmi, P.R. Bellamy, M. Bienz, A. Blumer,
E. Fenner, M. Gollin, T. Ramp, C. Reinhardt, A. Dübendorfer and B. Cotton

Institute of Zoology
University of Zurich
Switzerland

* To our deep sorrow Ilan Ivan Deak died on July 29, 1979. Since his death, his group has continued working on his muscle development project. This paper represents the joint efforts of all the members of his group.

INTRODUCTION

These studies were undertaken with the aim of investigating myogenesis in *Drosophila* with the special emphasis on the embryonic origin of precursor cells and the establishment of the ultrastructural organization during muscle differentiation.

Before taking up specific questions of muscle development, we will briefly describe the anatomy and organization of *Drosophila* muscles. Then we shall concentrate on the two main approaches of our work: the search for the primordia of the adult indirect flight muscles, and the detailed analysis of flightless mutants which should lead to an understanding of the genes involved in muscle development.

ANATOMY AND ORGANIZATION

Larval muscles are characterized by their large nuclei placed at the periphery of the fibers (Figure 1a). Further, each myosin filament is surrounded by 10-12 actin filaments arranged in a rather irregular fashion (Osborne, 1967; Crossley, 1968; Dübendorfer et al., 1978). All larval muscles are highly contractile, shrinking to 25% of their resting length in the contracted state. This threefold change in length is attributed to perforations in the Z-discs

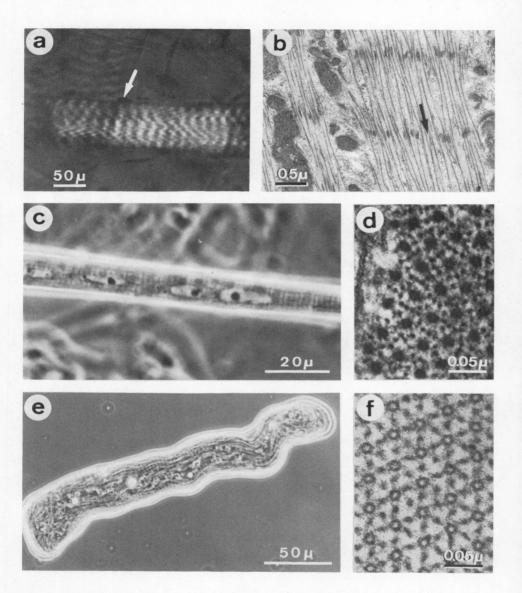

Figure 1: Different muscle types of larvae and flies. Third instar larval muscle: a) cross striation, nuclei in the periphery (arrow), b) electron micrograph showing perforated Z-bands (arrow). Tubular adult muscle from *in vitro* culture: c) sharp cross striation, central canal with nuclei, d) e.m. cross section. Adult fibrillar muscle: e) one muscle in tissue culture, f) e.m. cross section (Modified from Dübendorfer et al., 1978; Dübendorfer and Eichenberger-Glinz, 1980).

(Figure 1b) that allow the myofilaments to slide into the neighboring sarcomeres (Goldstein, 1971; Dübendorfer et al., 1978).

Among the muscle types of the adult fly we chose to concentrate on the large and well organized tubular and fibrillar muscles. Tubular muscles can easily be recognized since their nuclei are always in an axial core surrounded by the myofibrils (Figure 1c). The sarcomeres lie exactly in register, the nuclei are small and elongated, 3-6 µm long. Each myosin filament is surrounded by 10-12 actin filaments forming a more regular pattern than in larval muscle (Figure 1d) (Pasquali-Ronchetti, 1970; Dübendorfer et al., 1978).

Sarcomeres of fibrillar muscles are smaller than those of tubular muscles and are not exactly in register, so that cross-striation is not easily seen (Figure 1e). The nuclei lie scattered throughout the fiber. The most characteristic feature of fibrillar muscle is the strictly regular array of their filaments, six actin filaments surrounding each myosin filament (Figure 1f) (Shafiq, 1963; Elder, 1975; Dübendorfer et al., 1978).

The indirect flight muscles, which are fibrillar, constitute the bulk of the muscle mass in the thorax of a fly (Figure 2a,b). Six dorsolongitudinal muscles run from the anterior to the posterior margin of each half thorax (Figure 2a). All other fibrillar muscles of the thorax run dorsoventrally (Figure 2b). Tubular muscles in the thorax are the small direct flight muscles and the leg muscles; among the latter are the large tergal depressor of the trochanter of the second leg (Figure 2b), which enables the fly to jump (Behrendt, 1940).

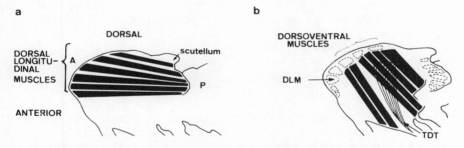

Figure 2: Diagrammatic sagittal sections (a and b) through the thorax of a fly. TDT = tergal depressor of the trochanter of the second leg. (Modified from Deak, 1978).

MUSCLE PRIMORDIA

Based on analyses of gynandromorphs Deak (1977) published blastoderm fate maps of the mutants heldup (hdp), flapwing (flw), vertical wing (vtw), upheld^2 (up^2), and indented thorax (int). Scoring for ability to jump, morphological defects in the tergal depressor of the trochanter, and leg weakness, he mapped the foci of all these tubular muscle defects (with the exception of *vtw*) to the site of the prospective midventral thoracic mesoderm on the blastoderm. The mutations upheld (*up*) and heldup^2 (*hdp^2*) (Hotta and Benzer, 1972), flightless H (*flt H*), and flightless O (*flt O*) (Koana and Hotta, 1978), as well as *int*, and *flw* (Deak, 1977), all affecting the fibrillar muscles of the thorax, mapped to the same region, as did nine more flightless mutations isolated in our laboratory. From Poulson's analysis (1950) it is known that the midventral mesoderm of the embryo contains the cells of the prospective muscle tissue of the larva. We conclude therefore that those mutations which have their foci in this region affect the muscle precursor cells directly and draw the further conclusion that the precursors of the adult thoracic tubular and fibrillar muscles are located in the narrow stripe of thoracic ventral mesoderm of the embryo at the blastoderm stage (for further discussion see Deak, 1977).

In addition to the early embryo, we also studied the anlagen of adult muscle in third instar larvae. From transplantation and extirpation experiments by Zalokar (1947), and Poodry and Schneiderman (1970), and from surface transplantations by Bhaskaran and Sivasubramanian (1970) it became evident that there are precursors of tubular muscles in the leg discs of the third instar larva. The possibility that the muscles surrounding the metamorphosed implanted leg discs had migrated there from the host was excluded by Ursprung et al. (1972). In their experiment they marked donors and hosts with different enzyme markers, and found that the muscles were always of the donor genotype.

We too have transplanted leg and wing discs of third instar larvae into metamorphosing hosts, and we always found tubular muscles around the implants. Even eye, antenna, leg and wing discs, which Dr. M. Milner cultured for us *in vitro*, began twitching autonomously after one day in culture; and the muscles were seen to be tubular with the phase contrast microscope.

Whether or not imaginal discs of the last larval instar contain any progenitor cells of fibrillar thoracic muscles, proved more difficult to investigate. We analyzed the transplanted leg and wing discs with both electron and light microscopy, using interference contrast afer mechanical dissociation of the implants and digestion of the cuticular parts by chitinase. Yet we only found fibrillar muscles in three cases out of 100 implanted second leg discs, and none at all in wing disc implants. This finding agrees with the results of Zalokar (1947). When he extirpated wing discs, all the fibrillar muscles of the thorax developed normally, but after removing second leg discs, the dorsoventral fibrillar muscles were absent in the adult flies.

More evidence for the absence of the precursors of fibrillar muscles in the wing discs came from studies of the mutation wingless ($_{wg}$) (Deak, 1978). These mutant adult flies often have extremely large duplications in the mesothorax and at the same time lack other parts. If muscle precursor cells were located in a particular region of the wing discs and this region was either duplicated or absent in the mutant animal, one would expect to find either an increase or decrease in the amount of that muscle. However, no change in the amount of fibrillar muscle was observed in adult flies with nota duplicated or absent, suggesting that the fibrillar muscle precursors are not part of the wing disc. However, the mass of tubular muscle varied proportionately with the size of the notum, thus demonstrating that *wg* affects the muscles as well as the epidermis, and that the only muscle precursors in the wing disc are of the tubular type.

Our results and data found in the literature seem to point to the conclusion that the wing disc does not contain precursor cells of fibrillar muscle, but that the second leg disc may well contain the precursors of the dorsoventral fibrillar muscles. This may suggest that in the third larval instar the anlagen for dorsoventral and dorsolongitudinal muscles are separate. These two types of adult muscles are morphologically identical and proved indistinguishable in the biochemical comparison done in our laboratory. Interestingly, some of our EMS-induced X-chromosomal mutations, which will be described below, affect only the dorsolongitudinal muscles.

Assuming that there are two anlagen for fibrillar thoracic muscles, the question arises, where are the precursors of the dorsolongitudinal fibers located in the third larval instar? Tiegs suggested in 1922, based on histological work, that precursors of fibrillar muscles are associated with larval muscles. Using the BAO staining technique we found small (i.e. adult) nuclei in close association with larval mesothoracic muscles. Naturally we would need much more data in order to ascertain whether these cells represent developmental stages of dorsolongitudinal fibrillar muscle.

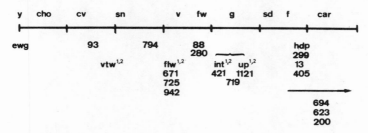

Figure 3: Distribution of flightless mutations along the X-chromosome. Above the chromosome: markers used for mapping, below the chromosome: mutations isolated in our laboratory (numbered) and mutations from other laboratories (lettered).

ANALYSIS OF FLIGHTLESS MUTANTS

This study is based on the idea that the components involved in myofibrillogenesis might possibly be uncovered by genetically altering particular structures and correlating the changes to any modifications in the protein pattern.

Most of the mutations analyzed were induced in our laboratory through the use of ethylmethane sulfonate. We selected non-flying animals with normal wing morphology, but abnormal wing position and/or identations of the thorax and/or gross morphological defects in their thoracic musculature as seen in the living fly under polarized light. First, we roughly mapped all sex-linked mutations using a number of markers evenly distributed along the X-chromosome (Figure 3). Then we crossed the mutants with each other and scored the trans-heterozygous flies for abnormal wing position and inability to fly. The preliminary data show nine complementation units on the X-chromosome affecting adult musculature (Figure 3). The mutations were analyzed for morphology, ultrastructure, and protein composition. Since some of the observed muscle defects were common to several mutants, we divided them up into four arbitrary groups.

We assigned the mutations erect wing (*ewg*) and 93 to the first group. In mutations belonging to this group the dorsolongitudinal

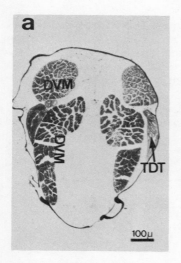 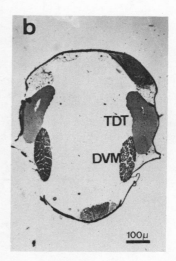

Figure 4: Longitudinal sections through the thoraces of mutant flies. (a) *ewg*: the dorsolongitudinals are missing; (b) *vtw*: all dorsolongitudinal and most dorsoventral muscles are missing. DVM = dorsoventral muscles, TDT = tergal depressor of the trochanter of the second leg.

muscles are missing completely (*ewg*, Figure 4a) or in part (*93*). The dorsoventral muscles are normal in *93* but swollen in *ewg*. The ultrastructure of the tergal depressor of the trochanter, however, is strongly disturbed in *93* and unaffected in *ewg* animals. Since in these two mutations the dorsolongitudinal muscles are affected selectively, we assume that there is at least one genetically controlled step that is different in the development of these two types of morphologically and biochemically identical fibrillar muscles.

Mutations of *vtw* (two alleles) and *flw* (five alleles) belong to the second group of mutations, in which both types of fibrillar muscles can be affected. In this group we expect to find both mutations primarily affecting muscle development, as well as mutations that primarily affect other organs such as trachaea or nerves. The mutants *vtw* and *flw* exhibit a highly variable expressivity. In some flies there is a large amount of completely disorganized fibrillar muscle material in the thorax, whereas in others most fibrillar muscles are missing (Figure 4b).

The third group comprises the genes *794 and 694*, the latter with three alleles. The muscle fibers of these mutants are thin and their organization is loose. There are small gaps between the fibers (Figure 5). These mutations apparently do not eliminate the muscle precursors, but may instead affect the assembly of the fibers.

The fourth group consists of the complementation units *88* with two alleles, *hdp* with four alleles, and *int-up* with seven alleles

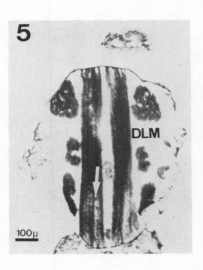

Figure 5: Cryotome section through the thorax of the mutant *794*. Arrow = gaps between the fibers, DLM = dorsolongitudinal muscles.

Figure 6: Electron micrograph of *int*. Arrows = disrupted Z-bands

(Figure 3). The mutations *int* and *up* may represent alleles of a single gene. Though phenotypically different, they do not complement (*up* phenotype), and no recombinants could be found in 15,000 flies scored. In *up* flies most of the muscle mass is present, but its organization is grossly abnormal. In *int* and hdp^2 animal muscles are normally formed but become disorganized some days after eclosion. Disrupted sarcomeres and ragged or totally absent Z-bands (Figure 6) are characteristic ultrastructural features of all the mutations of this group.

We have compared the protein patterns of the mutations with defective Z-bands (fourth group) on SDS-polyacrylamide gels, and have seen that most of them exhibit absence, reduction or modification of the same three protein bands (Figure 7). In fact, our data suggest that at least two of these bands represent Z-band proteins, since they co-migrate with purified Z-bands. Such material was obtained by treatment of a clean preparation of undegraded myofibrils with concentrated salt solution (Gollin, unpublsihed).

In summary, we now have a variety of muscle mutations at our disposal, which seem to interfere with muscle development at various levels, thus providing a basis for further research towards a better understanding of muscle formation and muscle function in insects.

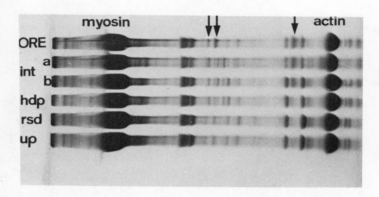

Figure 7: Sodium dodecylsulfate-polyacrylamide gels of flightless mutants. Arrows = modified protein bands. *int* (a) = young flies, *int* (b) = old flies.

REFERENCES

Bhaskaran, G., and Sivasubramanian, P., 1969, Metamorphosis of imaginal discs of the housefly: evagination of transplanted disks, J. Exp. Zool., 171: 385.

Behrendt, R., 1940, Untersuchungen über die Wirkungen erblichen und nicht-erblichen Fehlens bzw. Nichtgebrauchs der Flügel auf die Flugmuskulatur von *Drosophila melanogaster*, Z. wiss. Zool., 152: 129.

Crossley, A.C., 1968, The fine structure and mechanisms of breakdown of larval intersegmental muscles in the blowfly, *Calliphora erythrocephala*, J. Insect Physiol., 14: 1389.

Deak, I.I., 1977, Mutations of *Drosophila melanogaster* that affect muscles, J. Embryol. exp. Morph., 40: 35.

Deak, I.I., 1978, Thoracic duplications in the mutant *wingless* of *Drosophila* and their effect on muscles and nerves, Develop. Biol., 66: 422.

Dübendorfer, A., Blumer, A., and Deak, I.I., 1978, Differentiation in vitro of larval and adult muscles from embryonic cells of *Drosophila*, Wilhelm Roux Arch., 184: 233.

Dübendorfer, A., and Eichenberger-Glinz, S., 1980, Development and metamorphosis of larval and adult tissues of *Drosophila* in vitro, in: "Part IV: Use of Invertebrate Tissue Cultures in Endocrinology, Physiology, Genetics and Biochemistry", Kurstak, Maramorosch, and Dübendorfer, eds., Elsevier, Amsterdam.

Elder, H.J., 1975, Muscle structure, in: "Insect Muscle", P.N.R. Usherwood, ed., Academic Press, New York.

Goldstein, M.A., 1971, An ultrastructural study of supercontraction in the body wall muscles of *Drosophila melanogaster* larvae, Anat. Rec., 169: 326.

Hotta, Y., and Benzer, S., 1972, Mapping of behaviour in *Drosophila* mosaics, Nature, Lond., 240: 527.

Koana, T., and Hotta, Y., 1978, Isolation and characterization of flightless mutants in *Drosophila melanogaster*, J. Embryol. exp. Morph., 45: 123.

Osborne, M.P., 1967, Supercontraction in the muscles of the blowfly larva: an ultrastructural study, J. Insect Physiol., 13: 1471.

Pasquali-Ronchetti, J., 1970, The ultrastructural organization of femoral muscles in *Musca domestica* (Diptera), Cell Tiss. Res., 2: 339.

Poodry, C.A., and Schneiderman, H.A., 1970, The ultrastructure of the developing leg of *Drosophila melanogaster*, Wilhelm Roux Arch., 166: 1.

Poulson, D.F., 1950, Histogenesis, organogenesis, and differentiation in the embryo of *Drosophila melanogaster*, in: "Biology of *Drosophila*", M. Demerec, ed., Hafner, New York.

Shafiq, S.A., 1963, Electron microscopic studies on the indirect flight muscles of *Drosophila melanogaster*. I. Structure of the myofibrils, J. Cell Biol., 17: 351.

Tiegs, O.W., 1922, Research on insect metamorphosis. Part I. On the structure and post embryonic development of a Chalcid wasp, *Nasonia*. Part II. On the physiology and interpretation of the insect metamorphosis, Trans. Proc. Roy. Soc., S. Australia, 46: 319.

Ursprung, H., Conscience-Egli, M., Fox, D.J., and Willimann, T., 1972, Origin of leg musculature during *Drosophila* metamorphosis, Proc. Natl. Acad. Sci. USA, 69: 2812.

Zalokar, M., 1947, Anatomie du thorax de *Drosophila melanogaster*, Rev. Suisse Zool., 54: 17.

MONOCLONAL ANTIBODIES IN THE ANALYSIS OF *DROSOPHILA* DEVELOPMENT

Michael Wilcox, Danny Brower and R.J. Smith

M.R.C. Laboratory of Molecular Biology
Hills Road
Cambridge, England

We are attempting to approach the molecular mechanisms involved in *Drosophila* development by using monoclonal antibodies raised against the surfaces of imaginal discs. It is generally acknowledged that the cell surface plays a major role in development. Surface components have been implicated, for example, in tissue induction (Saxen, 1977) and in cell-cell recognition (Moscona, 1975; Marchase et al., 1975). However, with one or two exceptions (Burger, 1977), little is known of the molecular interactions involved in these processes. It is not clear whether all, or only some, classes of surface molecules - proteins, glycoproteins, glycolipids and so on - participate. Nor do we know whether cellular discrimination stems from qualitative or quantitative changes in these components, or even whether the molecules are likely to be major or minor constituents of the surface. Possessing a set of specific "tags" for surface molecules should enable us to approach these questions. One advantage of the monoclonal antibody technique over conventional immunological approaches is that in principle even minor surface components can be specifically labelled.

The advantage of *Drosophila* for such a study comes from the combination of a number of factors. For example, there is the sharp sub-division of the fly into compartments, units of clonal restriction which are thought to contribute to developmental control (Garcia-Bellido et al., 1976). Compartments have been detected, at this time, only in insects. There is the co-existence, in the larva, of fully differentiated structures with the relatively undifferentiated tissues which will give rise to the adult. Some of these latter structures, the imaginal discs, exist as discrete, morphologically simple, autonomously developing units. In addition, a wealth of information is available on the fates and on the level of determination (and hence the regulative potential) of the different regions of the imaginal discs (Bryant, 1978). While there is no direct evidence that the cell surface plays a critical role in *Drosophila* development, findings from the genetic analysis of the various homeotic mutants have led to the suggestion that compartmentalization is maintained via differential cell affinities (Crick and Lawrence, 1975). It has also been suggested that the complex interactions involved in the spatial organization of imaginal discs and in their morphogenesis require differences in cell surface properties (Reinhardt et al., 1977; Poodry and Schneiderman, 1971). Such differences may occur in a graded fashion within the discs (Nardi and Kafatos, 1976).

We describe here the early results of our study. The method, with some modifications which will be discussed later, is similar to that described by others (Galfré et al., 1973). Initially, mice are immunized with several hundred thoracic imaginal discs. We are currently trying several different immunization procedures in an attempt to optimize the regime. Lymphocytes from the spleen of an immunized mouse are fused with cells of a non-antibody producing mouse myeloma cell line and the resulting mixture of cells is plated out into 3-400 small wells and allowed to grow up in a selective medium (HAT). Aminopterin in this medium blocks an enzyme in one of the two biochemical pathways the cell uses for the synthesis of guanosine monophosphate; the remaining myeloma cells, which lack the function of an enzyme, hypoxanthine-guanine-phosphoribosyltransferase (HGPRT), in the alternative pathway, therefore die, while the hybrid cells, hybridomas, which have received wild-type HGPRT genes from the lymphocytes, are able to grow. The hybridomas are transferred as soon as possible to new wells, thus separating them from spleen fibroblasts which also grow in HAT medium (unfused lymphocytes, however, cannot proliferate *in vitro*). After further growth, the excreted products of each well are screened for their ability to bind to intact imaginal discs, and the positive lines then rigorously cloned by conventional techniques (Cotton et al., 1973).

For screening, we use an indirect immunofluorescence technique (cf. legend to Figs. 1-4). By this method we are able to detect antisera with spatially specific patterns of binding. Examples of such patterns are shown in Figs. 1-4. Before describing these it is pertinent to describe briefly the morphology of the discs. Within the basal lamina, the disc consists of a hollow sac of cells of two types, a columnar epithelial layer on one side and a cellular peripodial membrane on the other. The former, which includes the vast majority of the disc cells, and probably at least part of the latter (Sprey and Oldenhave, 1974) gives rise to adult cuticular structures. At various places between the columnar epithelium and the basal lamina are groups of adepithelial cells; while the fate of these cells is less certain, it has been suggested that they are muscle precursor cells (Poodry and Schneiderman, 1970; Reed et al., 1975).

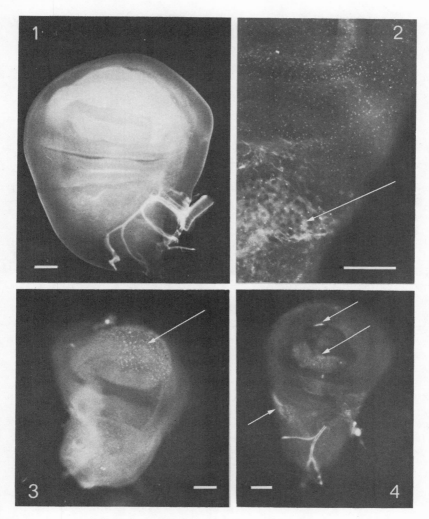

Figures 1-4: Fluorescence micrographs showing the binding of various antisera to D. melanogaster wing discs. Bar is 50 microns. Antisera were screened by the indirect immunofluorescence technique. Imaginal discs (or other tissues), either intact or sectioned, were treated first with the hybridoma antiserum (usually for 45' at 37º), then washed to remove unbound antibody. Retained antibody was then labelled by subsequent incubation (for 30' at 37º) with fluorescein-conjugated anti-mouse Ig (immunoglobulin) G. Figures 1 and 2: Binding of two antisera to intact wing discs. In Figure 2, only the lower part of the disc is shown. Figures 3 and 4: Binding pattern of a third antiserum on the peripodial membrane and epithelial sides, respectively, of the disc. Figure 3 is overexposed to enhance the relatively faint specific antibody binding.

Figure 1 shows an antiserum which seems to bind to the basal lamina but not to the disc cells themselves. Another antiserum (Figure 2) binds to the disc in very small patches, producing a pattern of dots; binding is also observed (arrow) either to the adepithelial cells or to an extracellular matrix in the same region. The binding of a third antiserum is restricted to particular regions of both the peripodial and epithelial layers of the disc (Figures 3-4).

Although our initial screening procedure can detect such potentially interesting antisera, it has drawbacks. Most obviously, it really requires mono-specific antisera, for the binding of an antibody whose specificity is restricted to one disc or to a region of a disc would be masked if the antiserum contained other antibodies which bound strongly, for example, to the whole surface of all the discs. We have overcome this problem to some extent by including an early isolation or "precloning" step in our procedure; small groups of hybridoma cells from the initial post-fusion wells are transferred early to new wells and it is the antisera from these second wells which we screen. Of course, positive hybridomas still require subsequent cloning. A second and potentially less tractable problem is the inability of some antibodies to reach their respective antigens. In general, the antibodies do not penetrate far into the intercellular spaces between the cells of the columnar epithelium. More importantly, we have evidence that some have difficulty in penetrating the surrounding basal lamina. This may reflect differences in the size of **the** antibodies; some may be IgG's and others the larger IgM's. With antibodies that do show some activity in the initial screen such penetration problems can be dealt with by manually removing the peripodial membrane or by using sections of frozen discs (cf. Figure 6). Such methods are too time-consuming, however, to be used during the routine screening of large numbers of antisera, and so we may fail to detect some positive clones.

We have examined the binding of the antibody produced by one clone, DA.1B6, to a variety of *Drosophila* tissues and organs. The antibody binds to all of the imaginal discs of 3rd instar larvae, and to those we have examined from 2nd instar larvae and early prepupae. Figures 5 and 6 demonstrate the pattern of binding to the wing disc. The cells of both epithelial and peripodial membrane layers are clearly fluorescent, as is the ring of imaginal cells (Bodenstein, 1965) surrounding the larval tracheae attached to the disc. The adepithelial cells, however, are not recognized by the antibody. The bulk of the other larval tissues show no antibody binding. But the rings of imaginal cells (Bodenstein, 1965) in some of these tissues, particularly those of the salivary gland and gut, are clearly recognized and stand out against the dark background of the larval cells of these organs (Figure 7). The brain, although largely negative, also shows patches of antibody binding which seem to be associated with extensions of the eye-antennal disc.

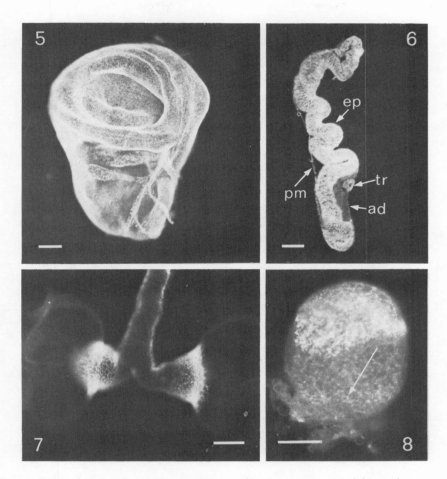

Figures 5-8: Binding of DA.1B6 antibody to *Drosophila* tissues. Bar is 50 microns. Figure 5: Binding to the intact wing disc. Figure 6: In frozen disc sections, the antibody binds to cells of both epithelial (ep) and peripodial membrane (pm) layers and to the tracheal imaginal cells (tr), but not to the adepithelial cells (ad). Figure 7: The imaginal cells in the base of each salivary gland show strong antigenic activity. Figure 8: The antibody binds to a cap of cells in the female gonad. A second line of binding activity can just be seen in the lower part of the organ (arrow). Again, the background of nonbinding gonad cells is enhanced by the printing procedure needed to demonstrate this latter binding.

The specificity of binding of the antibody to the larval ovary is particularly interesting. The ovary shows quite large but localized areas of binding (Figure 8). This pattern of binding carries over to the adult structure, where there is strong binding to the follicle cells, particularly those surrounding the early öogonia. In contrast, the developing egg and nurse cells are not recognized by the antibody.

Several other adult epithelia, those of the epidermis and male genitalia, for example, also bind the DA.1B6 antibody. At the other end of the ontological spectrum, cells of the cellular blastoderm lack the antigen. By embryonic stage 12 (post-segmentation, during dorsal closure; Bownes, 1975), however, several of the embryonic tissues, including the salivary gland, gut and the entire epidermis, bind the antibody. During later embyrogenesis the binding activity gradually disappears until, by the end of the 1st larval instar, it is restricted to the imaginal tissue.

The precise pattern of antigen acquisition and loss during embryogenesis remains to be elucidated. Nor do we know whether the abdominal histoblasts carry the antigen. At the present time, too, we know nothing of the molecular nature of the antigen, or of its function; nevertheless, we can reach certain conclusions. It is clear that the antigen is not entirely restricted to imaginal and adult cells - it is present also on (some) presumptive larval cells during embryogenesis. It is not specific for proliferating cells, since the antibody does not bind to the dividing cells of the early embryo, or to germ cells or adepithelial cells during their proliferative phases. Additionally, no binding can be detected to cells of the K_C *Drosophila* cell line. Rather, the antigen seems to be restricted to the cells of epithelial sheets. It begins to disappear from presumptive larval epithelia late in embryogenesis, but remains in the imaginal cells and persists in the differentiated cells of many adult tissues after metamorphosis. It is possible then that loss of the antigen is associated with the onset of polyteny in a cell. In this respect it is interesting to note that antigenic activity diminishes in the adult follicle epithelium at approximately the time when its cells become polytene and function in vitellogenesis (King, 1970).

The isolation of this antibody highlights the usefulness of *Drosophila* for the investigation of the cell surface and its role in development. Without the unique developmental history of holometabolous insects, the characterization of this antibody would have been more difficult and its original identification as something potentially interesting much more unlikely. All this encourages us to think that the technique will lead to the discovery of other molecules of developmental signficance.

REFERENCES

Bodenstein, D., 1965, The postembryonic development of *Drosophila*, in: "Biology of *Drosophila*", M. Demerec, ed., Hafner, New York.

Bownes, M., 1975, A photographic study of development in the living embryo of *D. melanogaster*, J. Embryol. exp. Morph., 33:789.

Bryant, P.J., 1978, Pattern formation in imaginal discs, in: "The Genetics and Biology of *Drosophila*", Vol. 2C, M. Ashburner and T.R.F. Wright, eds., Academic Press, London, pp. 229-335.

Burger, M.M., 1977, Mechanisms of cell-cell recognition: some comparisons between lower organisms and vertebrates, in: "Cell Interactions in Differentiation", M. Karkinen-Jääskeläinen, L. Saxen, and L. Weiss, eds., Academic Press, New York.

Cotton, R.G.H., Secher, D.S., and Milstein, C., 1973, Somatic mutations and the origin of antibody diversity: Clonal variability of the immunoglobulin produced by MOPC 21 cells in culture, Eur. J. Immun., 3:135.

Crick, F.H.C. and Lawrence, P.A., 1975, Compartments and polycones in insect development, Science, 189:340.

Galfé, G., Howe, S.C., Milstein, C., Butcher, G.W. and Howard, J.C., 1977, Antibodies to major histocompatibility antigens produced by hybrid cell lines, Develop. Biol., 48:132.

Garcia-Bellido, A., Ripoll, P., and Morata, G., 1976, Developmental compartmentalization in the dorsal mesothoracic disc of *Drosophila*, Develop. Biol., 48:132.

King, R.C., 1970, "Ovarian Development in *Drosophila melanogaster*", Academic Press, New York.

Marchase, R.B., Barbera, A.J., and Roth S., 1975, A molecular approach to retinotectal specificity, in: "Cell Patterning", Ciba Foundation Symposium 29, Elsevier, Amsterdam.

Moscona, A.A., 1975, Embryonic cell surfaces: Mechanisms of cell recognition and morphogenetic cell adhesion, in: "Developmental Biology - Pattern Formation - Gene Regulation", D. McMahon and C.F. Fox, eds., W.H. Benjamin, Reading Mass.

Nardi, J.B. and Kafatos, F.C., 1976, Polarity and gradients in lepidopteran wing epidermis. II. The differential adhesion model: gradient of a non-diffusible cell surface parameter, J. Embryol. exp. Morph., 36:489.

Poodry, C.A., and Schneiderman, H.A., 1970, The ultrastructure of the developing leg of *D. melanogaster*, Wilhelm Roux Arch., 166:1.

Poodry, C.A. and Schneiderman, H.A., 1971, Intercellular adhesivity and pupal morphogenesis in *D. melanogaster*, Wilhelm Roux Arch., 168:1.

Reed, C.T., Murphy, C., and Fristrom, D., 1975, The ultrastructure of the differentiating pupal leg of *D. melanogaster*, Wilhelm Roux Arch., 178:285.

Reinhardt, C.A., Hodgkin, N.M., and Bryant, P.J., 1977, Wound healing in the imaginal discs of *Drosophila*. I. Scanning electron microscopy of normal and healing wing discs, Develop. Biol., 60:238

Saxen, L., 1977, Morphogenetic tissue interactions: An introduction, in: "Cell Interactions in Differentiation", M. Markinan-Jääskeläinen, L. Saxen and L. Weiss, eds., Academic Press, London.

DISCUSSION

A. Shearn: Have you found antibodies which bind to all cells?

M. Wilcox: I don't know that we have any that bind to all cells; but our screening procedure selects for those which bind to imaginal discs. There are some which bind to all basal laminae that we have tested.

K. Kumar: Do the larval abdominal histoblasts bind the DA.186 antibody?

M. Wilcox: We tried very hard to find it. However, there is a problem caused by the autofluorescence of larval cuticle which we have not yet solved. So far we do not know the answer although the adult epidermis deriving from the histoblasts does possess the antigen.

A. Sarabhai: After having antibodies bound to these sites, is larval development affected?

M. Wilcox: This is something we are interested in; but we have not done any experiments yet.

A. Fodor: After adding ecdysone so that the disc starts to differentiate, do you find any differences in the binding capacity?

M. Wilcox: I don't know. We have not done many studies yet, although we do know that the evaginating wing and leg discs still bind the antibody strongly.

P.J. Bryant: Are the discs sonicated or are they intact when used as immunogens?

M. Wilcox: We have tried both and it does not seem to make any difference. Of course, selection for surface-specific antibodies comes at the screening stage.

Y. Hotta: How many clones have you isolated so far?

M. Wilcox: We have screened about 1500 supernatants so far from 5 fusions. Some 300 supernatants have shown binding activity of one kind or another and we have picked and frozen the 40-50 of these which seem most interesting.

THE EFFECT OF X-CHROMOSOME DEFICIENCIES ON NEUROGENESIS IN *DROSOPHILA*

J.A. Campos-Ortega and F. Jiménez

Institut für Biologie III
7800 Freiburg
Federal Republic of Germany

INTRODUCTION

   Complex biological processes, such as the development of the central nervous system (CNS) of *Drosophila*, can be dissected into logical steps by introducing genetic variations. In the beginning the major difficulty encountered in this process is that of defining the elements involved, which in turn is dependent on obtaining mutants suitable for analysis. The reason for this difficult is twofold: on the one hand, mutants affecting essential aspects of neurogenesis are expected to be lethal; on the other hand, the CNS is an internal structure, and the appraisal of mutant variations requires the use of special techniques. *Drosophila* goes through different developmental stages, embryonic, larval and pupal, before reaching the final imaginal stage. Assuming a low probability for neurogenic mutants to develop to the imaginal stage, a reasonable way of screening for such mutants is to concentrate on the study of embryos. For neurobiological studies *Drosophila* embryos are inconvenient because of their small size, which makes selective staining of neurones extremely difficult. However, they offer several important advantages for our purposes: the CNS of mature embryos is roughly comparable to that of third instar larvae, disregarding the proliferation within the imaginal anlagen; techniques have been developed which allow the simultaneous study of dozens of embryos as whole-mounts, in which the CNS can be visualized "in toto"; embryonic lethality can be used as a previous selection criterium. Wright (1970) proposed selecting for embryonic lethal point mutants when searching for mutants affecting embryogenesis; the same strategy could be pursued for studying neurogenic variations since they are probably found among embryonic lethals. However, a few arguments make clear that this approach may be very time consuming. Mutants with embryonic pheno-effective periods have been found in 30% of all point lethals (Hadorn, 1961); furthermore, mutagenesis studies have demonstrated that about 90% of all genes can mutate to lethal conditions (Judd et al., 1972). Accepting that the total number of genes in the chromosomal genome of *Drosophila* is about 5000 (Lefevre, 1974) a simple calculation indicates that about 1350 genes might show an embryonic lethal phenotype when mutated.

Another way of screening for genes concerned with neurogenesis is to use relatively small chromosome deletions and see whether the absence of a given part of the genome affects the development of a given part of the embryonic CNS (Poulson, 1940). The use of this approach should permit a faster screening of the genome than using embryonic point lethals; moreover, it overcomes automatically the difficulties posed by any possible hypomorphism present in many EMS-induced point mutants, and thus facilitates establishment of causal relationships between the lack of genetic functions, as opposed to a modified genetic function, and mutant phenotype. Some important difficulties are, however, inherent to this type of analysis; we will return to this point in the last section of this paper. Once any region producing an interesting neural phenotype is found, it can be further dissected by means of smaller deficiencies, or even point mutations mapping within it until a more precise localization of the gene(s) responsible is attained. At this stage of the work we have no prejudices as to the kinds of phenotypes expected to appear in embryos carrying deficiencies. Our main focus is on the genes responsible for specification of the developing CNS, since our final aim is the analysis of the genetic elements that enable nerve cell precursors to arise. Therefore, we are in principle interested in any kinds of modifications encountered in the CNS of embryos lacking given chromosomal bands, especially when these are specific modifications of the CNS and not of the remaining organs.

A prerequisite for any extensive screening of the genome for relevant neurogenic loci is a technique which enables rapid and easy examination of the CNS in a large number of embryos from each stock considered. Zalokar and Erk (1977) have devised a technique which permits simultaneous staining of dozens of embryos as whole-mounts The technique utilizes acidic fuchsin for exclusive staining of cell nuclei, suppressing that of cytoplasm, so that focusing of well stained and differentiated internal structures is possible (Figure 1). It must, however, be emphasized that the level of resolution provided by this technique is relatively low. It only allows the major parts of the CNS, brain lobes, subesophageal, thoracic and abdominal ganglia to be distinguished. It further shows single cell nuclei, but does not allow distinctions among them. Finally, it shows neither segmental nerves nor neural prolongations, but just a mass of neuropile. This itself restricts the kinds of detectable phenotypes to the level of rough organ changes, while many discrete modifications, for example of individual cell types, would escape our observation.

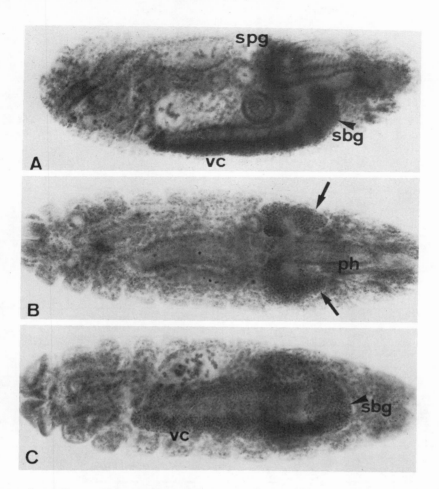

Figure 1: Shows the pattern of the CNS in lateral (A) and dorsoventral (B and C) orientation of fuchsin stained whole-mounts of wildtype embryos. Supraesophageal (spg), subesophageal (sbg) and neuromeric ganglia within the ventral cord (vc) can be distinguished. The neuropile is visible as a grayish homogeneous mass surrounded by the darkly stained cortex, B and C are two different focusing planes through the same embryos. Arrows in B point to the bilateral protocerebral lobes; ph: pharynx. In this, and the following figures (except in Figure 6) anterior is at the right hand side.

We have been using the approach outlined above for screening embryonic modifications produced by several chromosome deficiencies during development. In this report we present the results of our analysis of X-chromosome deficiencies.

THE EFFECT OF X-CHROMOSOME DEFICIENCIES ON NEUROGENESIS

Figure 2 shows Bridges' revised map (1938) of the X-chromosome on which the extent and location of the major deficiencies studied has been drawn. Most of these deficiencies have been isolated by G. Lefevre; they have been kindly provided by L. Cramer of the CalTech Stock Center at Pasadena, California. Deficiencies and duplications of the subdivisions 1A-B were supplied by A. Garcia-Bellido. Some deficiencies at 3A-C are from the Stock Center of the University of Texas at Austin; some of those at 7A-C from Bowling Green Stock Center. Table 1 shows the cytogenetic data of breakpoints of all deficiencies used for the present study (see Lindsley and Grell, 1968).

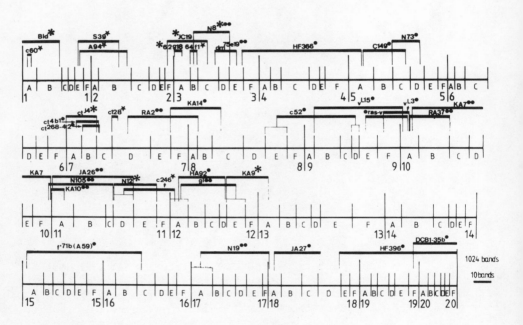

Figure 2: This figure shows the major X-chromosome deficiencies used in the present study. A simplified scale view of Bridges' revised map of the first chromosome is given, in which the actual banding pattern has been omitted. Deficiencies have been drawn as thick lines and the approximate location of their breakpoints as thin vertical lines. The mutant phenotypes found are indicated with different symbols. stars: no morphogenic defect stated; one dot: general morphogenic defects; two dots: very marked morphogenic defects; asterisks: neural defects without any important general change.

CHROMOSOME DEFICIENCIES AND NEUROGENESIS

Table 1: Breakpoints of deficiencies used in the present study

Deficiency	Breakpoints	Deficiency	Breakpoints
Df(1)c60	1A4 ; 1A6	Df(1)c128	7D1-2
In(1)y^{3PL}sc^{9R}	1A ; 1B2-3	Df(1)RA2	7D10 ; 8A4-5
Df(1)sc^8	1B2-3	Df(1)KA14	7F1-2 ; 8C6
In(1)sc^{8L}sc^{4R}	1B2-3 ; 1B3-4	Df(1)c52	8E ; 9D
In(1)sc^{4L}sc^{9R}	1B2-3 ; 1B3-4	Df(1)v^{L15}	9B1 ; 10A1
Df(1)sc^{19}	1B1-2 ; 1B4-7	Df(1)ras-v^{17Gc8}	9E3-4 ; 10A4-5
Df(1)sc^{V1}	1A8-C3	Df(1)v^{L3}	9F6-7 ; 10A6-7
Df(1)sc^{S2}	1B4-7	Df(1)RA37	10A7 ; 10B17
Df(1)sc^H	1B4-C3	Df(1)KA7	10A9 ; 10F10
Df(1)260-1	1B5-6	Df(1)N105	10F7-8 ; 11C4-D1
Df(1)y$^{-74k24.1}$	1B9-10	Df(1)KA10	11A1 ; 11A7
Df(1)svr	1B9-10	Df(1)JA26	11A1 ; 11D-E
Df(1)Bld	1B11-14	Df(1)N12	11D1-2 ; 11F1-2
Df(1)S39	1E4 ; 2B11-12	Df(1)c246	11D ; 12A1-2
Df(1)A94	1E5 ; 2B15	Df(1)g^1	12A ; 12E
Df(1)JC19	2F3 ; 3C5	Df(1)HA92	12A6-7 ; 12D3
Df(1)62g18	3A1-2 ; 3A4-6	Df(1)KA9	12E1 ; 13A5
Df(1)64c4	3A4-6 ; 3C3-5	Df(1)f$^+$71b(A59)	15A4 ; 16C2-3
Df(1)64j4	3A8-9 ; 3B1-2	Df(1)N19	17A ; 18A2
Df(1)64f4	3A9-B1 ; 3B3-4	Df(1)JA27	18A5 ; 18D1-2
Df(1)N^8	3B4-C1 ; 3D6-E1	Df(1)HF396	18E1-2 ; 20
Df(2)N^{71}	3C4 ; 3D-E	Df(1)DCB1-35b	19F1 ; 20F
Df(1)dm^{75e19}	3C12 ; 3E4	Df(2L)al	21B8-C1 ; 21C8-D1
Df(1)HF366	3E8 ; 5A7	Df(2R)Px^2	60C5-6 ; 60D9-10
Df(1)C149	5A8-9 ; 5C5-6	Df(3L)Asc	78D1-2-E1-2; 79A3-4-C1-2
Df(1)N73	5C2 ; 5D5-6	Df(3R)126c	87E1-2 ; 87F11-12
Df(1)ct^{J4}	7A2 ; 7C1	Df(3R)kar^{3L}	87B15-C2 ; 87C9-D2
Df(1)ct$^{268-42}$	7A5-6 ; 7B8-C1	Df(3R)P9	89D ; 89E
Df(1)ct^{4b1}	7B2-4 ; 7C2-4		

Up to now the effect on neurogenesis of a total of 51 different, partially overlapping deficiencies, which cover about 65% of the first chromosome, has been studied with fuchsin stained 18+2h old embryos. Histological observations of developing and mature embryos were confined to those cases which had previously shown interesting phenotypes in the fuchsin staining. Unless otherwise stated, Df/+ females were crossed to wildtype males, so that 25% of the progeny, i.e. the hemizygous Df/Y animals, were expected to show the mutant phenotype produced by the deficiency. In several cases the deficiency chromosome was provided by the males, using an appropriate duplication to counteract male lethality, and $\hat{X}\hat{X}/Y$ females. Mutant phenotypes were defined as morphological deviations from the embryonic pattern of the wildtype, which itself is present in about 75% of the siblings in any cross considered. As a rule the mutant phenotype observed in each case is fairly constant among the remaining 25% individuals, as well as reproducible when the same cross is repeated. However, with a few deficiencies two different mutant phenotypes were found, which seem to represent two different pheno-effective periods during embryonic development(see Figure 3). The first one is almost identical within different deficiencies producing it, although the frequency of such embryos varies among crosses. Particularly high proportions of embryos of this class are found within the progeny of $Df(1)g^1/+$ females. Embryos showing this phenotype contain non-differentiated cells in which no organogenesis has taken place. The second phenotype is characteristic of each deficiency considered, consisting of more or less marked morphogenic defects (see below). The finding of two different mutant phenotypes in fuchsin-stained whole-mounts of deficiency-carrying embryos might be related to recent results concerning maternal information in some of the eggs derived from $Df/+$ females, reported by Garcia-Bellido and Moscoso del Prado (1979). If this interpretation is correct, the early phenotype found in our material might correspond to insufficiency of maternal gene products, and the late one to lack of products expressed by the zygotic genome. Interesting in this context is the observation of a few embryos from some of the crosses yielding two classes of mutant phenotypes in which a mixture of features from both early and late phenotypes is clearly distinguishable within the same embryo (Figure 3C), a phenomenon which only occurs in those stocks. Since such embryos exhibit parts that have reached different stages of development, they suggest that some maternal products might be unevenly laid down in the egg cytoplasm, or that cells have different needs.

The embryonic phenotypes found in the current analysis can be classified into three different categories: (i) deficiencies which apparently do not modify the overall morphogenic pattern shown by the wildtype; (ii) deficiencies which produce rather general morphogenic defects; and (iii) those which principally affect neurogenesis. A further category of its own is represented by $Df(1)Notch^8$, and by smaller deficiencies or amorphic point mutants at 3C7 producing the N^8 phenotype. Poulson (1940, 1968) described modifications of the N^8 embryo, in which there is an initial alteration of the neurogenic layer leading to a complex syndrome of defects of the remaining embryonic organs. Thus N shares features of deficiencies from both of the second and the third groups. We are currently studying neurogenesis of N embryos at the cellular level (U. Dietrich and J. Campos-Ortega, unpublished), attempting to understand the cellular basis of the Notch syndrome. However, in this communication we shall not present any detailed account of these results, which up to now fully confirm Poulson's original description (see Wright, 1970).

A total of 10 different deficiencies, from which only six are contained in Figure 2, do not produce any apparent change in the embryonic pattern. Three of these deficiencies, $Df(1)S39$, $Df(1)A94$ and $Df(1)c246$ are relatively large, about 22, 24 and 21-32 bands respectively, the remaining seven being smaller ones.

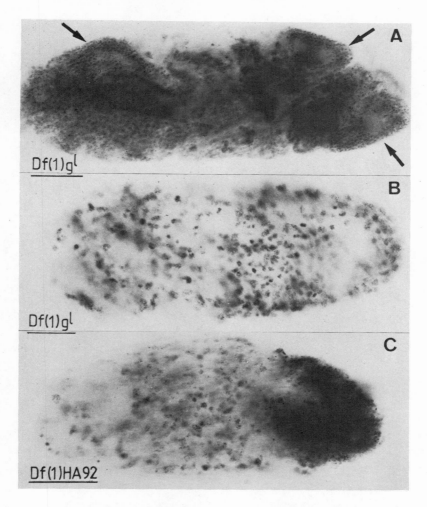

Figure 3: A and B show two different mutant phenotypes found within the progeny of the same cross. A is an example of a late phenotype, in which an abnormal organogenesis has been accomplished. Arrows point to masses of neural tissue. B is an example of an early phenotype where no organogenesis has taken place. C shows an embryo sharing characteristics of both early, posterior, and late, anterior, phenotypes.

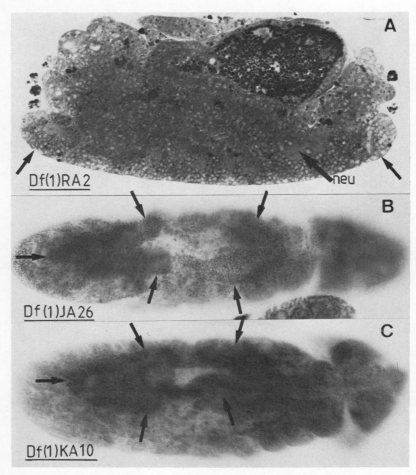

Figure 4: A is from a parasagittal section through a *Df(1)RA2* embryo showing the lack of hypoderm and of several other structures. neu: neuropile. The CNS of this embryo directly contacts the vitelline membrane without intervening hypoderm. B and C demonstrate that the size of the deficiency does not influence the expression of a given phenotype. *Df(1)JA26* lacks about 39 bands; *Df(1)KA10,* which is included in the former, only seven. In spite of this difference the phenotype produced by each is almost identical. The arrows point to the duplicated ventral cord. The pictures are ventral views of fuchsin stained embryos.

Most of the deficiencies studied belong to the second group, producing rather general morphogenic defects. Embryos hemizygous for these deficiencies show obvious anomalies in their CNS; but these anomalies appear to be secondary modifications of the CNS, the result of alterations in other structures. A full range of morphogenic defects is encountered among the embryos carrying deficiencies from this group, although development invariably proceeds long enough in all of them to readily permit identification of germ layer derivatives, as well as of the diverse embryonic organs. One of the modification patterns found is quite common within the members of this group, and consists of hypodermal defects, frequently holes, for example, in the dorsal cephalic hypoderm (through which the brain lobes eventually protrude) or in ventral regions of the embryo. Since the primary aim of our work was to locate regions of the genome primarily concerned with neurogenesis, we will not describe those phenotypes at this moment. However, interesting modifications of the CNS are visible in embryos hemizygous for Df(1)RA2 and Df(1)JA26 associated with many other embryonic pattern abnormalities. Embryos carrying Df(1)RA2 lack several of the organs found in the wildtype, including the hypoderm; in these embryos the CNS consists of a sheet of tissue organized in cortex and neuropile which lies in direct contact with the vitelline membrane (Figure 4a). This phenotype is due to failure of the missing parts to develop and leads to the death of the corresponding anlagen at 6-7h of development. Since embryos hemizygous for Df(1)KA14, which partially overlaps Df(1)RA2, do not show this phenotype, the defect is most likely to be due to the absence of bands comprised between 7D10 and 7F1-2. A very peculiar anomaly of the CNS is that produced by Df(1)JA26; the same phenotype is also produced by the much smaller Df(1)KA10 (Figures 4b and c, refer to Figure 2), and can therefore be referred to genes within 11A1 and 11A7. It consists of two almost completely separated ventral cords, which might well have resulted from either incomplete fusion of both halves or from extra proliferation due to duplication of the anlage. We do not know what causes this anomaly of the CNS to appear.

The third group includes deficiencies which affect the development of the CNS without significantly modifying the remaining structures of the embryo. This group comprised five major deficiencies Df(1)svr-included in Df(1)Bld (Figure 2), Df(1)JC19, Df(1)ct^{J4}, Df(1)N12 and Df(1)KA9, some of which have been further dissected by means of overlapping deficiencies. The phenotype of hemizygous Df(1)svr will be discussed in the next section in some detail. The CNS of Df(1)ct^{J4}, Df(1)N12 and Df(1)KA9 embryos has a common feature, namely, incomplete condensation of the ventral cord; in addition the ventral cord of Df(1)KA9 is split, showing two clearly unfused halves (Figures 5a, c and d). Histological analysis of Df(1)KA9 embryos has not provided any explanation for the failure of the ventral cord halves to fuse; nevertheless, it has shown that the neuropile is less developed in each of the unfused halves than in the corresponding structures of the wildtype, and that neuromeric commissures crossing the midline to link segments from both halves are present

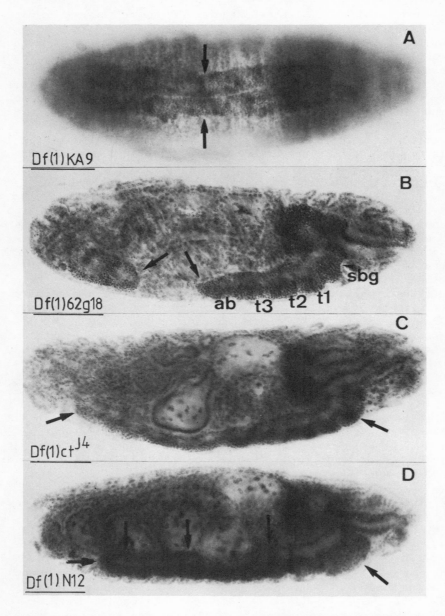

Figure 5: These pictures show the neural anomalies produced by four different deficiencies. Notice the abnormalities in the head of the embryos depicted in B, C and D. Arrows point to CNS structures. sbg: subesophageal ganglion; t1, t2, t3, ab: thoracic and abdominal neuromeres.

in 14h old mutant embryos. The defect found in *Df(1)JC19* is also visible in the smaller *Df(1)62g18*, which is included in the former (Figures 2 and 5B). The most prominent feature of this phenotype is an interruption in the ventral cord due to the absence of several neuromeres, which have been tentatively identified as abdominal ganglia 2nd to 6th (Figure 5B). In addition there is an evident shortening of the subesophageal ganglion. There is no evidence concerning the origin of this neural defect. Besides the neural defects, all four deficiencies produce to a variable extent an arrest in the development of head segments, with an incompletely involuted head and diverse pharynx anomalies. It is indeed very striking that the same kind of anomalies are present in *Df(1)svr* embryos as well (see below).

DEFICIENCIES AT THE SUBDIVISION 1B. A DISSECTION OF THE *Df(1)svr* PHENOTYPE

These include deficiencies which in hemizygous embryos produce CNS defects of varying severity depending on the deficiency considered. The most severe phenotype is that produced by *Df(1)B1d*, or by the deletion that produces a phenotypic equivalent, *Df(1)svr*. In the latter all bands from the tip of the chromosome to 1B9-10 are missing. The mature *Df(1)svr* embryo lacks any organized ventral cord and shows protocerebral lobes reduced to two clusters of nerve cells. Besides these CNS defects, the *Df(1)svr* embryo lacks the cuticular mechanoreceptors called Keilin organs and ventral pits. Finally, the mutant embryo shows a non-neural phenotypic component of variable extent, which consists of cell death localized in the lateral hypoderm at thoracic and anterior abdominal segments of developing embryos, in an incomplete involution of the head segments, accompanied by defects of the pharyngeal roof, and in imperfectly differentiated cuticular denticle belts, both features visible in the mature embryo. A similar description of the phenotype of *Df(1)svr* embryos has been recently provided by White (1980). Using several other chromosomal aberrations of this region we have dissected the complex phenotype of *Df(1)svr* embryos (see Jiménez and Campos-Ortega, 1979). The detailed dissection of the *Df(1)svr* phenotype described here can be taken as an example of how to proceed when the screening for neural defects produced by relatively large chromosomal deletions eventually uncovers an interesting phenotype. We have been very fortunate that a whole series of chromosomal aberrations of this region has already been prepared by previous workers. These chromosomes were used by Muller and associates for the dissection of the achaete-scute locus (see Muller, 1955; Garcia-Bellido and Santamaria, 1978; Garcia-Bellido, 1979). The following is a summary of the main results of our analysis.

1. There is a correlation between the severity of the neural defect and the size of the deficiency considered (Figure 6). In $(1)sc^{4L}$ sc^{9R}, a small synthetic deletion at 1B2-3 (Muller, 1935), produces the less pronounced neural defect, consisting of both a

main constriction at the middle of the ventral cord and a lack of condensation of the CNS. A similar phenotype is found in $In(1)y^{3PL} sc^{9R}$ and, slightly more pronounced in $Df(1)sc^{V1}$ hemizygous embryos. $Df(1)sc^{19}$ embryos, which lack the bands between 1B1-2 and 1B4-7 show a more serious defect, characterized by six or seven constrictions and incomplete condensation of the ventral cord, and by a visible reduction of the volume of the brain lobes. The phenotype produced by the terminal deletion $Df(1)sc^H$ is very similar to that of $Df(1)sc^{19}$ embryos; this indicates that the subdivision 1A, which is present in $Df(1)sc^{19}$ but absent in $Df(1)sc^H$ embryos, does not contribute in any important manner to the neural phenotype (see below). $Df(1)sc^{S2}$ and $Df(1)260-1$ are larger terminal deficiencies which result in the lack of large parts of the ventral cord and considerable defects of the brain lobes. Finally, $Df(1)y^{-74K}$, $Df(1)svr$ and $Df(1)Bld$ show a similar neural phenotype characterized by the virtual

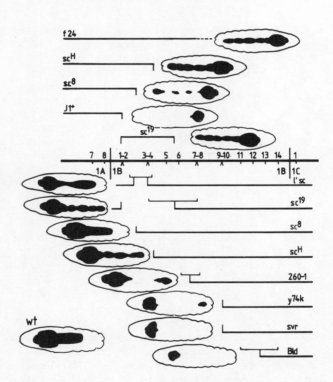

Figure 6: The effect on neurogenesis of deficiencies at 1B. The lower half illustrates the extent of some interstitial and terminal deletions studied; the neural phenotype found in each case is shown in drawings in ventral views of mutant embryos. The upper half shows the result of covering parts of the $Df(1)svr$ with some duplications. The wildtype pattern is given at the lower left hand corner of the figure.

absence of ventral cord structures and a very severe reduction in the volume of the brain lobes. We would like to emphasize that, despite the difference in size of the two terminal deficiencies $Df(1)svr$ and $Df(1)Bld$, some remnants of the CNS are still present in those embryos, i.e. a complete absence of CNS structures is not found in any of the deficiencies studied.

2. The neural phenotype of embryos carrying deletions at 1B is due to death of developing anlage cells. Histological analysis of increasingly aged mutant embryos has shown that early neurogenesis is normal in all cases. However, 7h old embryos already display extensive cell death throughout both ventral cord and brain anlagen, its extent being largely dependent on the deficiency considered. Since cells showing the characteristics of dividing neuroblasts are concomitantly present with dead cells, cell death in the CNS anlage is most likely to affect ganglion mother cells or differentiating neurones (see Bauer, 1906; Poulson, 1950).

3. Genes absent from the region 1B1-2 and 1B9-10 contribute with qualitatively similar effects to the achievement of the neural phenotype. This is indicated by two pieces of evidence: first of all by the described progression in the extent of the neural defect with increasing size of the deficiency; secondly, by the analysis of embryos carrying diverse deficiency-duplication combinations. Duplications of the tip of the X-chromosome, by translocation of some wild-type alleles of this region to other chromosomes, permit the synthesis of interstitial deletions (Figure 6). Using this approach it has been found that the left border of the region of interest is defined by the breakpoint of $Df(1;Y)1(1)J1^+$ at 1B1-2, since it is the largest terminal duplication which is unable to partially rescue the phenotype of $Df(1)svr$. Its right border is included within $Df(1)svr$, since this deficiency gives rise to the most severe phenotype, which does not differ from that produced by the larger $Df(1)Bld$. An interstitial deletion of 1B4-7 to 1B9-10 can be obtained by combining $Df(1)svr$ with $Dp(1;4)sc^H$; interstitial deletions of other bands are obtained combining other duplications with $Df(1)svr$, for example $Dp(1;2)sc^{S2}$ or $Dp(1;f)24$. In all these cases the phenotype found is comparable to that of $Df(1)sc^{19}$. Although slightly different in extent, these phenotypes are formally attained by the same mechanism, namely by the death of a given amount of cells in the neural anlage.

4. As a rule, deficiencies within the subdivision 1B show a more extreme phenotype in homozygosis than in hemizygosis; this affects both the neural and the non-neural components of the mutant phenotype. There is also a difference in the phenotype of embryos carrying deletions of the region 1B4 to 1B9-10, depending on the paternal or maternal origin of the deficiency chromosome, the latter being slightly more extreme than the former.

5. The neural phenotype found in $Df(1)svr$ embryos is produced by the absence of specific genetic information included within the

bands 1B1-2 to 1B9-10 and not by any unspecific effects resulting from the large size of the deficiency. This is supported by the study of hemizygous $Df(1)sc^{19}$ embryos which also carry homozygous autosomal chromosome deletions somewhere in the genome. With this experiment we intended to study embryos which were deficient in a number of bands similar to those of $Df(1)svr$. We combined $Df(1)sc^{19}$ with each of a total of six other deficiencies; some of those do not by themselves affect the morphogenic pattern of the embryo $(Df(2L)a1$, $Df(2R)Px^2$ and $Df(3R)kar^{3L}$, whereas others produce a characteristic embryonic phenotype $(Df(3L)Asc, Df(3R)P9$ and $Df(3R)126c)$. In both cases, double deficiency embryos did not show any augmentation of the neural phenotype encountered in hemizygous $Df(1)sc^{19}$ embryos. Particularly interesting in this context is the phentype of $Df(1)sc^{19}/Y; Df(3R)P9/Df(3R)P9$ and of $Df(3L)Asc$ ha-e been shown to include embryos, since both $Df(3R)P9$ and $Df(3L)Asc$ have been shown to include loci involved in the genetic control of embryonic segmentation (Lewis 1978; G. Jürgens, in preparation). The cuticular phenotype of $Df(3R)P9$ embryos has been previously described by Lewis (1978) and some of its features, for instance the unfused tracheal segments, are visible in fuchsin stained embryos as well. The CNS of 18h old $Df(3R)P9$ embryos (Figure 7C) shows a characteristic elongation of the ventral cord, which seems to contain more nerve cells and a more easily distinguishable neuromeric pattern than the wildtype; no modifications can be seen in the protocerebral lobes of $Df(3R)P9$ homozygotes. $Df(3L)Asc$ is a 20 band deficiency isolated by Gerd Jürgens that includes the Pc locus. The CNS of homozygous 18h old $Df(3L)Asc$ embryos shows a phenotype which is identical to that of Pc^3/Pc^3 embryos (unpublished observations). Its ventral cord is very elongated and has fewer cell bodies than the wildtype; the subesophageal ganglion is characteristically shortened (Figure 7A). The phenotype of double deficiency embryos $(Df(1)sc^{19}; Df(3R)P9$ and $Df(1)sc^{19}; Df(3L)Asc$ (Figure 7B and D)) can be described as resulting from addition of the phenotypes produced by the two deficiencies. Actually, the phenotype due to $Df(1)sc^{19}$ is more severe in combination with $Df(3L)Asc$ than with $Df(3R)P9$: in the latter case the total amount of nervous tissue left by the $Df(1)sc^{19}$ is even larger than in wildtype background. Whether or not neuromeres in $Df(3L)Asc$ are transformed into abdominal eight, and in $Df(3R)P9$ into mesothorax cannot be decided with the present material (see Lewis, 1978). Therefore presently the only possible interpretation for the different behavior of $Df(1)sc^{19}$ in both cases is that this is due to the different amounts of nerve cells present in $Df(3L)Asc$ and in $Df(3R)P9$ rather than to any qualitative differences in the neurons. This interpretation is supported by the results of the next series of experiments.

6. The severity of the neural defect found in $Df(1)svr$ embryos is not modified by increasing the amount of neural tissue in the mutant. Deficiencies at 3C6-8 (Poulson, 1968; Whelshons, 1965) give rise to the characteristic N^8 phenotype described by Poulson (1940).

CHROMOSOME DEFICIENCIES AND NEUROGENESIS 215

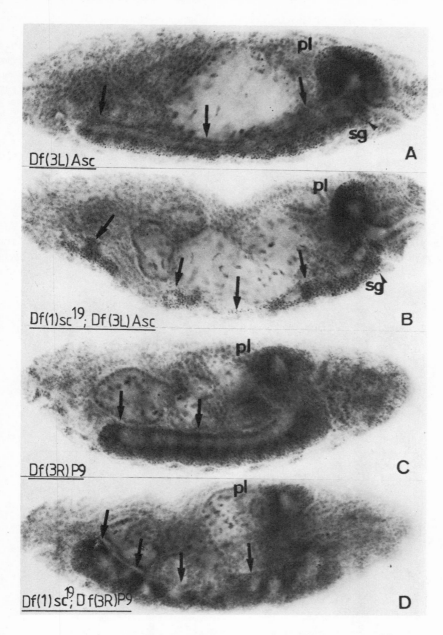

Figure 7: Illustrates the modifications of the $Df(1)sc^{19}$ phenotype when combined with two other deficiencies. pl: protocerebral lobes; sg: subesophageal ganglion. Arrows point to neural structures. Refer to text for further details.

According to this author, N embryos have a neurogenic layer which is considerably enlarged at the expense of lateral blastoderm cell derivatives as compared to the wildtype. As a consequence the mature N embryo completely lacks anterior and ventral hypoderm and some other organs, but has a correspondingly enlarged nervous system (Figure 8A). This consists of an anterior cerebral mass that surrounds the stomodeal opening, which continues caudally as a bilateral symmetrical sheet of neural tissue organized in cortex and neuropile regions. We have synthesized a $Df(1)svr$ $N264-39$ chromosome and studied the phenotype of double mutant hemizygotes (Figure 8B). The N transformation takes place in such embryos, but their CNS also clearly shows the effects due to $Df(1)svr$, consisting of the elimination of a large fraction of the neural tissue. $Df(1)svrN^{264-39}$ embryos lack virtually all the ventrally located nerve cells, have anterior regions which are seriously reduced in volume, but still show large clusters of neurons and neuropile in the lateral and most posterior part of the embryo. In fact, our impression is that $Df(1)svr$ does not eliminate many more neurons in the N background than it does in the N^+ back-

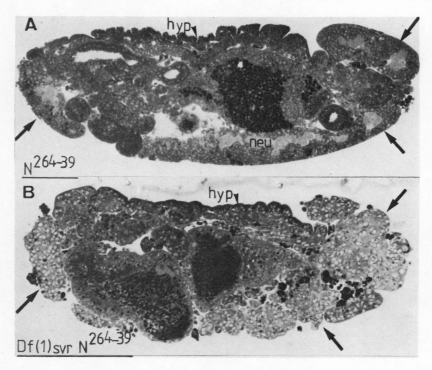

Figure 8: Parasagittal sections of mutant embryos. A shows the phenotype of N^{264-39}. B shows the expression of the $Df(1)svr$ in cis with N^{264-39}. This embryo is two hours younger than the one depicted in A. hyp: dorsal hypoderm. Arrows point to neural structures.

ground. This behavior of double mutant embryos appears to be in line with the finding of a correlation between the severity of a neural defect and the size of the deficiency described above. We would like to put forward the hypothesis that the amount of nerve cells lacking in each of the deficiencies under discussion is largely determined by the deficiency itself and not by the number of neurons available.

7. The absence of Keilin organs and ventral pits in the cuticle of mature *Df(1)svr* embryos derives from deletions including the scute^B system (Garcia-Bellido, 1979) and/or genes to its right up to 1B9-10. Using the technique of Van der Meer (1976) we have prepared the cuticle of embryos carrying different deletions of the region of interest. We found that any deficiency including bands to the right of the breakpoint of *Df(1)sc^{V1}* up to the breakpoint of *Df(1)svr* leads to the absence of mechanoreceptors. Neither *In(1)sc^{4L}sc^{9R}* nor *Df(1)sc^{V1}* hemizygous embryos lack them.

8. There is a variable non-neural phenotype in *Df(1)svr* embryos, already mentioned above (see Figure 9), which has two main components. One is cell death localized at the lateral hypoderm of thoracic and abdominal segments in developing 10-12h embryos (refer to Jiménez and Campos-Ortega, 1979); the other is a defect of variable extent in the involution of head segments visible in mature mutant embryos. In our previous account of the *Df(1)svr* phenotype (Jiménez and Campos-Ortega, 1979) we did not give any attention to this phenomenon, which seemed to be somewhat erratic and variable. White (1980) has carefully described this variation in head morphology of *Df(1)svr* embryos for the first time. The extent of hypodermal cell death increases from *Df(1)260-1* to *Df(1)Bld* hemizygous embryos, whereas *Df(1)sc^H* or *Df(1)sc^{19}* hemizygous embryos do not show any striking difference compared to the wildtype as far as this phenomenon is concerned. *Df(1)svr; Dp(1;2)sc^{19}* embryos, on the other hand, show fewer dead cells than those found in *Df(1)svr* embryos; thus a strict correlation between hypodermal cell death and the bands comprised between the breakpoints of *Df(1)sc^{19}* and *Df(1)svr* is not justified by the present evidence. It is unknown whether or not this phenomenon accounts for the lack of Keilin organs and ventral pits from the cuticle of embryos deficient for the same bands. Similar to hypodermal cell death, the incomplete head involution of some mutant embryos cannot be correlated with certitude with any of the bands included in the 1B region. Initially we thought that the pharyngeal defect is associated with deficiencies 1B4 to 1B9-10. However, a few homozygous *Df(1)sc^{19}* embryos which are deficient for 1B1-4 have been recently found to show this defect as well (Figure 9B). The intensity of its expression largely depends on several factors, among them the age of the embryos, the paternal and maternal origin of the deficiency chromosome and the homo- or hemizygosity of the individuals. Strikingly enough, the same incompleteness of head involution is found in other deficiencies of the X-chromosome which also affect the condensation of the ventral cord (see above and Figure 5). Finally, imperfectly differentiated denticle belts within abdominal seg-

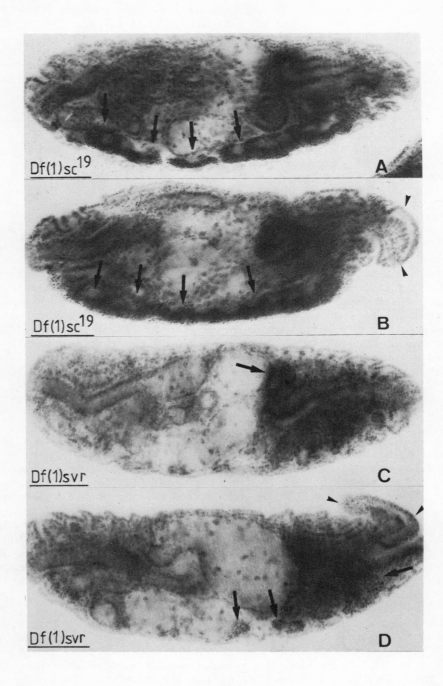

Figure 9: Two extreme cases of phenotypic expression. A shows a hemizygous $Df(1)sc^{19}$ with normal mouth parts and the characteristic neural defect; B shows a homozygous $Df(1)sc^{19}$ embryo in which the roof of the pharynx is completely everted (arrow heads). C shows a $Df(1)svr$ embryo with normally patterned head whereas the embryo at D has pharyngeal defects (arrow heads) very similar to those shown at B. Arrows point to neural structures.

---

ments can be seen in the cuticle of mature embryos carrying deficiencies of any gene from 1B1 to 1B9-10. This defect is particularly pronounced in homozygous embryos.

CONCLUDING REMARKS

The results from our analysis of embryos carrying deficiencies of the first chromosome outlined above permit a few conclusions of general interest. In our opinion the most important, albeit still tentative, conclusion is the apparent validity of this approach as a tool for screening the genome in search of neurogenically relevant genes. The reproducibility of mutant phenotypes and the specificity of the expression encountered in most of them support this view. Additivity of effects resulting from the absence of high numbers of genes, which would eventually lower the general physiological conditions of deficiency embryos, is not supported by the present material. Relatively large deletions were found to produce slight or no morphogenic defects in hemizygous embryos. Nevertheless, two factors are worth discussing as possible sources of interference with the expression of mutant phenotypes in deficiency embryos. One is the possible occurence of maternal information laid down in the unfertilized egg, which could eventually mask, or modify, the expression of a phenotype due to the absence of gene functions in the zygotic genome. Actually only two among 51 deficiencies of the X-chromosome, $Df(1)g^1$ and $Df(1)HA92$, which is included in the former have been found to be good candidates for uncovering haplo-insufficient loci, as judged from the occurence of high numbers of mutant embryos with early and with late pheno-effective periods. Other deletions have been found to produce the same phenomenon, but the number of embryos with early pheno-effective periods is very low in these cases. There is a simple way of testing the importance of maternally conditioned effects, and this is to compare the expression of the deficiency when provided by the female and by the male. Unfortunately, appropriate duplications, necessary for compensating for male lethality, are not available for most of the regions. The few cases in which we were able to perform this test did not show any great difference. However, more information is necessary before generalizations can be made. The other factor to be taken into account is the possibility of genes included in a deficiency with earlier pheno-effective periods than those of presumptive neurogenic genes. One could imagine a situation in which the earlier pheno-effective period of one gene might hinder the expression of another gene. This eventuality is perhaps improbable, but there is no proof of the contrary.

The sample studied in our current research has uncovered the effects produced by the absence in stages of about 65% of the first chromosome. Considering also the results of analysis of the effects of several other autosomal deletions, not included in this report, we have actually studied the effect of deletions in about 25% of the whole genome. Referring now to our primary concern, which is neurogenesis, the only interesting loci found as the result of this work are two complex loci on the X-chromosome. One is well known at 3C7 i.e. Notch; the other is represented by some, or all, of the bands from 1B1 to 1B9-10. In this paper we have not dealt with Notch in detail, since we could not make any important addition to, or correction of, Poulson's original descriptions (1940, 1945, 1968; Wright, 1970); we shall return below to the effects produced by the other deficiencies. Assuming a statistical distribution of genes relevant for neurogenesis throughout the genome and the reliability of our approach, the results of our analysis indicate that the number of genetic elements which specify the development of CNS, as opposed to the remaining ectodermal derivatives, cannot be very high. Certainly, many chromosomal deletions studied were found to produce neural malformations; it might even be that some of them include loci directly involved in neurogenesis. This we do not know. However, if genes within them are directly concerned with neurogenesis, they are most likely involved in very particular aspects in this process rather than in a differential activation of genes that specify CNS precursors. Another possibility might be that the number of neurogenic genes is very high but that they are clustered and not statistically distributed through the genome. In this case it could be that our search has not yet touched such regions other than Notch and the 1B region. It is obviously still premature to decide between these alternatives. A last word of caution is necessary at this point: we must be aware that we are dealing with a working hypothesis which, although plausible, still lacks any conclusive experimental evidence, namely that there is a process of differential gene activity specificying neuronal precursors and that this process requires the zygotic genome for its realization. We will be able to answer these questions at the latest after having completed the analysis on the neurogenic effects of chromosomal deficiencies of all loci in the *Drosophila* genome.

A very interesting phenotype indeed is the one found in embryos lacking bands at 1B1 to 1B9-10. This region has been further dissected and the development of mutant embryos has been followed in some cases to phenomenologically ascertain the origin of the mutant phenotype. This case also demonstrates the kind of difficulties inherent to our approach, since quite a considerable effort has been put into this analysis without yielding any satisfactory answer to a series of questions. The current results indicate that the integrity of most, if not of all, bands within this region is indispensable for normal neurogenesis; deletions within it result in the death of a given amount of developing anlage cells. These results further suggest that some of the genetic functions included within this region selectively act upon both central (Jiménez and Campos-Ortega, 1979) and peripheral (Garcia-Bellido and Santamaria, 1978) neurogenesis whereas others might be concerned with rather general morphogenetic processes. However, there is no conclusive experimental evidence that this is so. The phenotype produced by the largest terminal deficiency, $Df(1)svr$ is predominantly neural albeit some non-neural components of this phenotype have been found as well. Precisely this last group of effects, including non-viable hypodermal cells and incomplete head involution, render the interpretation of the experimental results very difficult. The difficulty is further enforced by the apparent inability to correlate either of both effects with absence of any of the bands included in the region. It has already been mentioned that four other deficiencies $Df(1)62g18$, $Df(1)ct^{J4}$, $Df(1)N12$ and $Df(1)KA9$ which show discrete alterations of the CNS in hemizygotic conditions, have been found to produce a similar cephalopharyngeal syndrome. This observation suggests that the embryonic head defects might be secondary to the neural defects in all these cases. Thus, this aspect of the $Df(1)svr$ phenotype still requires further investigation. It is from the results of such work that we hope to be able to determine whether genes from this region are exclusively involved in neural development.

ACKNOWLEDGEMENTS

We thank Sigrid Krien for patient and expert technical assistance, A. Garcia-Bellido and Gerd Jürgens for stimulating discussions and the Deutsche Forschungsgemeinschaft (SFB 46) for financial support. F.J. holds an EMBO long-term fellowship.

REFERENCES

Bridges, C.B., 1938, A revised map of the salivary gland X-chromosome, J. Hered., 29:11.
Bauer, V., 1906, Zur inneren Metamorphose des Centralnervensystems der Insecten, Zool. Jb., Abt. Anat. Ontog. Tiere, 20:123.
Garcia-Bellido, A., 1979, Genetic analysis of the achaete-scute system of Drosophila melanogaster, Genetics, 91: 491.
Garcia-Bellido, A., and Moscoso del Prado, J., 1979, Genetic analysis of maternal information in Drosophila, Nature, 278: 346.
Garcia-Bellido, A., and Santamaria, P., 1978, Developmental analysis of the achaete-scute system of Drosophila melanogaster, Genetics, 88: 469.
Hadorn, E., 1961,"Developmental Genetics and Lethal Factors", Methuen, London.
Jiménez, F., and Campos-Ortega, J.A., 1979, A region of the Drosophila genome necessary for CNS development, Nature, 282: 310.

Judd, B.H., Shen, M.W., and Kaufman, T.C., 1972, The anatomy and function of a segment of the X-chromosome of *Drosophila melanogaster*, Genetics, 71: 139.

Lefevre, G., 1974, The relationship between genes and polytene chromosome bands, Annu. Rev. Genet., 8: 51.

Lindsley, D.L., and Grell, E.H., 1968, "Genetic Variations of *Drosophila melanogaster*", Carnegie Inst. Wash. Publ., 627.

Muller, H.J., 1935, The origination of chromatin deficiences as minute deletions subject to insertion elsewhere, Genetica, 17: 237.

Muller, H.J., 1955, On the relation between chromosome changes and gene mutations, Brookhaven Symp., 8: 126.

Poulson, D.F., 1940, The effects of certain X-chromosome deficiencies on the embryonic development of *Drosophila melanogaster*, J. Expl. Zool., 83: 271.

Poulson, D.F., 1945, Chromosomal control of embryogenesis in *Drosophila*, Am. Naturalist, 79: 340.

Poulson, D.F., 1950, Histogenesis, organogenesis and differentiation in the embryo of *Drosophila melanogaster* Meigen, in: "Biology of *Drosophila*", M. Demerec, ed., Hafner, New York.

Poulson, D.F., 1968, The embryogenetic function of the Notch locus in *Drosophila melanogaster*, Proc. 12th Int. Congr. Genet., Tokyo, 1: 143.

Van der Meer, J., 1977, Optical clean and permanent wholemount preparation for phase-contrast microscopy of cuticular structures of insect larvae, Dros. Inf. Serv., 52: 160.

Welshons, W.J., 1965, Analysis of a gene in *Drosophila*, Science, 150: 1122.

White, K., 1980, Defective neural development in *Drosophila melanogaster* embryos deficient for the tip of the X-chromosome, Devel. Biol., in press.

Wright, T.R.F., 1970, The genetics of embryogenesis in *Drosophila*, Adv. Genet., 15: 262.

Zalokar, M., and Erk, I., 1977, Phase-partition fixation and staining of *Drosophila* eggs, Stain Technol., 52: 89.

FORMATION OF CENTRAL PATTERNS BY RECEPTOR CELL AXONS IN *DROSOPHILA*

John Palka and Margrit Schubiger

Department of Zoology
University of Washington
Seattle, Washington   98195   U.S.A.

"SEQUENTIAL POLYMORPHISM" AND THE CENTRAL NERVOUS SYSTEM (CNS)

   The larva of a fly possesses a complex nervous system which enables it to seek and recognize food, eat, avoid obstacles and dangers, regulate the release of hormones, respond to hormonal changes - in other words, to behave in an integrated fashion. It is in every sense a fully functioning organism.

   However, throughout the larval period important changes are occurring in the CNS. Neuroblasts divide, ultimately to give rise to neurons, even while the nervous system continues to carry out functions appropriate to larval life (White and Kankel, 1978). The differentiation of the compound eyes and the optic lobes, for example, starts in the third larval instar, well before pupariation and metamorphosis (Ready et al., 1976; White and Kankel, 1978). The general theme of the progressive transformation of the larval to the adult nervous system in holometabolous insects has been elaborated by Edwards (1969).

   Late in this prolonged and complex metamorphosis of the nervous system from its larval to its adult architecture occurs the arrival of the axons of newly differentiated sensory cells. The first signs we have seen of the formation of the triple row of bristles on the leading edge of the wing by scanning electron microscopy have been at around 15 hours after pupariation in animals maintained at 25°C (E.S. Cole, unpublished). By comparison, the neuroblast divisions that occur throughout larval life cease at around 24 hours after pupariation (White and Kankel, 1978).

   We do not yet know when sensory axons reach the CNS and what is happening to motor- and interneurons at this time, but certainly by the time the axons arrive, the CNS is highly structured, has most if not all of its cells, and its metamorphosis to the adult form is well under way. This complex organ, then, is the substrate through which the arriving axons grow.

CELLULAR DEVELOPMENT AND CELL-CELL INTERACTIONS

Before presenting the details of some of our work on neural development in *Drosophila*, it seems appropriate to set the scene a little more concretely by summarizing a few recent studies that give an indication of the kinds of neuronal events that might be occurring as sensory fibers are growing into the CNS.

Matsumoto (1979) has examined part of the development of interneurons in the antennal lobes of moths during the metamorphosis from caterpillar to adult. These cells first form the outline of their adult shape, the pattern of primary dendrites, and then fill this in with increasing secondary and tertiary branching at the time of arrival of antennal afferents. The post-synaptic cells in this case develop well even if their afferents never reach them, but a quantitative effect of pre- upon post-synaptic cells is indicated (Hildebrand, Hall and Osmond, 1979).

In contrast, a long literature (reviewed recently by Bate, 1978; Palka, 1979a; Anderson et al., 1980) shows that the first-order interneurons of the visual system are critically dependent upon the axons of the primary afferents. The most detailed description of this relationship in insects is by Anderson (1978) in locusts. She has shown, for example, that interneurons produced in excess of the number required by the arriving sensory axons die without differentiating. The most detailed description of the actual sequence of events is for the crustacean *Daphnia* (LoPresti et al., 1973) based on reconstructions from long series of electron micrographs. It is found that the recently formed lamina cell bodies lie dormant until contacted by a retinal axon. They respond to this contact by forming a complex but temporary junction with it (LoPresti et al., 1974) and then by spinning out their neurite. Again, if no retinal axon arrives, they die; if the wrong number arrive, their morphology may be altered (Macagno, 1977, 1979).

Some larval motor neurons in metamorphosing moths die and some new ones are formed from neuroblasts. A third group persist into the adult where they serve new and different functions (Taylor and Truman, 1974). This functional transformation is accompanied by major alterations in their branching patterns (Truman and Reiss, 1976). There are no studies on experimental manipulations of this process, but it seems likely that such regrowing cells would show some reaction to changes in their inputs.

The tympanic organs of crickets develop post-embryonically. If they are never permitted to form, a qualitative change in the dendritic branching pattern of auditory interneurons occurs (Hoy et al., 1978; Hoy and Moiseff, 1979). This is in contrast to the merely quantitative response of abdominal giant interneurons to chronic removal of cercal sensory neurons described by Murphey et al. (1976), possibly because in the latter case the first set of receptors was formed normally in the embryo and the cerci were only removed post-embryonically.

The cercal system has produced the first account of how the sensory fibers themselves might develop in a system other than the compound eye. This analysis is important, since in the eye the receptors and the first interneurons develop synchronously, whereas in most other systems the interneurons develop early and are innervated by a long succession of sensory cells developing later.

Murphey (Murphey, 1979; Murphey et al., 1980) has found that the first-formed hairs of a particular modality (clavate hairs) on the cerci have the most extensive, most central and least dense projections into the CNS. The axons of hairs formed later terminate nearer the edge of the ganglion, cover a smaller territory but do so more densely. Furthermore, receptors arrayed in different rows on the cerci form qualitatively different patterns centrally. The picture, then, is one of pattern classes related to circumferential location on the surface, coupled with quantitative variation related to longitudinal location and time of differentiation and of arrival in the CNS. Strikingly similar conclusions have been drawn by Ghysen (1980) on the basis of the projections of thoracic bristles in adult *Drosophila*.

Reviewing these glimpses of the whole course of development, we wish to emphasize that: (1) there is substantial interaction between pre- and post-synaptic elements; (2) there can be qualitative differences in the detailed branching patterns of receptors that look similar externally but differ in location; and (3) quantitative differences, perhaps suggesting competitive interactions of some kind, also occur. The formation of axonal projections is a dynamic, interactive process even late in post-embryonic life.

THE IMPORTANCE OF CELL SURFACE CONTACTS

In the previous sections we tried to establish the general point that the post-embryonic growth of axons within the CNS occurs in a substrate that is complex and highly structured, and that this growth is interactive with pre- and post-synaptic cells influencing each other. We want now to summarize briefly some observations that emphasize the pervasive importance of cell surface contacts in axonal growth. Again, the best studied case is that of *Daphnia*, already briefly described above. We add here several pertinent studies on insects.

Many have observed that peripheral sensory axons always reach the CNS by growing in contact with pre-existing axons (some recent reviews: Edwards and Palka, 1976; Edwards, 1977; Bate, 1978; Palka, 1979; Anderson et al., 1980). But what of the very first axons? The complete answer is not yet clear, but there are now several demonstrations of the existence of pioneer fibers, axons formed by peripheral cell bodies that generally have no known function (they do not form a sensory dendrite, for example) other than to serve as guiding surfaces for regular sensory axons that differentiate much later (Sanes and Hildebrand, 1975; Bate, 1976; Edwards, 1977; Edwards and Chen, 1979). The cell bodies of the pioneer fibers lie on the inner surface of the epidermis of all the appendages, and their differentiation as well as the growth of their axons has been observed with time lapse photography under Nomarski optics and with the help of intracellular injection of the dye Lucifer Yellow (Keshishian, 1979).

How the pioneer axons themselves come to lie in their characteristic locations is not known, but since they grow over the surface of non-neural cells and do not always follow a straight line, it seems certain that the spatial factors controlling their growth are related to the overall morphogenesis of the appendages. In *Daphnia*, glial cells appear to provide at least some of the surface required for the orderly growth of visual receptor axons (LoPresti et al., 1973).

The pioneer fibers thus appear to link spatial information encoded in non-neural tissues with the growth of axons that have a strong tendency to adhere to one another. Their importance has recently been demonstrated in a study of the embryonic development of cricket cerci by Edwards, Berns and Chen (1979 and personal communication). When pioneer fibers were eliminated by a laser lesion of the cercus, the usual pair of nerves within the appendage, one dorsal and one ventral, did not form. Rather, multiple fascicles were seen, presumably reflecting the usual tendency of axons to follow each other but without the ability to select particular paths over the non-neural substrate. In at least one case these fascicles failed to reach the CNS. Thus, pioneer fibers not only occur in the course of normal development, but they also appear to be necessary. The axons seen in the stalks of imaginal discs in dipterans have been thought to be the equivalents of pioneer fibers (Edwards, 1977), but their characteristics have not been worked out.

But what of the CNS? Are there special fibers within ganglia that can be identified as serving a guiding role for later ones? The ramifications of the peripheral pioneer fibers of locust cerci have been filled with cobalt, and it appears that the later-formed regular sensory fibers follow them even within the CNS, at least to reach the right regions of the neuropile (Shankland, 1979). The pioneer fibers reach the CNS after a number of intrinsic neurons have already formed a gridwork of processes. Some of the very first of the intrinsic processes persist into later life but some die, apparently after their guidance function has been completed (Spitzer et al., 1979). The earliest cells of the CNS differentiate from a distinctive group of precursors (Goodman et al., 1979). Thus, throughout the development of the nervous system, both peripheral and central, we see that nerve cells follow one another's surfaces and in the beginning special nerve cells develop that seem to bridge the gap between neuron-neuron surface interactions and neuronal growth over non-neural substrates.

Our overall picture of neural development, then, is a series of interactive steps in which pre-existing cells guide the processes of newly formed ones and these in turn modify the properties of the older cells. The interactions that we know about require cell-cell contact. The nature of the interactions is assumed to change as maturation of the nervous system proceeds.

NEURAL PATHWAYS IN HOMEOTIC MUTANTS

Having laid a general groundwork of recent studies on neural development, both embryonic and postembryonic, we will proceed to describe a particular experimental system in *Drosophila* in which we have used both genetic and surgical approaches to perturb normal development and thus to try to gain some insight into the factors that generate sensory projection patterns.

Our working assumptions in designing and interpreting such experiments are as follows: (1) Every sensory neuron has a developmental program. Just as sensory neurons differ from one another in peripheral anatomy and physiological characteristics, so are their central axons guided by different sets of instructions. (2) The CNS is far advanced when sensory axons enter it, and most if not all of the postsynaptic target cells are already present. These, and possibly glial cells as well, provide spatial information of more or less detailed character which the arriving axons will select or respond to according to their own programs. (3) These interactions may be time-dependent. The goal of our work is to attach more specific meaning to these assumptions. This can be done by studying the response of nervous systems to spatial and/or temporal perturbations of normal development.

There is evidence that homeotic genes at the bithorax locus in *Drosophila* (Figure 1) start to be effective very early in development, during or before the initial formation of imaginal discs in the first instar larva (Morata and Garcia-Bellido, 1976) and perhaps as early as blastoderm (Capdevila and Garcia-Bellido, 1974, 1978). The time of origin of the axons in the stalks of the imaginal discs is not known, but it seems likely that these fibers, whose cell bodies appear to lie in the discs rather than the CNS (Edwards, 1977), express the homeotic transformation right from the time of their first differentiation.

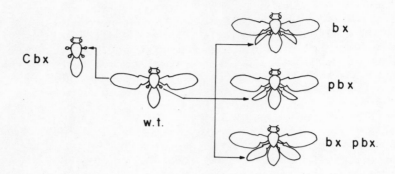

Figure 1: Geometry of dorsal appendages in mutants of the bithorax series. In bithorax (*bx*) animals, the anterior compartment of the haltere is transformed into anterior wing; this includes all of the sensilla of both appendages except for a few sensilla trichodea on the capitellum of the haltere. In postbithorax *(pbx)*, the posterior compartment is transformed; this involves only a trivial number of sensilla. Thus both *bx* and *bx pbx* mutants have essentially no haltere sensilla but a double set of wing sensilla. In *Cbx* both compartments of the wing are transformed into haltere, but the expressivity is variable and flies with no wing sensilla but a double set of haltere sensilla are very rare. (From Palka, 1979b).

Thus, if we study the projection pattern of axons from a wing that forms in the place of a haltere, it is as if we had removed the haltere and grafted a wing in its place prior to the formation of any neural guide paths at all - either those native to the graft tissue, or those native to the host site. To the extent that the projection patterns of the wing and haltere are distinguishable from one another, we can investigate how axons behave when they grow into a region of the CNS which is not their own so that the pathway-determining factors of the axons and the substrate are not matched in the usual way. Such studies have been carried out independently by Ghysen (1978, and this volume), by our own laboratory (Palka, 1977; Palka et al., 1979; Schubiger and Palka, 1979) and recently by Strausfeld and Singh (this volume). The results from different types of receptors have proved to be different, thus providing a rich material for investigation.

Wild-type Projections

The wing carries several hundred bristles (both mechanosensory and mixed chemo-mechanosensory) as the triple row along its anterior margin; some dozen large sensillar campaniformia (large s.c.) on several of the anterior veins; about 50 small sensilla campaniformia (small s.c.), all on the radial vein; and two chordotonal organs within the radius. The sensory supply of the haltere is simpler, consisting of only 15-20 sensilla trichodea on the capitellum, about 200 small s.c. on the pedicellus and scabellum, and two chordotonal organs in the interiors.

Both Ghysen (1978) and Palka et al. (1979) have offered fairly detailed descriptions of the projection patterns formed within the thoracico-abdominal nerve mass by the axons of these sensilla. The two studies are based on different filling methods, horse radish peroxidase (HRP) and cobalt respectively, and agree in most respects. The recent studies of Strausfeld and Singh (this volume) based on cobalt fills analyzed in sectioned material have largely confirmed the earlier work and provided new details. Figure 2 summarizes the principal features. The wing bristles project ventrally into the accessory mesothoracic neuromere just anterior to the main mesothoracic neuromere; the large s.c. send thick axons into the ventral parts of all three neuromeres; the axons of the small s.c. mainly form a dorsal, bifurcating tract that reaches termination sites in the metathorax and in the head; and the projection of the chordotonal organs is not identified.

The axons of the haltere (Figures 2 and 5a) form a dorsal bundle that gives off short tufts of fibers medially in the anterior metathorax and laterally in the mesothorax; it forms a very short twig of a few fibers between the arms of the dorsal bifurcating tract from the wing; and it sends a small bundle dorsolaterally to culminate in an extremely dorsal group of axons crossing the midline. These latter are not consistently filled by cobalt. Our whole-mount cobalt preparations have shown no projection at all into the ventral regions of the ganglion but a slender one has been seen with HRP and in sectioned cobalt material.

Mutant Projections

Figure 3 illustrates the central projections from sensilla on homeotic wings. The bristle axons accumulate in the ventral, anterior metathorax, generally curving around the anterior and medial face of the neuropile in bx^3/Ubx^{130} flies but often forming a discrete ovoidal projection in $bx^3\ pbx/Ubx^{130}$ double mutants. They also send fibers anteriorly into their "native" segment, the mesothorax, but the majority cannot be described as reaching their home destinations. The thick axons of the large s.c. also project ventrally, and form a pattern in all three neuromeres very much like they would have formed had they entered in the mesothorax. Unlike the bristle axons, therefore, they can fairly be described as reaching home destinations and doing so by the usual routes. Finally, the axons of the small s.c. form a dorsal tract indistinguishable from that which would have been formed by haltere axons, ex-

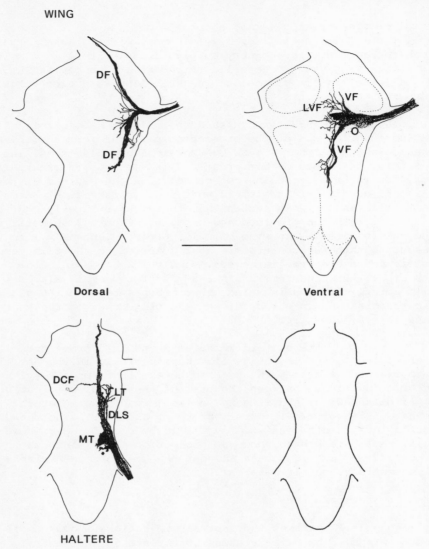

Figure 2: The central projections of receptor cells in the wing and haltere of wild type *Drosophila*. DF - dorsal fork, containing axons from wing small s.c.; O - ovoid, formed by axons from bristles; VF - ventral fork, containing fine fibers probably also originating from bristles; LVF - large ventral fibers from large s.c., some of which also travel with the fine fibers of the ventral fork; MT - medial tuft of the haltere projection; LT - lateral tuft; DLS - dorsolateral strand; DCF - dorsal crossing fibers. Cobalt whole mounts show no ventral projection from the haltere. Scale bar is 100 μm. (After Palka et al., 1979).

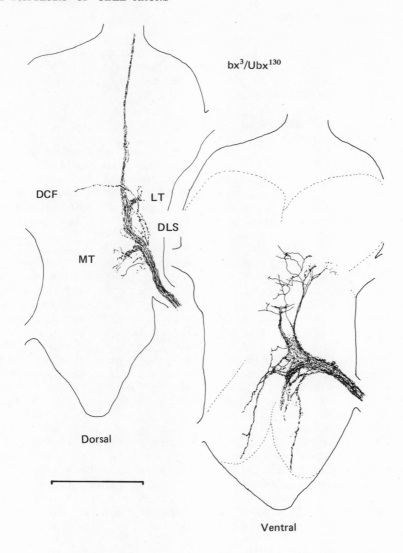

Figure 3: The central projections of receptors on the homeotic appendages of mutant flies of the genotype bx^3/Ubx^{130}. The dorsal component has all the landmarks of a normal haltere projection, including the MT, LT, DLS and DCF (compare with Figure 2). The ventral component has fine fibers from bristles concentrated in the anterior metathorax but also extending both backwards and forwards. It also has thick fibers from large s.c., whose distribution seems to overlap that of the corresponding large s.c. axons of the normal mesothoracic wing. Note that a normal haltere shows no projection at all in those regions in our wholemount preparations. Scale bar is 100 μm. (From Palka et al., 1979).

cept perhaps for subtle differences in the form of the medial tuft. It must be remembered that a quantitative difference is to be expected, since the haltere has around 200 small s.c. and the wing only about 50. Again, we do not know the destinations of fibers from the chordotonal organs.

EXPERIMENTAL ANALYSIS OF HOMEOTIC MUTANTS

This mutant material provides a rather complex natural experiment in transplantation, with a number of features that distinguish it from a grafting experiment. As pointed out above, bithorax genes start to act very early, almost certainly before any nerve fibers associated with the discs are formed, and this constitutes a significant conceptual simplification compared with most transplantation experiments. However, at least three complications exist.

First, we must be concerned about possible pleiotropic effects of the mutations on the CNS. We have approached this question primarily by creating genetic mosaics in which homeotic wing fibers entered a wild type CNS. Second, a fly with metathoracic homeotic wings retains the normal mesothoracic ones (Figure 1), creating a situation in which axons from the two sets of wings could interact with each other or with a common substrate in the CNS. To explore possible interaction effects we have been studying the effects of removing the normal wings on the distribution of fibers from the homeotic wings. And third, we cannot interpret the projection patterns we see unless we have evidence that the nerve cells associated with the homeotic sensilla have a wing character, just as the cuticular elements have. For bristle and large s.c. axons we can make a circular but probably reasonable argument: we will assume that these nerve cells have differentiated as true wing cells because their projection patterns are consistent with a wing identity. But the axons of the homeotic wing small s.c. distribute like those of haltere small s.c., and we propose to raise quite seriously the possibility that these particular nerve cells have in fact <u>not been transformed</u>, even though the cuticular structures clearly have been.

<u>A Mosaic Experiment</u>

Homeotic mutations affect the outside of the adult fly conspicuously. Effects on internal tissues have rarely been described, though there are a few reports such as that of Rizki and Rizki (1978) on changes in the morphology of the larval fat body and of Lewis (1978 on the ovaries and the embryonic CNS. Even though these internal changes may be difficult to detect, one cannot assume that they do not exist. Therefore, we have made a comparison between the projections of homeotic sensilla into mutant CNS, in homeotic flies, and into wild type CNS, in mosaic flies. We accomplished this by using the method of mitotic recombination to generate clones of wing tissue in the halteres of otherwise wild type animals (Palka et al., 1979).

The projections from such clones were generally similar to those seen in mutant flies. Ventral components of fine fibers (presumably from bristles) and of thick fibers (presumably from large s.c.) were found regularly in regions to which the normal haltere does not project. However, there were a number of differences in details, leading us to suggest that a mutant ganglion is, after all, a somewhat different substrate than a wild type ganglion. This agreed with the difference mentioned above in the shape of the homeotic bristle projection in bx^3/Ubx^{130} flies (arcuate and/or branching) and in the double mutants $bx^3\ pbx/Ubx^{130}$ (ovoid). It also agreed with some unpublished anatomical evidence on the double mutants obtained by David King (personal communication).

The mosaic experiment is subject to at least one alternative interpretation, pointed out by J.C. Hall (personal communication). We used the Minute method (Morata and Ripoll, 1975) to produce large clones. Inasmuch as Minute has an effect on the rate of cell division and produces a significant slowing of adult development, it could have resulted in a change in the time of arrival of sensory axons (Minute+) relative to the maturation of the CNS (Minute). This possibility and its potential influence on axon distribution patterns has not yet been tested.

Effects of Deafferentation

Figure 4 illustrates the point that wing and haltere fibers from the small s.c. of wild type flies travel very close to each other though on present evidence they do not intermingle; the same is true for fibers from the normal and the homeotic wings of mutant flies. At the points marked by the arrows the two fiber bundles cross, and some details of this intersection can be seen in Figure 6a. Could the ascending homeotic axons be induced to grow towards the root of the wing nerve at this intersection point if the normal wing axons were lacking and their substrate were vacant? If yes, this would offer strong support for Ghysen's (1978) hypothesis that afferent fibers follow precise markers through the CNS and that these are not polarized, so that the afferents will trace out the pattern of the markers irrespective of the point at which they first encounter it (Schubiger and Palka, 1979, and in preparation).

Wing removal was accomplished in two ways: surgically, by extirpating most of the wing disc on one side of a homeotic prepupa; and genetically, by producing a double mutant stock generating flies of the genotype wg/wg; bx^3/Ubx^{130}. This stock gave flies which lacked 0, 1 or 2 normal wings and/or 1 or 2 homeotic wings. Thoracic duplications were seen as described previously (Sharma, 1973; Sharma and Chopra, 1976; Morata and Lawrence, 1977; Deak, 1978). Small wing stumps were fairly common, but only flies missing all of the mesothoracic wingspread were analyzed.

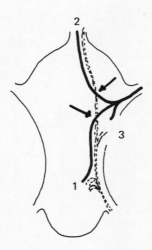

Figure 4: A schematic representation of the dorsal components of the wing and haltere projections of normal flies, both carrying axons from small s.c. The longitudinal fiber bundles are adjacent though apparently not intermingled. The termination areas in metathorax (1), mesothorax (3) and subesophageal ganglion (2; see Ghysen, and Strausfeld and Singh, this volume) are also close together but not overlapping. The arrows mark points where fiber bundles intersect. (See also Figure 6a.) (After Palka et al., 1979).

The results are illustrated in Figures 5 and 6b, and are easily summarized. We could not find any effect of wing removal by either method on the distribution of homeotic small s.c. axons. They continued to behave just like normal haltere small s.c. fibers and showed no detectable tendency to send even a few fibers out along the entry path of the wing small s.c. axons.

An unexpected byproduct of this experiment was the occurrence, in about 10% of cases, of homeotic wing nerves rerouted forward so that they entered the CNS in the mesothorax where the normal wing axons would have entered had they not been removed. These fibers did not distribute as normal wing small s.c. do. In every case we have seen, they have formed one or several supernumerary bundles within the CNS. The normal wing projection pattern and one example of a rerouted homeotic pattern are compared in Figure 6c and d.

The surprising outcome of the deafferentation experiment, therefore, is that the axons of homeotic wing small s.c. consistently behave like haltere small s.c. when they enter the metathorax, even

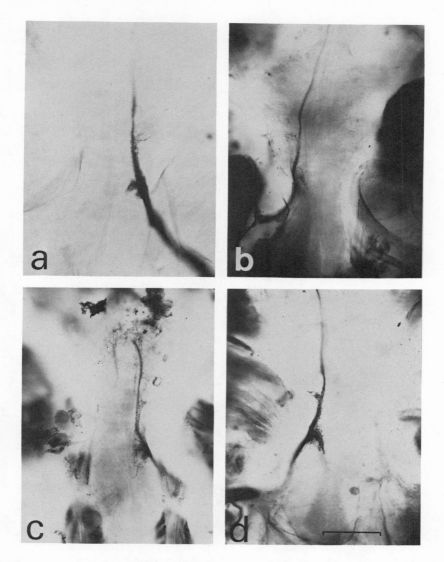

Figure 5: Lack of a normal mesothoracic wing has no detectable effect on the projection of the small s.c. of the homeotic wing in bx^3/Ubx^{130} mutants. (a) Normal haltere projection of a wild type fly. (b) Dorsal component of the homeotic wing projection of an intact $bx^3\ Ubx^{130}$ mutant. (c) Dorsal component of a surgically de-winged fly. (d) Dorsal component of a double mutant, $wg/wg; bx^3/Ubx^{130}$ in which all the dorsal appendages except a single homeotic wing were lacking. The MT and LT are evident in all cases, and in no case did we find any homeotic small s.c. axons deviating from the ascending tract towards the root of the wing nerve. Scale bar is 100 µm.

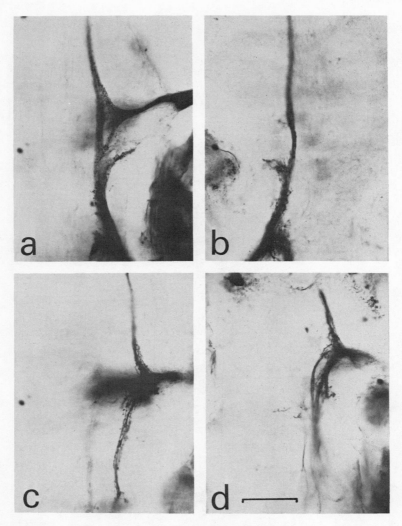

Figure 6: (a) Intersection of wing and haltere small s.c. tract in a wild type fly. Note the streams of wing axons (entering from the right) crossing the bundle of haltere axons ascending from below. The LT is seen behind the posterior arm of the wing dorsal fork (DF). (b) Same preparation as Figure 5d, at higher power. Note that the ascending bundle is intact, with no sign of laterally directed fibers anterior to the LT. Part (a) shows where fibers following a wing path backwards would lie if they were present. (c) The DF of a wild type fly. Note that a single bundle projects posteriorly. (d) The dorsal component of a re-routed homeotic nerve of a wg/wg; bx^3/Ubx^{130} double mutant which entered the mesothoracic neuromere at the point where wing fibers normally enter. Note the multiple strands projecting posteriorly. These have been seen in all cases of re-routed homeotic nerves. Scale bar is 40 µm

when potentially competing normal wing small s.c. are removed; however, when these same axons, arising from tissue that externally is wing-like and not haltere-like, enter the mesothorax, they do not behave like perfect wing axons. The simplest explanation of this behavior would seem to be that the homeotic axons still carry a haltere identity, i.e., that they have not been transformed. Very similar results have been obtained by A. Ghysen and Janson (personal communication) in an independent experiment using wingless.

Our analysis of bristle axons in this experiment is not yet complete. However, present indications are that they respond to deafferentation quite differently. Unlike the small s.c. axons, the bristle axons do make a normal wing projection when they enter the mesothorax through rerouted nerves. Also in contrast to the axons of the small s.c., when they enter the metathorax their distribution does seem to be affected by removal of the normal wings. Our analysis of the large fibers from the large s.c. is just beginning, but Ghysen and Janson have described this projection as being unaffected by the absence of normal wings.

WHAT HAVE WE LEARNED?

The various homeotic mutants provide fascinating material for study but, as emphasized above, each case needs to be approached in detail and with caution. We often think of mutants as Nature's experiments, but the problem is that we do not have control over the design and execution of these experiments.

More specifically, what besides caution have we learned by studying homeotic mutants? Here is a partial list:

(1) Different receptor cell classes are affected differently by any given mutation. We have seen these differences in the patterns formed by their axons in the CNS, in their response to deafferentation, and in the evidence that the neurons of the small s.c. are not transformed by bx^3 whereas the cuticular elements are.

(2) No major pleiotropic effects of bx^3 or pbx on the CNS have been detected by any investigator. A recent study of mosaics in spineless-aristapedia by R.F. Stocker and P.A. Lawrence (in preparation) leads to the same conclusion for that mutation. However, there is a significant tendency for bristle projection patterns to be different in bx^3/Ubx^{130} single mutants and in $bx^3 \ pbx/Ubx130$ double muttants. All receptors in the wing are in the anterior compartment, and there is at present no reason to suspect that pbx affects cells in the anterior compartment. Thus, we view the differences in the projection patterns as evidence for a subtle effect of pbx on the CNS. The mosaic results suggest weak effects of bx^3 on the CNS, and the study of Stocker and Lawrence also reports minor differences be-

tween the projections in mutant and mosaic animals in spineless-aristapedia. Thus, even though the evidence is clearly against any major effects of those alleles of homeotic genes that have been studied upon the adult CNS, indications of minor effects crop up repeatedly and the possibility of such effects upon the projection of any particular class of sensilla from a homeotic region must always be considered.

(3) The evidence that the external morphology of sensilla may not be a reliable indicator of the differentiative state of the associated neurons is fairly strong. We discuss our evidence for this hypothesis below. Further evidence is given by A. Ghysen and R. Janson (this volume); the anatomical data of Strausfeld and Singh (this volume) are also compatible with this view.

YOUR CAN'T TELL A BOOK BY ITS COVER

We summarize here our reasons for advancing the suggestion that there can be a disparity in differentiative state between the cuticle-secreting cells of a sensillum and its sensory neuron.

(1) Axons of homeotic small s.c. follow a path which matches that of haltere small s.c. in all details that we have examined carefully. There is no major exception to this statement either in our data or that of Ghysen and Janson and of Strausfeld and Singh.

(2) This pathway is stable in the face of both genetic and surgical removal of the normal wing, i.e. there is at present no evidence of interactions between the axons of normal and homeotic small s.c. We have some preliminary evidence that the bristle projection is influenced by the same experimental manipulations (Schubiger and Palka, unpublished), suggesting that populations of axons from apparently identical sensilla on these adjacent appendages can interact. The lack of interaction among small s.c. axons, then, could be due to the fact that the neurons are not identical even though the cuticular elements are.

(3) Rerouted homeotic nerves carry both small s.c. and bristle axons to the mesothorax. The axons of the homeotic small s.c. do not project like native wing small s.c. in this situation, whereas they do project like native haltere axons when they enter the metathorax. In contrast, we have seen a number of cases in which rerouted homeotic bristle axons formed perfectly wing-like projections in the mesothorax.

These observations are economically explained by the supposition that the small s.c. neurons of homeotic wings carry a haltere label, a possibility already discussed by Palka et al. (1979). We would add to the observations the following "arguments of plausibility".

(1) In other alleles of bithorax, notably bx^{34e}, some of the small s.c. are not transformed, even when an excellent homeotic wing blade and a distinct radial vein have been formed. Even in bx^3, the most extreme allele, Adler (1978) reported that the homeotic radius carries a mixture of wing and haltere small s.c. We have recently started a systematic examination of this region with the scanning electron microscope (Cole and Palka, unpublished), and we too find at least occasional haltere-like small s.c. present on bx^3/Ubx^{130} and bx^3 pbx/Ubx^{130} homeotic wings (Figure 7). The determinative state even of the cuticle-secreting cells seems to be somewhat unstable.

(2) A similar situation prevails in Cbx, where Morata (1975) found that the proximal anterior region containing the radius and the small s.c., was transformed from wing to haltere in only 11% of homozygotes, compared with 100% for most of the posterior compartment.

(3) Perhaps even more important, Morata also found that even when a given _area_ of the Cbx appendage was transformed, individual cells within it need not be. In particular, he found single wing sensory bristles apparently unaffected by the homeotic mutation but surrounded by a sea of ordinary epidermal cells secreting haltere trichomes and thus showing the homeotic effect.

(4) A trichogen cell secreting the shaft of a bristle or the dome of a campaniform sensillum is not the sister but the cousin of the nerve cell - one cell division intervenes (Lawrence, 1966). It seems entirely possible that a gene product required for haltereness (such as the S_1 of Lewis, 1978) might be present in different amounts in these cousins, or that their threshold to it might consistently be different. This is the kind of explanation proposed by Morata to account for regional and cellular differences in expression in the cuticle secreting cells of Cbx, and there is no reason why it should not apply equally well to the trichogen and nerve cells of bx^3.

The case is certainly not unambiguous, but we feel that it is sufficiently plausible to merit discussion. On the basis of interesting genetic evidence, Ghysen and Janson have proposed a different and quite specific model to account for the haltere-like behavior of homeotic small s.c. axons. Whatever the correct explanation proves to be, the aggregate evidence for the suggestion that you cannot tell a neuron by the cuticle that covers it seems quite strong.

AN EMERGING PICTURE

The phenomenology presented by the sensory projections from homeotic appendages is complex and diverse, but it is reasonably consistent if it is viewed as the outcome of the working assumptions we

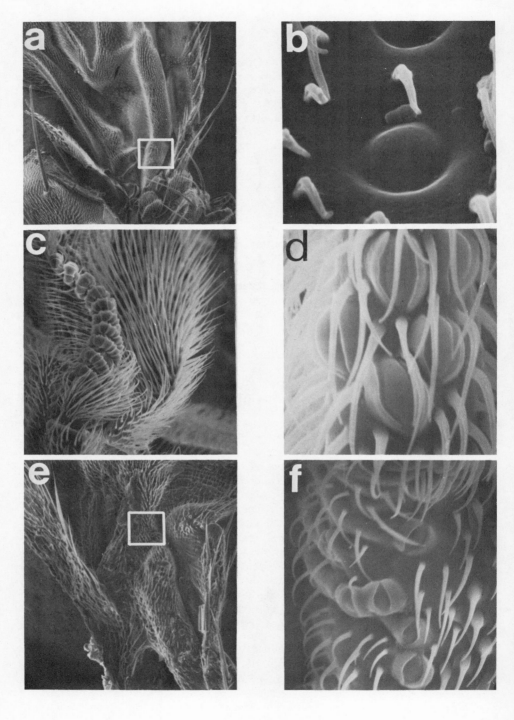

Figure 7: Examples of small campaniform sensilla. (a) Base of wild type wing showing location of small s.c. on dorsal aspect of proximal radius. 130x. (b) Small s.c. from field shown in (a). Note smooth contours. 9600x. (c) Ventral pedicellus of haltere. Note elaborately sculptured small s.c. fused into rows. 1000x. (d) Unfused small s.c. on scabellum of wild type haltere. Note lips. 5400x. (e) Base of homeotic wing from a $bx^3\ pbx/Ubx^{130}$ mutant, indicating the tip of the distal radius. 220x. (f) Supernumerary small s.c. from the field shown in (e). A wild type wing would have no small s.c. in this location. Note haltere-like lips. 2000x. (Figure kindly prepared by E.S. Cole.)

---

stated earlier. The picture we see at present can be described this way:

(1) We suppose that the axons of the homeotic large s.c. are so successful in establishing their normal projection pattern because their normal guide paths are present in all three neuromeres and because their own program directs them to recognize these paths and also permits them to grow far and branch widely.

(2) In contrast, the homeotic bristle axons may form variable projections but with a strong tendency towards metathoracic duplication of their normal mesothoracic pattern because (a) a path from meso- to metathorax is appropriate for only a small number of fibers from the normal wing and hence should only be followed in the opposite direction by a small number of axons from the homeotic wing; (b) the metathorax contains an adequately homologous area which promotes the branching and termination of bristle axons; and/or (c) their own program calls for branching soon after entry into the CNS.

(3) The homeotic small s.c., we suppose, encounter the normal haltere substrate and follow it because in spite of external appearances they still carry the normal haltere developmental program which is presumably optimally matched to that particular substrate. The essential feature of this picture is that projection patterns are the outcome of the interaction between the recognition and growth programs of axons and the spatial information provided (in greater or lesser detail) by the substrate they grow on or through.

Many other findings are well accomodated in this general picture. For example, Anderson and Bacon (1979) showed that certain groups of hairs on the head of locusts have distinctive projections in the CNS, and that these are re-established when one group of hairs is grafted in the place of another: the hairs project according to their cuticle of origin. H. Anderson (personal communication) has also grafted head hairs onto the pro- and mesothorax. Now some normal head hairs do travel through the CNS all the way to the prothoracic ganglion, and axons from the grafts form central patterns that are variable but resemble these normal long-distance projections in many ways. On the other hand, head hairs normally do not reach the mesothoracic ganglion at all, and correspondingly the projections of head hairs grafted to the mesothorax are extremely variable and difficult to relate to any normal projections. We suggest that, roughly speaking, the large s.c. in bithorax mutants behave like hairs grafted from one head location to another in locusts (because appropriate substrates in the CNS are within reach of the in growing axons); the bristles behave like head-prothorax grafts (because the metathoracic neuromere in *Drosophila* is similar though not identical to the mesothoracic one; i.e. it has at least some suitable substrates for bristle axons); and the small s.c. can be thought of as logically resembling control grafts where a patch of hairs is removed and replaced in the same location (because they retain a metathoracic identity).

Stocker et al. (1976) and R.F Stocker and P.A. Lawrence (in preparation) have found that sensory neurons from the homeotic legs of Antennapedia and spineless-aristapedia (ss^a) project to regions of the CNS appropriate to antenna, not leg, though with substantial differences in the details of the projection pattern. On the other hand, Deak (1976) and Stocker (1977) have shown that at least some of the homeotic sensory neurons function like leg receptors and can initiate a reflex response normally characteristic only of legs, not of antennae. These results, like those on bithorax, raise questions about whether particular homeotic sensory neurons have a leg or an antennal identity, whether antennal and leg neuropile areas of the CNS have some underlying similarities in spite of their very different histological appearance. Again, a full analysis will require consideration of both the developmental programs of the axons and the nature and spatial distribution of substrates or guide paths in the CNS. It may be useful to point out that in other systems evidence has been obtained for the development of cell-specific axonal branching patterns under *in vitro* conditions where the usual complex substrate cues provided by the CNS are not available (e.g. Guillery et al., 1968).

As long as we ask questions such as "Do bithorax sensory neurons project like wing neurons or like haltere neurons?", the empirical answer is complex: the large s.c. project like wing, the small s.c. like haltere, and bristles rather like wing but in an adjacent segment. This is because such a question focuses attention on the outcome of developmental processes. But if we ask, "What processes produce these projection patterns?", a unified picture emerges in which pre-existing substrate factors in the CNS and differentiation programs in the developing sensory neurons play equally important roles. The picture will continue to grow, especially as temporal factors are also examined (e.g., LoPresti et al., 1973; Macagno, 1979).

WHERE DO WE GO FROM HERE?

We introduced this paper with a selection of studies that emphasize the dynamic nature of developmental processes. The experimental results, however, were all based on studying the final outcome of development even though the various perturbations occurred very early. It seems appropriate at this point to start to bridge the time gap by examining neural development as it is happening. Fortunately, the rapidly growing literature on the development of the nervous system in other insects as well as a variety of other animals provides us with many indications of what to look for.

At the same time, we are gradually acquiring sufficient information about neural organization in *Drosophila* to be able to contribute to basic inquiries into the development and pattern formation of the whole organism. It seems likely that analyses of basic developmental processes such as segment determination and compartmentalization will gain in richness and scope from studies directed explicitly to the nervous system.

ACKNOWLEDGEMENTS

The original work reported here has been supported by grants No. NS-07778 from the NIH and No. BNS-7914111 from the NSF, and by the University of Washington Graduate School Research Fund. J.P. was introduced to the fascinations of the bithorax system while visiting Peter Lawrence as a Guggenheim Fellow.

REFERENCES

Adler, P.N., 1978, Positional information in imaginal discs transformed by homeotic mutations, Wilhelm Roux Arch., 185:271.
Anderson, H., 1978, Postembryonic development of the visual system of the locus, *Schistocerca gregaria*. I. Patterns of growth and developmental interactions in the retina and optic lobe, J. Embryol. exp. Morph., 45:55.
Anderson, H., and Bacon, J., 1979a, The form of the C.N.S. arborizations of sensory neurons is governed by their epidermis of origin, Abst. Soc. Neurosci., 5:152.
Anderson, H., and Bacon, J., 1979b, Developmental determination of neuronal projection patterns from wind-sensitive hairs in the locust, *Schistocerca gregaria*, Devel. Biol., 72:364.
Anderson, H., Edwards, J.S., and Palka, J., 1980, Developmental neurobiology of invertebrates, Ann. Rev. Neurosci., 3:97.
Bate, C.M., 1976, Pioneer neurons in an insect embryo, Nature, 260: 54.
Bate, C.M., 1978, Development of sensory systems in arthropods, in: "Handbook of Sensory Physiology, Vol. IX: Development of Sensory Systems", M. Jacobson, ed., Springer-Verlag, Berlin-Heidelberg-New York, p. 1.

Capdevila, M.P., and Garcia-Bellido, A., 1974, Developmental and genetic analysis of *bithorax* phenocopies in *Drosophila*, Nature, 250:500.

Capdevila, M.P., and Garcia-Bellido, A., 1978, Phenocopies of bithorax mutants, Wilhelm Roux Arch., 185:105.

Deak, I.I., 1976, Demonstration of sensory neurons in the ectopic cuticle of *spineless-aristapedia*, a homeotic mutant of *Drosophila*, Nature, 260:252.

Deak, I.I., 1978, Thoracic duplications in the mutant *wingless* of *Drosophila* and their effect on muscles and nerves, Devel. Biol., 66:422.

Edwards, J.S., 1969, Postembryonic development and regeneration of the insect nervous system, Adv. Insect Physiol., 6:97.

Edwards, J.S., 1977, Pathfinding by arthropod sensory nerves, in: "Identified Neurons and Behavior of Arthropods", G. Hoyle, ed., Plenum Press, New York, p. 483.

Edwards, J.S., Berns, N.W., and Chen, S.-W., 1979, Laser lesions of embryonic cricket cerci disrupt guidepath role of pioneer fibers, Abst. Soc. Neurosci., 5:158

Edwards, J.S., and Chen, S.-W., 1979, Embryonic development of an insect sensory system, the abdominal cerci of *Acheta domesticus*, Wilhelm Roux Arch., 186:151.

Edwards, J.S. and Palka, J., 1976, Neural generation and regeneration in insects, in: "Simpler Networks and Behavior", J. Fentress, ed., Sinauer, Sunderland, Massachusetts, p. 167.

Ghysen, A., 1978, Sensory neurones recognize defined pathways in *Drosophila* central nervous system, Nature, 274:869.

Ghysen, A., 1980, Choice of the right pathway in *Drosophila* central nervous system, Devel. Biol., in press.

Ghysen, A., and Deak, I.I., 1978, Experimental analysis of sensory pathways in *Drosophila*, Wilhelm Roux Arch., 184:273.

Goodman, C.S., Bate, C.M., and Spitzer, N.C., 1979, Origin, transformation, and death of neurons from an identified precursor during grasshopper embryogenesis, Abst. Soc. Neurosci., 5:161.

Guillery, R.W., Sobkowicz, H.M., and Scott, G.L., 1968, Light and electron microscopical observations of the ventral horn and ventral root in long term cultures of the spinal cord of the fetal mouse, J. Comp. Neurol., 134:433.

Hildebrand, J.G., Hall, L.M., and Osmond, B.C., 1979, Distribution of binding sites for ^{125}I-labeled α-bungarotoxin in normal and deafferented antennal lobes of *Manduca sexta*, Proc. Natl. Acad. Sci. U.S.A., 76:499.

Hoy, R., Casaday, G., and Rollins, S., 1978, Absence of auditory afferents alters the growth pattern of an identified auditory interneuron, Abst. Soc. Neurosci., 4:115.

Hoy, R., and Moiseff, A., 1979, Aberrant dendritic projections from an identified auditory interneuron are innervated by contralateral auditory afferents, Abst. Soc. Neurosci., 5:163.

Keshishian, H., 1979, Origin and differentiation of pioneer neurons in the embryonic grasshopper, Abst. Soc. Neurosci., 5:166.

Lawrence, P.A., 1966, Development and determination of hairs and bristles in the milkweed bug, Oncopeltus fasciatus (Lygaeidae, Hemiptera), J. Cell Sci., 1:475.
Lewis, E.B., 1978, A gene complex controlling segmentation in Drosophila, Nature, 276:565.
LoPresti, V., Macagno, E.R., and Levinthal, C., 1973, Structure and development of neuronal connections in isogenic organisms: Cellular interactions in the development of the optic lamina of Daphnia, Proc. Natl. Acad. Sci. U.S.A., 70:433.
LoPresti, V., Macagno, E.R., and Levinthal, C., 1974, Structure and development of neuronal connections in isogenic organisms: Transient gap junctions between growing optic axons and lamina neuroblasts, Proc. Natl. Acad. Sci. U.S.A., 71:1098.
Macagno, E.R., 1977, Abnormal synaptic connectivity following UV-induced cell death during Daphnia development, in: "Cell and Tissue Interactions", J. W. Lash and M.M. Burges, eds., Raven Press, New York, p. 293.
Macagno, E.R., 1979, Cellular interactions and pattern formation in the development of the visual system of Daphnia magna (Crustacea, Branchiopoda). I. Interactions between embryonic retinular cell fibers and lamina neurons, Devel. Biol., 73:206.
Matsumoto, S.G., 1979, Physiology and morphology of neurons in the antennal lobes of mature and developing Manduca sexta, Abst. Soc. Neurosci., 5:170.
Morata, G., 1975, Analysis of gene expression during development in the homeotic mutant Contrabithorax in Drosophila melanogaster, J. Embryol. exp. Morph., 34:19.
Morata, G., and Garcia-Bellido, A., 1976, Developmental analysis of some mutants of the bithorax system of Drosophila, Wilhelm Roux Arch., 179:125.
Morata, G., and Lawrence, P.A., 1977, The development of wingless, a homeotic mutation of Drosophila, Devel. Biol., 56:227.
Morata, G., and Ripoll, P., 1975, Minutes: Mutants of Drosophila autonomously affecting cell division rate, Devel. Biol., 42:211.
Murphey, R.K., 1979, The development of distinct synaptic arborizations in the cricket nervous system is correlated with birthday and position in a receptor array, Abst. Soc. Neurosci., 5:256.
Murphey, R.K., Jacklet, A., and Schuster, L., 1980, A topographic map of sensory cell terminal arborizations in the cricket CNS: Correlation with birthday and position in a sensory array, J. comp. Neurol., in press.
Murphey, R.K., Mendenhall, B., Palka, J., and Edwards, J.S., 1976, Deafferentation slows the growth of specific dendrites on identified giant interneurons, J. comp. Neurol., 159:407.
Palka, J., 1977, Neurobiology of homeotic mutants in Drosophila, Abst. Soc. Neurosci., 3:187.
Palka, J., 1979a, Theories of pattern formation in insect neural development, Adv. Insect Physiol., 14:251.

Palka, J., 1979b, Mutants and Mosaics, Tools in Insect Developmental Neurobiology, Soc. Neurosci. Symp., 4:209.
Palka, J., Lawrence, P.A., and Hart, H.S., 1979, Neural projection patterns from homeotic tissue of Drosophila studied in bithorax mutants and mosaics, Devel. Biol., 69:549.
Ready, D.F., Hanson, T.E., and Benzer, S., 1976, Development of the Drosophila retina, a neurocrystalline lattic, Devel. Biol., 53:217.
Rizki, T.M., and Rizki, R.M., 1978, Larval adipose tissue of homeotic *bithorax* mutants of Drosophila, Devel. Biol., 65:476.
Sanes, J.R., and Hildebrand, J.G., 1975, Nerves in the antennae of pupal Manduca sexta Johanssen (Lepidoptera: Sphingidae), Wilhelm Roux Arch., 178:71.
Schubiger, M., and Palka, J., 1979, The haltere-like projection of segmentally translocated wing receptors in Drosophila homeotic mutants, Abst. Soc. Neurosci., 5:178.
Shankland, M., 1979, Development of pioneer and sensory afferent rpojections in the grasshopper embryo, Abst. Soc. Neurosci., 5:178.
Sharma, R.P., 1973, *wingless*, a new mutant in *D. melanogaster*, Dros. Inform. Ser., 50:134.
Sharma, R.P., and Chopra, V.L., 1976, Effect of *wingless(wg)* mutation on wing and haltere development in Drosophila melanogaster, Devel. Biol., 48:461.
Spitzer, N.C., Bate, C.M., and Goodman, C.S., 1979, Physiological development and segmental differences of neurons from an identified precursor during grasshopper embryogenesis, Abst. Soc. Neurosci., 5:181.
Stocker, R.F., 1977, Gustatory stimulation of a homeotic mutant appendage, *Antennapedia*, in Drosophila melanogaster, J. Comp. Physiol., 115:351.
Stocker, R.F., Edwards, J.S., Palka, J., and Schubiger, G., 1976, Projection of sensory neurons from a homeotic mutant appendage, *Antennapedia*, in Drosophila melanogaster, Devel. Biol., 52:210.
Taylor, H.M., and Truman, J.W., 1974, Metamorphosis of the abdominal ganglia of the tobacco hornworm, Manduca sexta, J. comp. Physiol., 90:367.
Truman, J.W., and Reiss, S.E., 1976, Dendritic reorganization of an identified motoneuron during metamorphosis of the tobacco hornworm moth, Science, 192:477.
White, K., and Kankel, D.R., 1978, Patterns of cell division and cell movement in the formation of the imaginal nervous system in Drosophila melanogaster, Devel. Biol., 65:296.

DISCUSSION

Discussion of this paper follows the chapter by Strausfeld and Singh.

SENSORY PATHWAYS IN *DROSOPHILA* CENTRAL NERVOUS SYSTEM

Alain Ghysen and Renaud Janson

Laboratoire de Génétique
Université Libre de Bruxelles
Belgium

SUMMARY

We have analyzed the central projections of identified sensory neurons in wild type *Drosophila* and in the homeotic mutant *bithorax postbithorax*. The results indicate that: (i) the choice of a given pathway in the central nervous system of the adult depends on the developmental history of the neuron, namely on the type of sense organ of which it is part, and on the developmental compartment to which it belongs; (ii) the establishment of the projection involves the specific recognition of a preexisting "trail" in the central nervous system. Here we examine and rule out alternative explanations of our results, and conclude that the directed growth of an axon relies at least in part on a physical guidance mechanism to which specificity is conferred by programming the growing axon to recognize the appropriate guide.

We speculate that the guides are preexisting, specifically marked nerve fibers and that the basic pattern of connectivity is largely laid down early during neurogenesis, at a time when a standard set of connections is relatively easy to specify.

INTRODUCTION

While it is probably true that the genetic control of the development of the nervous system and of its connectivity remains one of the most fascinating problems in the field of neurobiology (Garcia-Bellido, personal communication), there is little doubt that it is also a most challenging subject in the field of developmental genetics.

The explosive growth in the formal understanding of the mechanism of heredity which resulted from the work of Morgan, Sturtevant, Muller and Bridges on *Drosophila* soon led to the hope that this material might also prove valuable in the analysis of the nervous system, its function and its development. However, early attempts to analyze the nervous defects in some behavioral mutants were unsuccessful. More recently this approach was brilliantly revived by Benzer and coworkers (for an overview see reference 1). Large numbers of behavioral mutants were isolated and characterized in the early 1970's. In a few cases the analysis of such mutants allowed an explanation of the nervous defect in terms of specific steps of the conduction or synaptic transmission of the nerve impulse[2,3,4]. On the other hand in no case could the defect be ascribed to some perturbation in the development of the nervous system or of its connectivity. This disappointing result led to the somewhat pessimistic statement that "In general, ... behavioral geneticists have asked one set of questions and answered another".[5]

Simultaneously with the behavioral approach mentioned above, much work has been devoted to the genetic analysis of the development of the adult epidermis. This work has recently met with considerable success and has provided new insights on the mechanism by which the genome codes for the piecemeal construction of an organism.[6,7] The understanding of this mechanism has now reached a state where it becomes possible to ask meaningful questions about the regulatory mechanism which is responsible for the orderly expression of segment- and compartment-specific genes.[8]

Part of the progress made along this line stems from the existence of a special class of developmental mutations, the homoeotic mutations, which affect the process of segmental or compartmental determination and consequently transform one body part into another (reviewed in references 6,9,10). Independently of their importance for the analysis of the development of the external morphology of the fly, the existence of homeotic mutations suggests an alternative approach for the developmental analysis of the nervous system. Instead of searching for mutations which would specifically alter the connectivity, and trying to infer the rules used to establish the appropriate connections, why not take advantage of the known homeotic mutations to face the nervous system with modified sensory inputs? Assuming that the genome has no special provision for coping with such extraordinary situations, one would expect that the handling of the homeotic sensory input will follow the same general principles that govern the normal development of the nervous system. One may then hope that comparing the fate of normal and homeotic inputs will reveal the nature of some of these general principles.

This approach was pioneered by Deak.[11] He used the mutant aristapedia (which transforms part of the antenna into the development-tally homologous part of a leg) to show that some of the ectopic leg neurons are able to establish functional connections similar to those of the homologous neurons on the normal legs.

Parallel experiments along this line have attempted to analyze in more detail the fate of the ectopic projections by using cobalt backfill or degeneration stain to visualize the homeotic sensory projection in the central nervous system.[12,13] However, the complexity in the projection from whole appendages makes the interpretation of these results very difficult. Therefore, it seemed necessary to find a system where the projection of single identified sensory neurons could be analyzed in normal and homeotic situations. This can be done with some of the mesothoracic sensory structures.

Many mesothoracic bristles can be unambiguously recognized as they occupy defined and reproducible locations on the notum. The same is true for the campaniform sensilla on the distal part of the wing blade. The central projection of each of these sense organs can be individually visualized by horseradish peroxidase backfill Homeotic mutations are known which transform the metathorax into a mesothorax. This transformation is specially appropriate since the normal metathorax consists of a very small notum which is void of bristles, and a haltere on the distal part of which no campaniform sensilla arise, so that those mesothoracic sensory neurons which can be individually marked have no homologues on the metathorax. Furthermore, the homeotic mutations which induce this transformation belong to the bithorax locus, which is by far the most thoroughly studied and the best understood among the different available homeotic systems.

It is important to mention here that some of the mesothoracic sensory structures have metathoracic homologues (e.g., the "proximal" campaniform sensilla on the base of the wing and on the stalk of the haltere). The analysis of the projections and connections established by these homologous neurons has allowed us to analyze the mechanism which determines the segmental identity of sensory neurons (A. Ghysen, R. Janson, and P. Santamaria; P. Vandervorst and A. Ghysen, in preparation; see also Palka and Schubiger, this volume). On the other hand, the fact that the projections of these homologous sensory neurons run side by side over most of their course[14] makes them unsuitable when using the homeotic mutations as a means to generate "genetic grafts". For this reason they will not be considered further in this paper.

Choice of a Pathway by Normal Sensory Neurons

The description of the different pathways followed in the central nervous system by mesothoracic sensory neurons has been reported elsewhere.[13,14] A series of experiments aimed at understanding why a given neuron decides to establish one projection rather

Table 1. Determination of the pathway of a sensory neuron

Sensory Projection[a]	Sensory Structure	Developmental State					
		A	B	C	D	E	F
notum bristles	B	–	–	0	–	1	=
wing bristles	B	–	–	0	–	0	=
wing camp. sens., distal	C	–	–	0	=	0	1
wing camp. sens., prox.	C	0	–	0	–	0	0
haltere camp. sens., prox.	C	1	–	0	–	0	0
meso. leg camp. sens.[§]	C	0	?	1	–	–	–
meta. leg camp. sens.[§]	C	1	?	1	–	–	–
meso. leg bristles, ant[*]	B	0	0	1	–	–	–
meso. leg bristles, post[*]	B	0	1	1	–	–	–
meta. leg bristles, ant[*]	B	1	0	1	–	–	–
meta. leg bristles, post[*]	B	1	1	1	–	–	–

Legend to Table 1: The first column lists the different sensory projections which have been identified so far. The first have been described elsewhere. [*]: only femur bristles have been analyzed. We do not know whether the differences between the projections of the pro-, meso- and metathoracic leg bristles depend on their site of entry in the thoracic ganglion or on their segmental identity. [§]: the specific projection of these sensilla has not been analyzed; however, the complete leg projection (which includes the projection from the leg sensilla) contains no component which follows the same path as the projection from the wing or haltere campaniform sensilla. This suggests that the leg sensilla must follow their own pathway. The second column represents the type of sensory structure. B: mechanoreceptor bristles; C: campaniform sensilla. The third column represents the state of expression of a hypothetical set of genes (A-F) which is supposed to control the progressive restriction of the developmental potential of imaginal cells. According to the current views, each of these genes define two subgroups in an otherwise homogeneous population of cells, one in which the gene is "on" (symbolized by "1" in the Table) and the other in which it is "off" (symbolized by "0"). The genes A to F are, respectively, expressed in the meta- but not in the mesothoracic segment (A); in the posterior but not in the anterior compartments (B); in the ventral (leg) but not in the dorsal (wing, haltere) discs (C); in the dorsal but not in the ventral compartments (D); in the notum but not in the proximal compartments (F). Various simplifying assumptions have been made for the sake of the illustration: a single gene is assumed to be involved in the definition of a given compartment; the assignment of 0's and 1's is arbitrary except possibly in the cases of genes A and B; last but not least, the current view of the compartmental mode of development has been taken for granted. A dash is than another indicated that this choice depends on the developmental history of the neuron both in the locust[15] and in *Drosophila*[16]. In *Drosophila*, it was found that the choice of a pathway depends on the type of sensory structure that the neuron innervates, and on the developmental compartment to which the neuron belongs. For instance, the projection from campaniform sensilla and from bristles differ even when both types of sense organs belong to the same compartment, and the projection from structurally similar campaniform sensilla differs depending on whether they belong to the proximal or to the distal wing compartment. The application of this rule to the choice of the various projections identified so far is schematized in Table 1. It should be mentioned that besides this main choice, there may be consistent variations in the detail of the projection depending on the exact location of the neuron within a compartment. However, this aspect will not be considered here.

As for the first parameter (type of sensory structure) it is known that the neuron derives from the same mother cell as the cells which differentiate the cuticular elements of the sensory structure.[17] It is likely that the developmental decision of the mother cell to give rise to a bristle or to a campaniform sensillum remains registered in the genome of its daughter cells, so that all of them differentiate according to the bristle or sensillum blueprint. This would account for the fact that bristle and sensillum neurons will follow different pathways in the central nervous system, much as some of their epidermal sister cells will differentiate cuticular structures specific of a bristle or of a campaniform sensillum.

As for the second parameter (compartment), it is thought that all of the cells of a given developmental compartment share a common "state of determination" which is also registered and passed on to their descendents. In at least one case, the difference between cells in two compartments has been ascribed to the fact that a control gene is "on" in one compartment and "off" in the other.[18] A corresponding difference has been observed at the level of specific cell-cell affinities: the cells of a given compartment preferentially reaggregate together after they have been dissociated and mixed with cells of another compartment.[19] A similar process of differential

---

Legend to Table 1 (continued):

found in those cases where we do not know whether the state of expression of a given gene (0 or 1) is important for the choice of the corresponding sensory pathway; this is usually because the sensory structure is found only in one of the compartments. Most of these cases could be settled by mutations which add supernumerary sense organs in compartments where such structures would normally not develop (*Hairy-wing*, *Tuft*). A double dash (=) means that the state of expression of the gene has no influence on the pathway followed by the corresponding sensory neurons.

recognition may explain why the neurons of a given compartment follow the same pathway in the central nervous system, while neurons of another compartment may follow another pathway (see below).

Possible Mechanisms for the Establishment of a Given Projection

The results summarized above indicate that a combination of the different developmental decisions which are already known to take place during the normal development can account for the choice of a specific pathway by a given neuron. We are now faced with a second question: how is this choice implemented, i.e. what is the mechanism which ensures the establishment of the chosen specific projection?

By analogy with the mechanism used to explain to somebody how to get somewhere, one could imagine different solutions. One way would be to give a succession of topographical instructions (in our analogy, this would be "take the second turn right, go on for two km and then turn left") possibly combined with the use of specific clues ("turn left at the level of the ITT building"). Another way would be to mark the path with specific signs ("follow the green arrows, or the white stones"). Yet a third possibility is to take advantage of gradients to specify positional information (the generals of Napoleon used to "marcher au canon", i.e. to head towards the battlefield by relying on the sound of the guns). It is immediately obvious that if an axon reaches the central nervous system at a place far from its normal site of entry, the strict application of these three orienting systems will lead to completely different results. The analysis of the projection of homeotic sensory neurons might therefore allow us to distinguish among those three possibilities.

The Pathway of Homeotic Sensory Neurons

Homeotic mutations result in sensory neurons developing at ectopic locations. We will call "homeotic neurons" those neurons which develop on a body part which has been transformed to another by a homeotic mutation. Thus in the mutant *bx pbx* which transforms the metathorax to a mesothorax, homeotic neurons will differentiate in the transformed segment. It is known that growing axons are guided towards the central nervous system by preexisting nerve fibers.[20,21,22] In the case of *Drosophila*, it has been shown that adult axons follow the larval segmental nerves, and that no segmental specificity is involved in this guidance process.[23] Therefore the homeotic neurons send their axons along the metathoracic segmental nerve and join the thoracic ganglion at the place appropriate for the metathoracic axons, far posterior to the place of entry of the normal mesothoracic fibers. Thus the *bx pbx* mutation provides a situation where one can compare the central projection of identified sensory neurons which have entered the central nervous system at two different places.

The analysis of this situation has shown that the projection established by the homeotic axons is essentially identical to the normal projection.[14] An example of these results is shown in Figure 1b in the case of the campaniform sensilla in the distal compartment of the wing. Small differences may be observed between the normal (right) and the homeotic (left) projections in the region where the homeotic axons reach the prospective course of the normal projection. However, there is no doubt that the homeotic axons grow along the appropriate pathway and display the normal pattern of branching. Furthermore, it has been shown by backfilling individual neurons that each homeotic axon bifurcates when it reaches its prospective pathway and follows it in both directions. In order to do so, the homeotic axon has to follow some stretches of the path in a direction opposite to that followed by normal axons. This result is difficult to reconcile with the first and third hypothetical mechanisms outlined above. On the other hand, it is most easily explained by assuming that the growing axon recognizes a specific "trail" which preexisted in the central nervous system, and follows it no matter where the first contact was made. A similar conclusion was put forward in the case of the optic axons of *Xenopus* by Katz and Lasek[24], who propose the term "substrate pathway" for the preexisting route along which growing axons are guided. Before we discuss this conclusion, we have to examine two alternative explanations which are consistent with our results.

Do Homeotic Neurons Use a Normally Silent Program?

The Dipterans presumably derive from four-winged ancestors where the metathorax was essentially identical to the mesothorax. It is conceivable that in these primitive insects the distal sensilla of both pairs of wings had to establish a similar projection for functional reasons, much as in the present-day *Drosophila* the projection of the proximal campaniform sensilla of the haltere extensively overlaps the projection from the homologous sensilla of the wing.[14] Therefore, the metathoracic sensory neurons may have had their own program of projections, independent of the mesothoracic programs but eventually yielding overlapping projections. These programs could be written in terms of topological instructions or gradient reading (see above) or any other conceivable system. Then one could imagine that the metathoracic programs have survived in *Drosophila*, even though they are not used in the fly because the corresponding sense organs do not differentiate in the normal metathorax. These silent programs would then be revealed when homeotic mutations result in the development of the corresponding sense organs, and would allow the establishment of projections similar to the normal mesothoracic projections.

This explanation can be tested by using the same principle which was explained above: if the program is written in any terms but the following of the a preexisting trail, one does not expect the pro-

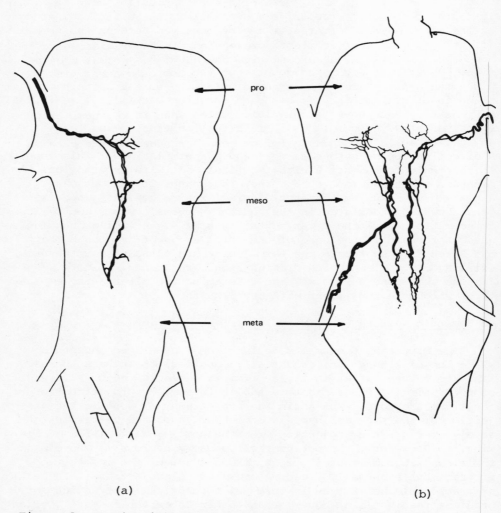

Figure 1: Projections from the distal campaniform sensilla in wild type (a) and $bx^3\ pbx/Ubx^{105}$ (b) flies. The projections were visualized by backfilling the neurons with horseradish peroxidase, as described elsewhere.[14] The thoracic ganglion is viewed from above; anterior is up. In (b), the distal sensilla of the right mesothoracic wing and of the left homeotic wing were backfilled simultaneously.

jection to remain unchanged if the axons enter the central nervous system at an abnormal place. Thus we examined the projection established by homeotic axons which are misrouted so that they now enter the ganglion at the site appropriate for the normal mesothoracic axons. This misrouting can be achieved by making the fly homozygous for the mutation *wingless* (*wg*, see appendix). The result of this experiment is exemplified in Figure 2a: the misrouted homeotic axons establish a projection which is identical to that of the normal mesothoracic neurons. We conclude that the establishment of the homeotic projection does not result from the expression of a silent metathoracic program, and that the mesothoracic program is such that the normal projection is established whatever the site of entry of the axons in the central nervous system.

Are the Homeotic Fibers Guided by the Normal Fibers?

The previous experiment strengthens the conclusion that homeotic axons recognize and follow a specific "trail" in the central nervous system. It could be that this trail is simply provided by the normal mesothoracic fibers. This hypothesis, suggested by Palka et al., [13], could be used to reconcile our results with any mechanism for the establishment of the projection. Indeed one simply has to assume that one among all the neurons which follow a given pathway is embodied with the ability to lay out this pathway. If all these neurons have a high affinity for each other, they will form a bundle and follow the leader. The homeotic axons may then follow the normal axons and establish an essentially normal projection. Thus, our results would be accounted for and yet the initial establishment of the projection could still rely on any conceivable mechanism such as these outlined above.

One way to test this explanation would be to suppress the normal sensory input. This should then result in the disorganization of the homeotic projection. This experiment has been done with the *wg* mutation. This mutation results in the absence of a wing or haltere with a penetrance of about 50% for each appendage.[25] Thus homozygous *wg* flies will have 0, 1 or 2 wings and 0, 1 or 2 halteres. When *wg* is combined with *bx pbx*, the resulting flies may have 0, 1, 2, 3 or 4 wings. In particular, flies will be found which have a metathoracic but no mesothoracic wings. We have examined the projection of the homeotic distal sensilla in flies of the latter phenotype where the homeotic fibers were not misrouted. The result is shown in Figure 2c. The projection remains essentially similar to that obtained when a mesothoracic wing is present (Figure 2b), which indicates that the homeotic fibers do not rely on the normal mesothoracic fibers to lay out the path they will follow. We conclude that whatever is specifically recognized by the adult axons was probably laid out earlier in the central nervous system.

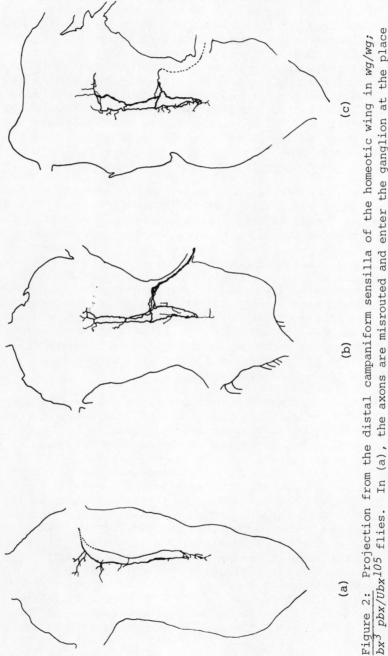

Figure 2: Projection from the distal campaniform sensilla of the homeotic wing in wg/wg; bx^3 pbx/Ubx105 flies. In (a), the axons are misrouted and enter the ganglion at the place appropriate for normal mesothoracic axons; in (b), the ipsilateral mesothoracic wing was present and in (c), the ipsilateral mesothoracic wing was absent and replaced by a duplicated heminotum.

How Do Axons Reach the Appropriate "Trail"?

Our results suggest that the "trail" which sensory axons recognize and follow does not extend into the cortex of the ganglion. If it did so, one would expect the homeotic fibers to follow it outwards up to the root of the dorsal mesothoracic nerves. Thus the mechanism whereby axons first reach the "trail" that they will eventually follow is probably not based on a specific recognition mechanism. It is conceivable that there is no specific mechanism at all, and that neurites extend randomly until one of them meets the appropriate "trail". The unsuccessful neurites would then regress. Alternatively it is possible that the program of axonal growth includes a topological instruction to orient the growth of the incoming axon towards the appropriate region of the ganglion.

The second possibility is supported by the generally similar appearance of normal and homeotic projections between their place of entry and the point where they reach the "trail". The fact that axons which follow different pathways usually separate as soon as they enter the ganglion is also consistent with this explanation. Furthermore, in the rare cases where homeotic wing fibers were misrouted through the metathoracic leg nerve, they nevertheless succeeded in establishing the appropriate projection. This implied in some cases that the ingrowing axon had to follow a dorsoventral direction very different from that followed by normal axons in order to reach the "trail". Thus if the initial course of the ingrowing axons is somehow programmed, it must be in terms of dorsoventral or mediolateral level to reach rather than in terms of dorsoventral or mediolateral directions to follow.

However, none of these circumstantial observations rules out the alternative possibility, if one assumes that exploratory neurites can extend over fairly long distances in their search for the "trail", and that existing fiber bundles or other texture properties of the ganglion limit the number of courses possibly followed by the exploratory processes. This would account for the relative homogeneity and reproducibility of the stretch between the entry and the main course of the pathway. It is possible that some special situation will be found where the two possibilities could be distinguished. Alternatively, it may be that this question will remain open until the dynamics of axonal growth have been elucidated in this system.

A Speculation on the Nature of the Recognized "Trail"

We would like to propose that the "trails" which are recognized by the growing axons are specifically marked nerve fibers which developed earlier, possibly during early neurogenesis. At this time a standard set of connections would be established between the first differentiating neurons on the basis of very simple directional cues such as the anteroposterior, mediolateral and possible dorsoventral axes of the body. Recent studies on the early neurogenesis of the locust central nervous system indicate that the very first neurites establish simple anteroposterior and contralateral connections within and between the segmental groups of neuroblasts (C.M. Bate and P.M. Whitington, personal communication). It is conceivable that these early connections are specifically labeled and serve as pioneering fibers within the central nervous system. Any subsequent fiber may then follow a given path in terms of these labeled guides. This hypothesis is schematized in Figure 3, where A shows hypothetical early connections and their labels, and B shows different possible adult pathways resulting from different sets of labels being recognized. It appears from the figure that different hypothetical projections may share common stretches. Such partial overlaps have been observed between actual projections: for example part of one component of the metathoracic leg projection coincides with part of the projection from the distal sensilla of the wing, while another part of the same projection coincides with part of the notum bristle projection, and yet a third part of the same projection coincides with no other identified thoracic projection in its extension to the suboesophageal ganglion (unpublished). This type of partial coincidence between different pathways also occurs between sensory and motor neurons, which raises the comforting possibility that the mechanism of axonal guidance proposed here is not limited to the establishment of sensory projections.

In this view, each of the major fiber tracts as they appear in the adult central nervous system[26] would be the result of an accretion process initiated by one or a few pioneering fibers.

Figure 3: A possible mechanism for the formation of adult sensory pathways. A represents a hypothetical set of connections established early during neurogenesis in the segmental ganglia a, b and c. The establishment of these connections would rely on simple cues such as the anteroposterior and mediolateral axis of the body. The same connections are assumed to be formed in all segments, but their label differs from one segment to the next. Within each segment different connections also bear different labels (1, 2, 3 ...). B represents three adult pathways, each of which would result from the recognition of a specific set of labeled pioneer fibers. Above are the recognized subsets of labeled fibers, below are the pathways followed respectively by the proximal campaniform sensilla of the wing, the homologous sensilla of the haltere, and the notum bristles. The development of the nervous system between the embryonic (A) and adult (B) stages distorts the initially simple stretches and increases the general complexity of the ganglion and the distances which have to be covered by connecting fibers.

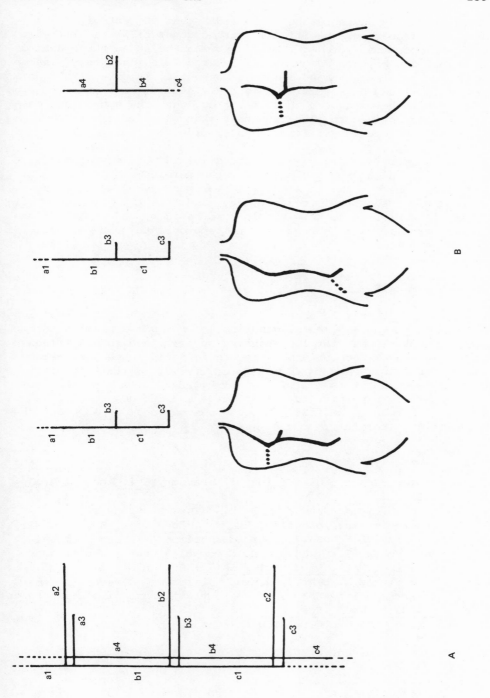

There is of course no reason to consider that the establishment of these pioneering, specifically marked connections is limited to the embryogenesis; more of these fibers can be added throughout development and be used as guides by later developing axons. The main point is that in each case a marked connection is established at a time when it can be specified relatively easily, so that it will be available to be used as a guide later on, when the system has developed and changed to such an extent that the same connection would have become very difficult to specify.

In summary, we propose that the simple hypothesis of contact guidance[20], which proved correct for the development of the peripheral nervous system, is also valid for the development of the central nervous system's connectivity, and that the required specificity is conferred to this simple hypothesis by assuming that the growing axon is able to recognize the appropriate guide or set of guides.

ACKNOWLEDGEMENTS

A.G. is greatly indebted to E. Lewis and to A. Garcia-Bellido for making the developmental genetics of *Drosophila* what it has become, to R. Thomas for longstanding support, and to his friends in the labs of S. Benzer and R. Thomas for help in many occasions. The profound influence of C.M. Bate on our way of thinking of the nervous development is gratefully acknowledged. Part of this work was submitted in partial fulfilment of the B.Sc. degree of R.J. A.G. is "chercheur qualifié" of the Fonds National de la Recherche Scientifique (Belgium).

REFERENCES

1. S. Benzer, Genetic dissection of behavior, Sci. Amer. 229(#6):24 (1973).
2. Y.N. Jan, L.Y. Jan and M.J. Dennis, Two mutations of synaptic transmission in *Drosophila*, Proc. R. Soc. Lond. B 198:87 (1977).
3. Y.N. Jan and L.Y. Jan, Genetic dissection of short term and long term facilitation at the *Drosophila* neuromuscular junction, Proc. Natl. Acad. Sci. USA 75:515 (1978).
4. O. Siddiqi and S. Benzer, Neurophysiological defects in temperature-sensitive paralytic mutants of *Drosophila melanogaster*, Proc. Natl. Acad. Sci. USA 73:3253 (1976).
5. W.G. Quinn and J.L. Gould, Nerves and genes, Nature 278:19 (1979).
6. E.B. Lewis, A gene complex controlling segmentation in *Drosophila* Nature 276:565 (1978).
7. A. Garcia-Bellido, P. Ripoll and G. Morata, Developmental compartmentalization of the wing disc of *Drosophila*, Nature New Biol., 245:251 (1973).

8. A. Garcia-Bellido and M.P. Capdevila, Initiation and maintenance of gene activity in a developmental pathway of Drosophila, in: "The Clonal Basis of Development", S. Sobtenly and I.M. Sussex, eds., Academic Press, New York, pp. 3-21 (1979).
9. W.J. Gehring and R. Nöthiger, The imaginal discs of Drosophila, in: "Developmental Systems: Insects", Vol. 2, S.J. Counce and C.H. Waddington, eds., Academic Press, New York, pp. 211-290, (1973).
10. G. Morata and P.A. Lawrence, Homeotic genes, compartments and cell determination in Drosophila, Nature 265:211 (1977).
11. I.I. Deak, Demonstration of sensory neurons in the ectopic cuticle of spineless-aristapedia, a homoeotic mutant of Drosophila, Nature 260:252 (1976).
12. R.F. Stocker, J.S. Edwards, J. Palka and G. Schubiger, Projections of sensory neurons from a homoeotic mutant appendage, Antennapedia, in Drosophila melanogaster, Develop. Biol. 52:210 (1976).
13. J. Palka, P.A. Lawrence and H.S. Hart, Neural projection patterns from homoeotic tissue of Drosophila studied in bithorax mutants and mosaics, Develop. Biol. 69:549 (1979).
14. A. Ghysen, Sensory neurons recognize defined pathways in Drosophila central nervous system, Nature 274:869 (1978).
15. H. Anderson and J. Bacon, Developmental determination of neuronal projection patterns from wind sensitive hairs in the locus, Schistocerca gregaria, Develop. Biol. 72:364 (1979).
16. A.Ghysen, The projection of sensory neurons in the central nervous system of Drosophila: Choice of the appropriate pathway, Develop. Biol. (in press) (1980).
17. Reviewed in: C.M. Bate, Development of sensory systems in arthropods, in: "Handbook of Sensory Physiology" Vol. IX, M. Jacobson ed., pp. 1-53 (1978).
18. G. Morata and P.A. Lawrence, Control of compartment development by the engrailed gene of Drosophila, Nature 255:614 (1975).
19. A. Garcia-Bellido, Pattern reconstruction by dissociated imaginal disc cells of Drosophila melanogaster, Develop. Biol. 14:278. (1966).
20. V.B. Wigglesworth, The origin of sensory neurons in an insect, Quart. J. Microsc. Sci. 94:93 (1953).
21. C.M. Bate, Pioneer neurons in an insect embryo, Nature 260:54 (1976).
22. J.R. Sanes and J.G. Hildebrand, Nerves in the antenna of pupal Manduca sexta, Wilhelm Roux Arch. 178:71 (1975).
23. A. Ghysen and I.I. Deak, Experimental analysis of sensory nerve pathways in Drosophila, Wilhelm Roux Arch. 184:273 (1978).
24. M.J. Katz and R.J. Lasek, Substrate pathways which guide growing axons in Xenopus embryos, J. Comp. Neurol. 183:817 (1979).
25. R.P. Sharma and V.L. Chopra, Effect of the wingless mutation on wing and haltere development in Drosophila melanogaster, Develop. Biol. 48:461 (1976).
26. M.E. Power, The thoracico-abdominal nervous system of an adult insect, Drosophila melanogaster, J. Comp. Neurol. 88:347 (1948).

APPENDIX: Nerve misrouting in the mutant *wingless*

The most prominent effect of the mutation *wingless*[1] is to replace the wings and halteres by mirror-symmetrical duplications of the meso- and metanotum respectively. The penetrance of this phenotype is incomplete, each appendage having a probability of about 50% of being absent. In the remaining cases the appendage appears completely normal.

Two explanations have been proposed for this phenotype. According to the first one,[2] *wg* would be responsible for a substantial amount of cell death in the central region of the wing and haltere discs. The remaining parts of the discs may then either regenerate the damaged region, or duplicate themselves, depending on the extent of the damage.[3] In the first case a normal appendage will be formed, while in the second case no wing will be formed but the notum (and possibly the pleura) will duplicate. This explanation accounts readily for the variability in phenotype, the incomplete penetrance and the all-or-none effect on the presence of the appendages. Furthermore, this explanation is supported by the observation that X-rays, which are known to induce cell death, can mimic the effect of *wg*.[4]

According to the second explanation,[5,6] *wg* would be a homeotic mutation affecting the process of compartmentalization. In normal flies, a compartment boundary separates the wing from the notum and pleura.[7] The mutation *wg* would act at this level so that both compartments follow the notum developmental pathway, much as in *engrailed (en)* flies the cells of both anterior and posterior compartments follow the anterior developmental pathway.[8,9] However, in the case of *en* the expressivity is partial (i.e., the transformation from posterior to anterior is far from complete) and the penetrance is complete, while in *wg* the penetrance is partial but the expressivity is complete (i.e. when the wing is "transformed", it is completely so). This explanation has been supported by the results of a clonal analysis which indicated that cell death during the larval development could not account for the *wg* phenotype.[6]

More recently it has been argued that the results of the clonal analysis did not exclude an explanation based on cell death if this occurred during embryogenesis, and/or if the cell death affects neither the presumptive notum region nor the presumptive pleura region.[10] If the latter is true, intercalary regeneration may occur and complicate the interpretation of the gynander analysis. The extreme variability in the location of the plane of symmetry in the cases of duplications, as well as the occasional presence in these cases of material which is typical of the wing compartment (costa, alula and other wing blade material) were taken as further support for the cell death hypothesis. Indeed the homeotic hypothesis requires that the line of symmetry coincides with the compartmental boundary, unless additional effects are postulated.

The analysis of haltere projections in homozygous *wg* flies shows that in some of the cases where an apparently normal haltere develops, the haltere nerve is misrouted and joins the ganglion at a place appropriate either for one of the dorsal mesothoracic nerves, or less frequently for the metathoracic leg nerve. This effect depends on the temperature: very few misroutings were observed in flies raised at 25°, while as many as 40 % of the haltere nerves may be misrouted in flies raised at 16° (this frequency is somewhat variable in different experiments). When the haltere is replaced by a homeotic wing in the mutant *wg/wg; bx^3 pbx/Ubx^{105}*, the homeotic nerve may also be misrouted, and the frequency of misrouting is also higher when the flies are raised at a lower temperature. In this appendix we show how this peculiar phenotype is easily accounted for by the cell death hypothesis.

It is known that a larval nerve is used as a guide by the wing and haltere sensory axons in their journey towards the central nervous system.[11] In the case of the wing disc, this nerve contacts the disc at a position corresponding to the prospective tegula.[12] Therefore if this region is affected by the process of cell death, the adult sensory axons will loose their normal guide. This cannot be easily observed in the case of the wing nerve, because a second larval nerve contacts the same disc in the region of the prospective notum.[12] Presumably the wing axons which have lost their normal guide will follow this second nerve, and enter the thoracic ganglion along the posterior dorsal mesothoracic nerve rather than along the anterior dorsal mesothoracic nerve. Both nerves enter the thoracic ganglion at about the same place, and therefore it is not easy to decide whether or not the wing axons have been misrouted.

In the case of the haltere disc there is only one larval nerve connecting the disc to the central nervous system. If one assumes that this nerve contacts the disc at a position homologous to the prospective tegula in the wing disc, then the contact will be lost whenever cell death encompasses this region. If regeneration takes place subsequently, so that a haltere develops, the haltere nerve will lack its normal guide and will be misrouted. Larval operations aimed at breaking the "haltere disc" larval nerve resulted in misroutings of the adult haltere nerve, which usually joined a dorsal mesothoracic nerve and occasionally a metathoracic leg nerve.[11] This is exactly what is observed in *wg* flies.

The effect of temperature on the frequency of misrouting is entirely consistent with this interpretation. Indeed in cases where the wing is replaced by a duplicated notum, the position of the plane of symmetry depends on the temperature at which the flies were raised (Figure 4). It can be seen that at 25° the region of cell death usually does not include the presumptive tegula, since this structure is usually present in the duplication. *A fortiori* one expects that this region has remained intact in those cases where a normal appendage develops, since regeneration would occur in the cases where cell death is less extensive. Thus at this temperature nerve misrouting should be rare. On the other hand, at 18° the tegula is usually missing in the cases of duplication, and therefore, the corresponding region of the disc may also have been affected in the case where regeneration occurred. Thus the frequency of misrouting should be higher at 18° than at 25°, in complete agreement with the experimental data.

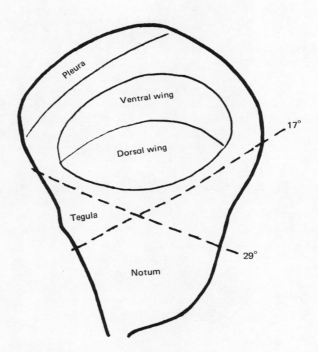

Figure 4: The position of the line of symmetry in cases of duplication of the mesonotum in *wg* flies is shown on the fate map of the wing discs. The position depends on the temperature at which the flies were raised (from ref. 10).

REFERENCES FOR APPENDIX ONLY

1. R.P. Sharma, *wingless*, a new mutant in *Drosophila melanogaster*, Droso. Inf. Serv. 50:134 (1973).
2. R.P. Sharma and V.L. Chopra, Effect of *wingless* mutation on wing and haltere development in *Drosophila melanogaster*, Develop. Biol. 28:461.
3. P.J. Bryant, Pattern formation in the imaginal wing disc of *Drosophila melanogaster*: fate map, regeneration and duplication, J. exp. Zool. 183:49 (1975).
4. J.H. Postlethwait, Pattern formation in the wing and haltere imaginal discs after irradiation of *Drosophila melanogaster* first instar larvae, Wilhelm Roux' Arch. 178:29 (1975).
5. P. Babu, Early developmental subdivisions of the wing disc in *Drosophila*, Molec. Gen. Genet. 151:289 (1977).
6. G. Morata and P.A. Lawrence, The development of *wingless*, a homeotic mutation of *Drosophila*, Develop. Biol. 56:227 (1977).
7. A. Garcia-Bellido, P. Ripoll and G. Morata, Developmental compartmentalization of the wing disc of *Drosophila*, Nature New Biol. 245:251 (1973).
8. A. Garcia-Bellido and P. Santamaria, Developmental analysis of the wing disc in the mutant *engrailed* of *Drosophila melanogaster*, Genetics 72:87 (1972).
9. G. Morata and P.A. Lawrence, Control of compartment development by the *engrailed* gene of *Drosophila*, Nature 255:614 (1975).
10. I.I. Deak, Thoracic duplications in the mutant *wingless* of *Drosophila* and their effect on muscles and nerves, Develop. Biol. 66:422 (1978).
11. A. Ghysen and I.I. Deak, Experimental analysis of sensory nerve pathways in *Drosophila*, Wilhelm Roux Arch. 184:273 (1978).
12. C.A. Reinhardt, N.M. Hodgkin and P.J. Bryant, Wound healing in the imaginal discs of *Drosophila*, I. Scanning electron microscopy of normal and healing wing discs, Develop. Biol. 60:238 (1977).

DISCUSSION

Discussion of this paper follows the chapter by Strausfeld and Singh.

PERIPHERAL AND CENTRAL NERVOUS SYSTEM PROJECTIONS IN NORMAL AND

MUTANT (BITHORAX) *DROSOPHILA MELANOGASTER*

N.J. Strausfeld' and Ram Naresh Singh"

'European Molecular Biology Laboratory
69 Heidelberg
Postfach 102209
Federal Republic of Germany

"Tata Institute of Fundamental Research
Homi Bhaba Road
Bombay 400 005
India

ABSTRACT

Projections of axons from the wing, haltere and the head are described from silver intensified cobalt preparations of normal and bithorax mutant *Drosophila*. Dorsal projections from the homeotic wing into thoracic ganglia and to the brain are haltere-like. Ventral projections are wing-like. In general these results are in complete agreement with previous studies by Palka et al.[1] and Ghysen[2]. However, considerable variation of projection patterns is seen between individual mutant animals some of which were previously described only from mosaics (clones of bithorax sensory fibers into normal ganglia).[1] Evidence for possible transformations of the third thoracic ganglion in mutants is therefore ambiguous.

Projections of antennal sensory neurons and relay neurons from the head also do not support the notion that the third thoracic ganglion is transformed in these mutants.

INTRODUCTION

Several bithorax mutants result in abnormalities of the thoracic segments.[3] The bithorax mutant *bx* gives rise to transformation of cuticular structures of the anterior metathorax compartment[4,5]

into structures like those of the anterior mesothoracic compartment. The postbithorax mutant *pbx* has cuticular structures on the posterior metathorax like those of posterior mesothorax. Double mutants (*bx pbx*) transform both compartments. In *bx* mutants the anterior compartment of the haltere is wing-like. In double mutants the entire haltere is a wing-like structure. As shown by Palka et al.[1] the metathoracic wing is smaller than the mesothoracic wing. But apart from a slight reduction in the number of some species of sensillum, most sensilla are typical of the normal wing and are properly located.

Previous accounts employed these mutants to follow pathways of sensory axons into the central nervous system. These pathways were compared with those from normal wings and halteres. Using comparative anatomy it was possible to determine if a certain sensory axon established an appropriate projection even if it entered into the wrong ganglion from a homeotic appendage. The accounts by Palka et al.[1] and Ghysen[2] agree with respect to descriptions of axons from normal and mutant appendages. Endings from homeotic wings are haltere-like if they are derived from sensilla that are homologous to those of the normal haltere. Axons from sensilla of the homeotic appendage that are special to wing (e.g. distal campaniform sensilla of the third vein and anterior cross vein) project in a fashion that is basically wing-like. Ghysen emphasized that the results indicate that a growing axon is typed and recognizes a specific route. It follows this irrespective of its point of entry into the route. Ghysen stated that there was no compelling evidence that the third neuromere was transformed in the mutant. Palka et al. also studied fiber projections from clones of wing sensilla on halteres into normal neuropil. They presented evidence that some subtle differences might be evident in the central anatomy of the third thoracic ganglion in mutants. However, they concluded that the ability of a clonal homeotic fiber to project into normal ventral neuropil did not require a homeotic transformation of the neuropil.

Evidence that supports the notion that the third ganglion is transformed in mutants is tenuous although in the present bithorax-postbithorax mutants the ganglion appears to be somewhat larger than normal, using the dimensions of the prothoracic ganglion and mesothoracic ganglion to standardize measurements. In $bx^3\ pbx/Ubx^{101}$ mutants the third thoracic ganglion is broader and deeper anteriorly, and slightly longer. The presence of ovoid and lateral ventral fibers from the homeotic wing into the third ganglion of $bx^3\ pbx/Ubx^{130}$ was also seen as significant.[1]

This account is a preliminary report of our results from fills into normal and homeotic wing and haltere and into central relay neurons from the brain using cobalt chloride. We used bithorax (bx^{34e}/bx^{34e}) and bithorax-postbithorax mutants (bx^3pbx/Ubx^{101}).

Over two hundred flies (normal and mutant) were studied. The results of our investigations indicate that all alterations of the thoracic ganglion in mutants can be related to the unusually large number of incoming fibers from the homeotic appendage. Although we have not yet used mosaics, many of the fiber distributions reported to be typical of mosaics[1] were seen by us in bithorax - postbithorax mutants.

Relay neurons and some antennal sensory fibers that project from the head into the thoracic ganglia also appeared to have normal distributions in mutant animals. The only exception to this relates to subtle differences of branching patterns of a multiganglionic neuron that normally has few branches in the third thoracic ganglion. The results are described and discussed in the succeeding sections.

MATERIALS AND METHODS

Normal ("wild-type") Oregon-S *Drosophila melanogaster* were obtained from stocks at the European Molecular Biology Laboratory. Dr. Campos-Ortega, University of Freiburg, kindly provided us with first and second generation bithorax mutants (TM3/TM6, homozygous for bx^{34e}) and the double bithorax mutant bx^3pbx/Ubx^{101} after the appropriate crosses (see refs.1,2,6). Mutants homozygous for Bar-eyed (B/B) were also obtained from him. These were used for studies of the dendritic morphology of the giant descending neuron[7] described later. Both sexes were used but the results described here compare females only. Double bithorax mutants were selected for anatomy which had well developed and flattened homeotic wings that showed the expected number and distribution of sensilla.[1] Bithorax mutants were selected for anatomy which had homeotic metathoracic appendages whose anterior margin showed a substantial anterior wing-like margin. We attempted to use individuals of each class that were obviously similar.

ANATOMICAL METHODS

Cobalt Diffusion into Appendages

Flies were mounted in plasticine, leaving their spiracles uncovered. The appendages were severed near to their attachment with the head or thorax. Antennae were severed across segment 2, two thirds along its length distally. The wing distal to the alula and costal vein was removed, damaging both the alula and costal vein. The pedicel and capitellum of the haltere was removed for haltere fills. Immediately after removing the appendage its stump was threaded into a fused fiberglass micropipette drawn to an appropriate diameter and mounted on a plasticine gimbel.[8] A 5% solution of $CoCl_2$ was backfilled into the pipette. Passive diffusion of cobalt ions into nerves took place during 1 hour at 4°C, in the dark. In the case of antennal fills, uptake of cobalt by relay neurons was achieved by removing the source of cobalt after 20 minutes and leaving the flies at 4°C for a further 1 hour.

Cobalt Injection into Neuropil (see ref. 9).

Double-barrelled fused fiberglass micropipettes were drawn to a tip diameter of about 0.2 µm (each tip). One tip was broken off between 0.1 and 0.5 mm shorter than the other. The intact barrel was backfilled with a 5% solution of $CoCl_2$ in distilled water. The pipette was mounted on a plasticine holder. Using reference serial sections of whole *Drosophila*, the tip was aimed at the desired region. The tip was advanced directly through the intact cuticle at the back of the head or ventral thorax. The tip usually broke to about 5 µm diameter.[8] The depth of penetration was gauged by the distance of the broken barrel from the cuticle surface. In about 50% of cases the unbroken tip penetrated the required area. Cobalt was allowed to diffuse from the pipette (at 4°C) for 15 minutes. The pipette was then removed and diffusion from the pool of cobalt in tissue continued for half an hour. The locations of peripheral fills (antennae, wing, haltere and homeotic wing) and injections (brains, 2nd and 3rd thoracic ganglia) are shown in Figure 1.

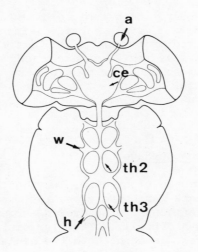

Figure 1: Locations of peripheral fills (large arrows) and injections into neuropil (small arrows). a = antenna; ce = cerebral ganglia; w = wing nerve; h = haltere or homeotic wing nerve; th2 and th3 = 2nd and 3rd thoracic ganglia.

Cobalt Precipitation and Intensification

Small openings were made in dorsal head cuticle and in the mesonotum. The animal was immersed for 2 minutes in 0.5% ammonium sulfide in distilled water containing 1 g sucrose and 0.1 ml glacial acetic acid/10 ml. The animal was then washed twice, two minutes, in an appropriate ringer solution and then submerged in fixative (acetic acid, alcohol, formaline:AAF). Fixation was 6 hours. Legs and proboscis were then removed and the hole in the mesonotum was enlarged. The animal was hydrated and CoS containing profiles were silver intensified according to the method of Bacon and Altman[10] for whole ganglia. In our case we intensified almost intact *Drosophila*. This treatment gave rise to negligible tissue distortion. Occasionally the head was slightly rotated with respect to the long axis of the body.

Reconstructions

After silver intensification the animal was dehydrated and embedded in Epon. Serial sections were made horizontally or sagittally at 15 µm. Photographs were made of each section every 2 µm depth using a Leitz Ortholux microscope equipped with a Zeiss plan-NEO-FLUAR x25 oil-immersion lens. Photographs were on 35mm Kodak Panatomic X or Ektachrome 64. Negatives or color transparencies were serially projected via a surface glazed mirror and serially reconstructed. Resolution is far superior to camera lucida visualization of whole mount preparations.

Cobalt Preparations

Optimal diffusion-injection timing was determined for wing projections. The criteria used were as follows: 1) Resolution of fibers as small as 0.2 µm diameter, against a transparent background. 2) Absence of cell bodies of interneurons (except in the case of antennal fills) excluding the possibility of transsynaptic fills.[11] 3) Reproducibility of projection patterns between one normal animal and the next. 4) Reproducibility of wing projections to the characteristic pattern in the brain (see Figure 3A).

Having established optimal injection times for wing fills, these were used for filling homeotic appendage axons. Slightly shorter periods were employed on haltere nerves. Any wing, haltere or homeotic appendage fill that showed the presence of cell bodies was not used for this analysis. Contralateral ascending fibers from normal haltere to the brain are assumed to be genuinely derived from haltere receptors. This pattern (Figure 3E) was also seen in reduced silver preparations (Figure 2B).

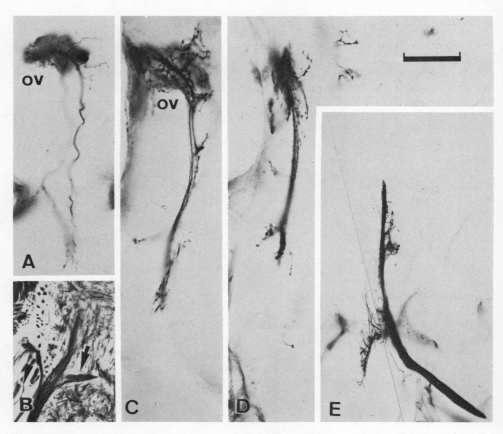

Figure 2: HRP, cobalt and reduced silver visualization of sensory fibers. A. Medial and lateral ventral fork fibers of the normal wing using HRP. Note the sinusoidal passage. B. Reduced silver preparation of the normal haltere nerve showing the root of the contralateral fibers (arrow). C. Lateral ventral fork fibers of the normal wing using silver intensified cobalt sulfide. D. Fibers of the dorsal fork of the normal wing (silver-cobalt sulfide). E. Dorsal anterior fibers and the medial tuft of the homeotic wing of bx^{34e}/bx^{34e} (silver-cobalt sulfide)

Horse Radish Peroxidase (HRP)

Ghysen[2] derived his results by filling nerves with HRP. Characteristically HRP filled fibers appear to have sinusoidal passages, compared with cobalt filled profiles (Figure 2A,C). Simultaneous fills with HRP on one side and cobalt on the other also show these differences as do combined HRP-cobalt fills into the same nerve.

Cobalt is first precipitated and then the tissue is treated with diaminobenzidine tetrachloride/H_2O_2 after fixation in phosphate buffered glutaraldehyde (see ref. 2). However, these discrepancies do not give rise to disagreement between our observations and Ghysen's. Probably using HRP the filled neuron shrinks less than does the surrounding tissue, as demonstrated when using PIPES buffer and postfixation in osmium. Then HRP nerves appear to be identical to cobalt sulfide nerves (Nässell, unpublished results).

"Transsynaptic Fills" (see ref. 11)

Prolonging cobalt diffusion results in Co^{++} migration into certain interneurons.[9,11] The patterns of contiguous neurons and sensory fibers are reproducible. We have used this phenomenon to reveal some relay neurons that descend from the posterior antennal centers to thoracic ganglia. Two types of neurons are specially well shown, the giant descending neuron[7] and a small antennal descending neuron (AD, Figure 7). Also the giant neuron gives up cobalt to peripheral neurons to muscle whereas the AD neuron donates cobalt to a small metathoracic interneuron (Figures 5,6,7). These are described later.

TERMINOLOGY AND FIGURES

We have used the basic terminology employed by Palka et al.[1], adding some new labels of our own. Abbreviations are explained with the accompanying figures. Line reconstructions show the brain to the right, in the case of peripheral fills into wing and haltere and to the left in the case of antennal fills. The scale bar on line drawings is 100 μm and on photomicrographs 50 μm

RESULTS

Passage of Fibers from Haltere, Wing and Homeotic Wing

Previous accounts[1,2] showed that sensory nerves from normal wing and haltere gave rise to two main fiber distributions, dorsal and ventral. Wings gave rise to prominent dorsal tracts and prominent ventral tracts, whereas the ventral component from the haltere was relatively minute. The sensory origins of the various fiber types were identified either by selectively filling tributaries of the wing nerve with cobalt or HRP[2], or by filling the wing nerve of mutants that lacked certain types of sensilla[1] or in which certain types of sensilla were exaggerated.[2] The reader is referred to the cited accounts for descriptions of sensilla and the shapes of the axonal arbors. Our studies included selective cobalt fillings of wing nerve tributaries and the findings do not differ from those of Palka et al.[1] and Ghysen.[2] The sensillary origins of fibers given on the following page are from references 1 and 2.

Dorsal Fibers of the Wing. Dorsal fibers in neuropil from the normal wing are derived from the following sensilla. Companiform sensilla at the base of the radius and chordotonal organs at the wing base[2] give rise to axons that project medio-dorsally between the pro- and mesothoracic ganglia and dorso-posteriorly and dorso-anteriorly through the ipsilateral neuropil of the meso- and metathoracic neuropil and the prothoracic neuropil, respectively. Posterior and anterior directed fibers are here termed the posterior and anterior dorsal fiber bundles (PDF and ADF; Figure 3A). A posterior tributary is seen to pass laterally around the anterior and lateral border of the metathoracic ganglion where it overlies a similar ventral tributary from ventral fibers (Figures 2C,D). Anterior dorsal fibers also give rise to occasional branches in the posterior mesothoracic ganglion and in the posterior prothoracic ganglion. Note that all dorsal fibers of the wing are ipsilateral except for the small medial bundle between pro- and mesothoracic ganglia. Note also that the projections of fibers to the brain via ADF (Figure 3A) end as a characteristic group of branched terminals in the posterior deuterocerebrum (the cerebral fork, CF). We have not seen other types of single ascending fibers as is typical of haltere nerves (Figure 3E). Also, the ADF bundle of the wing is slightly inferior to the homologous bundle from the haltere. The haltere ADF (Figure 3E) terminates further posteriorly in the deuterocerebrum as unbranched bush (cerebral bush, CB).

Dorsal Fibers of the Haltere. Sensilla trichoidea found on the capitellum are described by Ghysen[2] as ending as a bundle between the meso- and metathoracic ganglion. It appears from our preparations that these sensilla may also give rise to many short club-shaped terminals that reside superficially in the lateral mesothoracic ganglion and continue and merge with the median tuft (MT,Figure 3E). The short terminals are here termed the dorsal tuft (DT). HRP fills of the haltere stalk indicate that the great majority of sensory axons are derived from the pedicel and scabellum. These comprise the main parts of the haltere nerve projection. As described by Palka et al. the haltere nerve normally enters the posterior half of the metathoracic ganglion where it gives rise to the DT projection. The main trunks of the nerve give rise to bifurcations and terminals in the median tuft (MT) between the anterior margin of the metathoracic ganglion and the posterior mesothoracic ganglion. Palka et al.[1] describe two lateral components in the mesothoracic ganglion, the lateral tuft (LT) and a small group of fibers that arises at MT and passes along the edge of the mesothoracic ganglion anteriorly. We have not been able to clearly separate a discrete dorsolateral bundle from the LT and prefer to label the entire lateral complex as LT(Figure 3E). Our cobalt diffusions do, though, clearly reveal the small dorsal crossing fibers (DCF) which number between 1 and 3 elements. In about 60% of our preparations we can see unambiguously a contralateral dorsal fiber tract (CDF), derived from a root (RC) just anterior to the median tuft. Although it intensifies palely

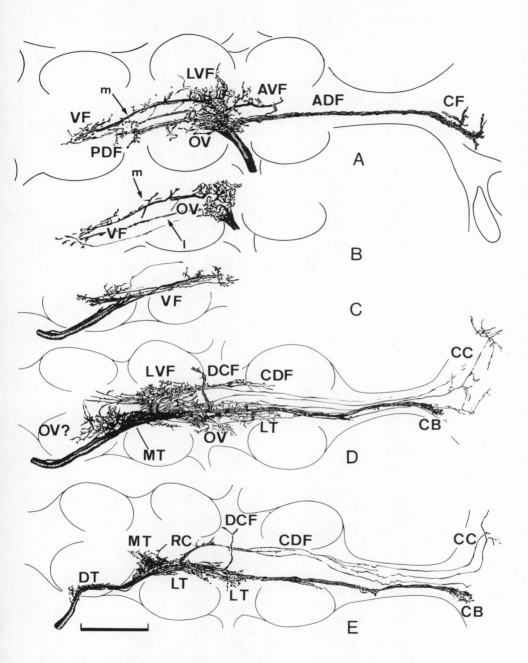

Figure 3: Cobalt fills into the normal wing (A,B), normal haltere (E) and a homeotic wing of the bithorax mutant bx^{34e}/bx^{34e} (C,D). For explanation see description in the text.

near the root, the CDF fibers are black. We have seen this tract unassociated with cell bodies. Typically it ramifies into between 5 and 8 single ascending axons that end diffusely in the deuterocerebrum (CC, the contralateral complex, Figure 3E).

Ventral Fibers of the Wing. Ghysen[2] has identified the ventral fibers that run between the meso- and metathoracic ganglion as being derived from campaniform sensilla, with the exception of those at the base of the radius. Particularly prominent are fibers derived from the sensilla of the third vein and the anterior cross vein. Palka et al.[1] filled large ventral fibers from axons of large campaniform sensilla at the tip of the costa. Short, multibranched and blebbed endings of the ovoid (OV) were shown to be derived from anterior bristles.[2] The interganglionic ventral fibers are distributed as two groups, one medial and one lateral (VFm and l, Figure 3B). Both have characteristic branching patterns into the posterior mesothoracic neuropil and the anterior and mid-metathoracic neuropil. Lateral ventral fibers (LVF) branch heterolaterally within the anterior mesothoracic neuropil. This feature is a crucial one when comparing projections from homeotic wings. The ovoid is restricted to ipsilateral mesothoracic neuropil. However, like LVF fibers these intra-ganglionic elements extend further towards the opposite side of the ganglion in males than in females. Lastly, there is a short anterior ventral projection (AVF) that gives rise to branches in the anterior mesothoracic ganglion and the posterior prothoracic ganglion.

Ventral Fibers of the Haltere. Between 1 and 4 ventral fibers have been seen which project from the haltere nerve beneath the RC, and forward into the posterior neuropil of the prothoracic ganglion. These underlie the dorsal ADF tract. Apart from this we have not seen any prominent ventral component, and are in agreement with the findings of Palka et al.[1] and Ghysen.[2]

Projections of Ventral Fibers of the Wing into the Metathoracic Ganglion of Normal and Mutant Flies. VF fibers normally extend three quarters of the way into the metathoracic neuropil (VFm) or as far as its posterior margin (VFl). In bithorax and bithorax-postbithorax animals the VFm fibers have sometimes been seen to project as far as the posterior margin of the metathoracic ganglion and the VFl fibers have been observed to invade the first abdominal ganglion. These features have been seen in 30% of our preparations and may possibly imply some minor and subtle transformation of the metathoracic ganglion in mutant animals.

Projections of Fibers from Homeotic Wings. Three examples are shown (Figures 3C, D and 4A-D). We find that projections are least variable amongst bithorax mutants and most variable amongst bithorax-postbithorax mutants with respect to dorsal projections. However, despite variations all have certain features in common.

Dorsal projections are usually haltere-like and include the root of the contralateral tract (RC), the dorsal contralateral fibers (DCF) and contralateral tract of dorsal fibers that ascend to the brain (CDF). Endings of the typically haltere-like ADF give rise to the cerebral bush (CB) superficially in the posterior deuterocerebrum. Thus these fibers do not usually occupy the homologous wing address with respect to brain endings although a mixture of haltere and wing-type endings have occasionally been resolved (Figure 4A). The contralateral complex (CC) is often much larger and more diffuse than in normal animals with fibers sometimes reaching into the protocerebrum and the optic lobes (Figures 3D,4A). As is the case for the DCF component, CC fibers appear to be more populous in mutants than in normal flies. Significantly the DCF branch is always observed to be in its proper mesothoracic location. Dorsal fibers also contribute to a lateral tract(LT in Figure 3D and Figure 4B). The two examples shown in Figure 4 also contained a dorsal lateral tract (DLT) and a set of superficial fibers that project near the edge of the anterior mesothoracic and prothoracic ganglia (DLF, dorso-lateral fibers). All the homeotic projections exhibit a lateral tuft (LT) which is sometimes situated slightly anterior to its normal position. The median tuft (MT) was also identified. However, this extends further posteriorly in mutants even though the homeotic nerve enters the ganglion more anteriorly than in normal flies. A dorsal tuft (DT) is not clearly distinguished in mutants though many short terminals typical of the normal dorsal tuft are seen to arise from the main trunk of the homeotic nerve at the point of its major divisions. Lateral ventral fibers (LVF in Figure 3D), and opposite to VF in Figure 4A) are observed at their proper location in the mesothoracic, not in the metathoracic. Likewise a metathoracic ovoid (OV?) has rarely been detected. It is commonly observed that small diameter ventral fibers project anteriorly and branch in a typically ovoid fashion in the proper mesothoracic location (OV).

In our preparations interganglionic ventral fibers usually have large diameters, typical of the VFm elements of normal flies. Their abnormal projections are, however, often seen mimicked by slender elements, possibly homologous to normal VFl fibers. Our preparations support Ghysen's contention that homeotic ventral fibers, derived from sensilla typical of the normal wing, project to highways that are occupied by homologous fibers from the wing. In the example shown in Figure 3C ventral fibers (VF) are distributed as a reverse projection of the normal situation. In double bithorax mutants ventral fibers tend to reach appropriate positions in both sides of the neuropil. In the two examples of homeotic projections in bx^3 pbx/Ubx^{101} (Figure 4), some large ventral projections occupy the area normal for the LVF fibers. In Figure 4B the majority of ventral fibers are contralateral. Unlike some figures of projections from the double bithorax mutant bx^3 pbx/Ubx^{130} posterior displacement of an LVF-type pattern was rarely seen.

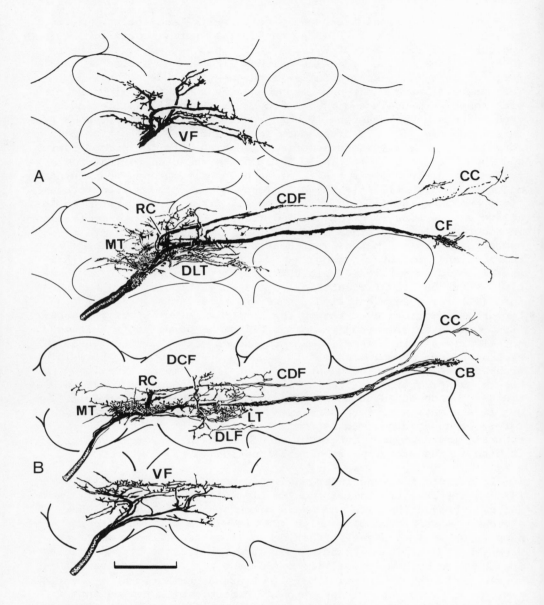

Figure 4: Projections of fibers from the homeotic nerve in double bithorax mutants. Note similarities between the dorsal haltere-like elements of Figure 4B with the normal haltere in Figure 3E. Ventral projections are characteristically variable between individual mutant flies. However, their fibers occupy locations that are typical for ventral projections from the normal wing. There is, however, one interesting and important variation amongst the $bx^3\ pbx/Ubx^{101}$ mutants. This is that in about 10% of our preparations the

Figure 4 (cont.)
dorsal ascending bundle, derived from small campaniform sensilla, shows a wing-like structure. The terminals in the brain are disposed as a forked structure (CF in Figure 4A) compared with a bush (Figure 4B). Also the profile of the dorsal bundle in the mesothoracic is reminiscent of a wing projection indicating that transformation of campaniform sensilla from haltere- to wing-like does sometimes occur. Note also the fine fibers accompanying the dorsal bundle: a feature reported to be shown by mosaics.[1]

Passage of Fibers from Brain to Thoracic Ganglia

There is little hard evidence from peripheral fills that the third thoracic ganglion is transformed to a mesothoracic structure in bithorax mutants and we agree with Ghysen that the third ganglion is essentially metathoracic in structure. Any difference in its volume may be due to the abnormally large number of sensory inputs and the additional branches that they make.

In order to test this further, we carried out a simple experiment. This consisted of injecting cobalt ions into meso- and metathoracic neuropil of normal and mutant flies. The cerebral dendrites of neurons that pass from these ganglia were revealed after silver intensification. Some ascending relay neurons originate in the mesothoracic ganglion of the normal fly, but not in the metathoracic ganglion. Injections into the third thoracic ganglion of mutants did not reveal supernumerary mesothoracic cells.

Another line of investigation was to trace primary sensory fibers from the antennae to the thoracic ganglia of mutant and normal flies. In addition, using appropriate injection-diffusion periods it was possible to fill certain prominent descending relay neurons from the brain. When these are revealed so too are certain intracerebral interneurons that project between the antennal lobes, and which contribute to the sub- and supraoesophageal tracts. Also, when cobalt migrates into the giant descending neuron (see for comparison, ref. 7) one of its ventral flanges (a dendritic branch) is seen to be contiguous with certain visual relay cells of the lobula. This pattern of connections was first resolved in larger flies (*Calliphora*) and seems to be ubiquitous.[9] Figure 5 shows a scheme of fills achieved by passage of cobalt into the antennae. Figure 6 illustrates various aspects of cobalt filled sensory fibers and interneurons as well as features that are typical of the giant descending neuron and its relationships with mesothoracic-derived motor neurons that terminate on dorsal longitudinal muscle (DLM) and muscle of the tergal depressor of the trochanter (TDT) (see ref. 12).

We have compared the projection patterns of two species of sensory fibers that pass from the lateral antennal nerve into the pro-

and mesothoracic ganglia of normal *Drosophila*. One of these is characteristically **wide-diametered; the other is slender.** The first type branches latero-medially into ipsi- and contralateral neuromeres of the pro- and mesothoracic ganglia. The second type gives rise to thin blebbed branches that invade ipsi- and contralateral neuromeres of these ganglia at a more dorsal location. Comparisons between invididuals shows that these branching patterns are quite variable. However, we have not seen them to penetrate the metathoracic ganglia. Amongst a population of normal *Calliphora* about 2% of the large diameter fibers send one or two branches into the anterior metathoracic ganglion.

We have also compared the shapes of giant neurons in normal and mutant flies. Both in normal and biothorax mutant flies the giant neuron axon projects only as far as the anterior mesothoracic ganglion. However, its dendrites do differ in different visual mutants. In Bar eyed, the ventral dendrite normally associated with lobula relay cells projects into the posterior antennal centers. In this mutant the number of ommatidia is greatly reduced[6] as is the number of retinotopic pathways and columnar interneurons. The example shown in Figure 8B demonstrates that a mutation that changes the peripheral organization of receptors may be reflected by changes of morphology of neurons deep within the brain.

Lastly we compared the form of the small antennal descending neuron (AD; Figures 7C,7D and 8). The axon of this cell branches

---

Figure 5: First and second order projections from antennae to the thoracic ganglia. The antennal nerve is roughly divided into a medial and a lateral component. The medial component terminates mainly in the ipsilateral antennal lobe and sends some fibers into the contralateral lobe. The lateral component terminates in the posterior deuterocerebrum where it forms the posterior antennal center. The lateral part of the antennal nerve also gives rise to two primary sensory axons (lower left) that branch in the pro- and mesothoracic ganglion.

Cobalt passes from antennal fibers in the posterior deuterocerebrum into relay neurons. Some of these comprise intracerebral interneurons (top right) that connect left and right antennal centers. Cobalt also enters the giant descending nerve (lower right, D) and a small antennal relay neuron (E). The giant neuron gives up cobalt to lobula neurons and to two fibers that project directly to musculature (G, motor). The small antennal neuron gives up cobalt into a small intraganglionic interneuron (F) that resides in the metathoracic and abdominal neuropil mass.

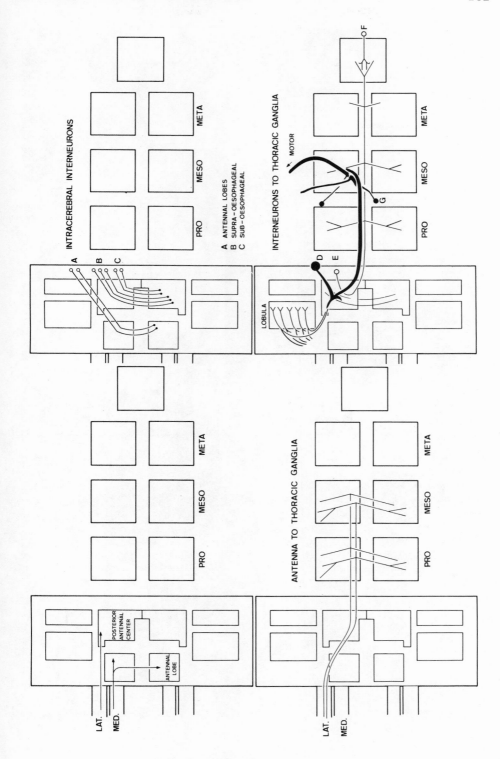

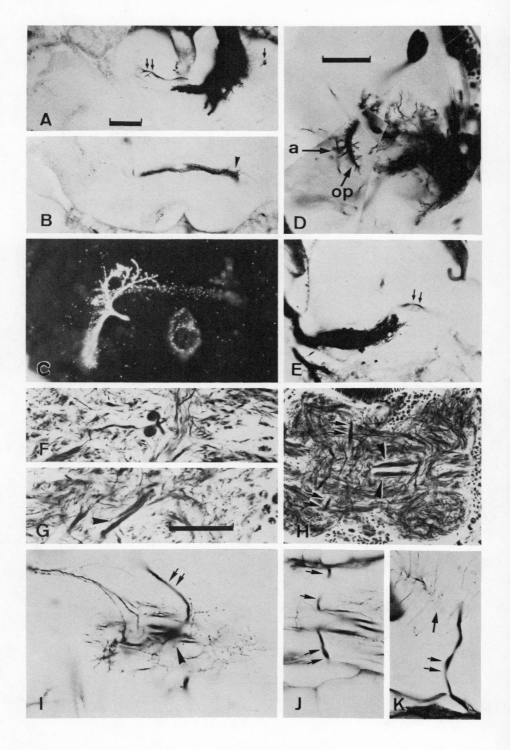

several times to ipsi- and contralateral neuromeres of the pro- and mesothoracic ganglia and sends a few slender fibers into the anterior metathoracic ganglion of normal flies. In the bithorax-postbithorax mutants the AD neuron gives rise to supernumerary branches in the metathoracic ganglion. However, the small interneuron associated with this cell (PI, in Figures 7,8) is present in the normal and mutant animal.

The differences between central morphology of normal and mutant *Drosophila* are adequately demonstrated in Figures 7,8,9 and 10. In all cases the mutant genotype was $bx^3\ pbx/Ubx^{101}$.

Figure 7 illustrates a giant descending neuron (in A) shown by Golgi impregnation of a normal fly. B illustrates the slender antennal projection into the 1st and 2nd thoracic ganglia (SAF) from the antennal lobe (AL) or posterior antennal center (PAC).

Figure 7C and D compare a normal and mutant projection of SAF and the large antennal fiber (LAF) with the small antennal descending neuron AD. The giant neuron (GD) is also shown in the mutant. Except for AD axon branches of both, normal and mutant projections are similar. The AD axon branches more profusely in the anterior metathoracic neuropil of the mutant (indicated by arrows). These projections are **shown** individually in Figure 8, where for each pair the

---

Figure 6: A. Mass filling of the lateral antennal nerve, showing its typical ramification in deuterocerebral neuropil. An axon of an intracerebral interneuron is also shown (double arrows) with its cell body to the right (single arrow). B. The supraoesophageal tract. The forward dendrite of the giant neuron (a in 6D) **inserts** into this tract at the arrow. C. A backfilled giant neuron revealed after injection of Co^{++} into the mesothoracic ganglion. Dark-field illumination shows its contiguity with the supraoesophageal tract. D. A **Golgi** impregnated giant neuron (montage). The lower dendritic flange is associated with visual neurons (op). E. The initial segment of an antennal sensory fiber to thoracic ganglia. F. Cross sections of the pair of giant fiber axons, G, the apposition between the axon and the TDT nerve. H. The pair of giant axons (single large arrows) and the DLM nerves. The cobalt fills in Figures 6I and J show the DLM nerves in the double bithorax mutant and in a normal fly after injection of cobalt into the brain from a micropipette and its uptake by the giant neuron. Figure 6 K illustrates the passage of the DLM nerve (double arrow) directly to dorsal longitudinal muscle where it gives rise to a huge ramifying termination (single arrow). K was revealed after injection of cobalt into the mesothoracic ganglion. F, G, H are reduced silver preparations.

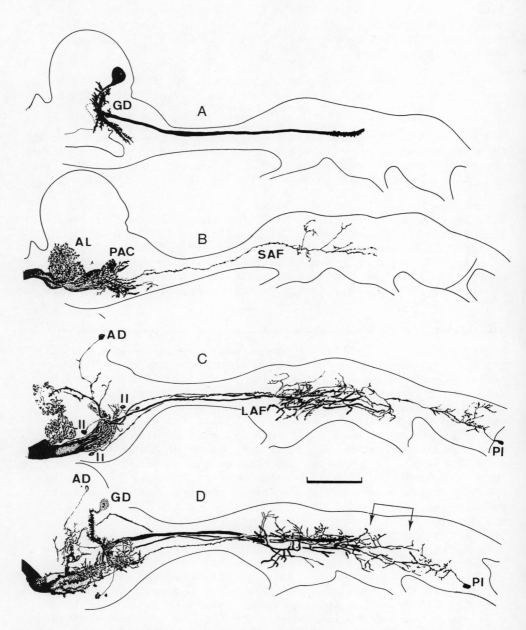

Figure 7: Comparisons between normal and mutant central projections. Sagittal sectons. For explanation see the text.

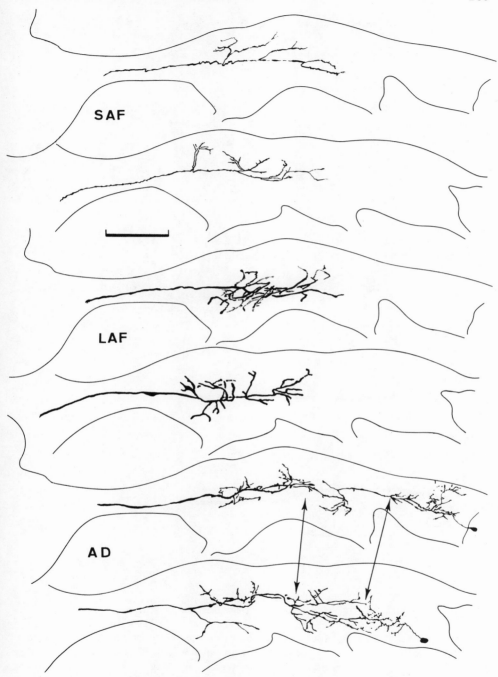

Figure 8: Comparisons of central projections of simple fibers to the thoracic ganglia. For explanation see the text.

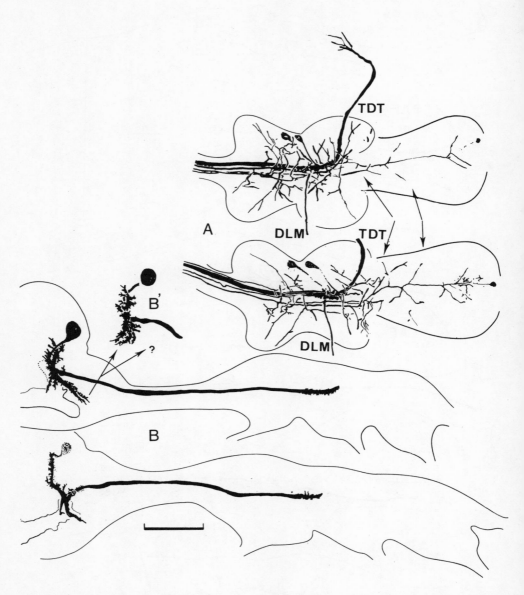

Figure 9: Central projections in normal and mutant *Drosophila*. A compares normal and mutant from horizontal sections and shows the origin of the TDT and DLM nerves. Increase in the number of metathoracic axon branches is indicated by arrows. B compares the forms of a normal giant descending neuron with the same element in the bithorax-postbithorax mutant (lower) and the dendritic pattern in Bar-eyed. For further explanation see the text.

upper reconstruction is from the normal animal, the lower reconstruction is from the mutant.

The giant descending neurons of normal and mutant flies are again compared singly in Figure 9B. B' shows the dendritic pattern of GD in a Bar-eyed *Drosophila*. Figure 9A compares another pair of flies, seen from the horizontal aspect, after injection of cobalt into the brain. The normal projection of fibers is seen at the top, with the mutant below. Again, there appear to be more branches from the AD neuron in the metathoracic ganglion of the mutant. In both the mutant and normal *Drosophila* the TDT and DLM neurons are identically located. Figure 10 shows a third comparison and also indicates supernumerary axon branches in the metathoracic ganglion (me).

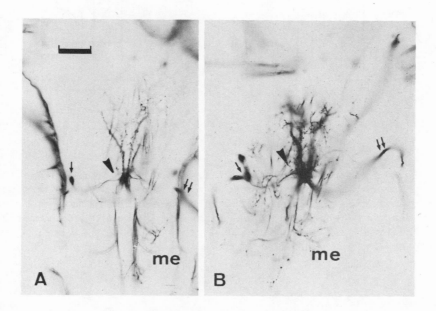

Figure 10: Two photomicrographs that compare the patterns of arborizations of certain neurons in the meso- and metathoracic ganglia of a normal *Drosophila* (left) and a bithorax-postbithorax mutant (right). The dense conglomerate (arrow) medially in the mesothoracic ganglia is the point of junction between the axon of the giant descending neuron and the TDT and DLM nerves. The TDT nerve cell body (contralateral to the axon) is indicated by a small single arrow, and its axon is indicated by double arrows. Both normal and mutant appear to be identical except for an apparent increase in the number of small-diameter axon branches in the anterior neuropil of the metathoracic ganglion (me).

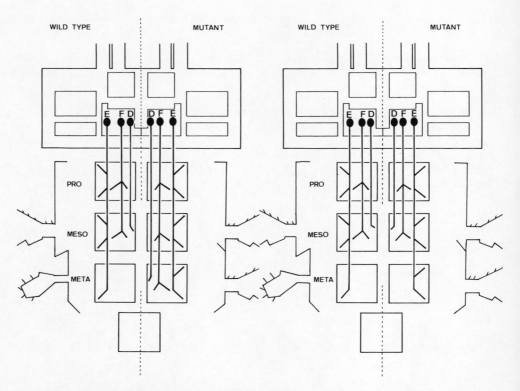

Figure 11: Diagram illustrating normal projection patterns (to the left in each case) and expected projections if metathoracic ganglion was transformed to mesothoracic in double-bithorax mutants (left hand diagram). The right hand diagram, to the right, summarizes the observed results (see text).

DISCUSSION

The majority of our findings do not support the notion of a general transformation of the 3rd ganglion to mesothoracic structure in the present bithorax mutants. However, some subtle differences are observed indicating a certain kind of transformation has occurred.

Using central projections from the head to the thorax as criteria it might be expected that if the metathoracic ganglion was generally transformed to a mesothoracic structure then it would contain neurons typical of the mesothoracic ganglion. This is, however, not the case.

In the extreme sense we might expect axon terminals that normally end in the mesothoracic neuropil to end in metathoracic if the 3rd ganglion was transformed and if the program for termination specified "grow to the border of the anterior metathoracic compartment". However, in no case do we see a repetition of an element, or branching pattern typical of mesothorax in metathorax, if the neuron normally ends in the mesothorax.

The only observed difference between normal and mutant projections is shown by the AD neuron which is usually multisegmental to all three ganglia. In the pro- and mesothoracic ganglia it branches several times and in normal metathoracic it branches sparsely. In mutants its metathoracic branches are more numerous.

We conclude that transformations of C.N.S. structures in metathoracic ganglia of bithorax mutants are subtle and possibly of a secondary nature. Extra branches may be due to additional sensory fibers from the homeotic appendage providing extra guides for postsynaptic dendrites. Therefore, descending interneurons that end on them would also show increased branches. If all the postsynaptic elements responded to homeotic projections in this way, then their supernumerary branches would significantly contribute to enlargements of the third thoracic ganglion in bithorax mutants.

ACKNOWLEDGEMENTS

We thank Dr. J.A. Campos-Ortega for providing mutants and for originally suggesting that these studies be undertaken. We also thank Malu Obermayer and Harjit Seyan for expert technical assistance.

REFERENCES

1. J. Palka, P.A. Lawrence, and H.S. Hart, Neural projection patterns from homeotic tissue of Drosophila studied in bithorax mutants and mosaics, Dev. Biol. 69:549 (1979).
2. A. Ghysen, Sensory neurons recognize defined pathways in Drosophila central nervous system, Nature (Lond.) 274:869 (1978).
3. E.B. Lewis, Genes and developmental pathways, Am. Zool. 3:33 (1963).
4. A. Garcia-Bellido, P. Ripoll, and G. Morata, Developmental compartmentalization of the wing disc of Drosophila, Nature New Biol. 245:251 (1973).
5. G. Morata and P.A. Lawrence, Homeotic genes, compartments and cell determination in Drosophila, Nature (Lond). 265:211 (1977).
6. D.L. Lindsley and E.H. Grell, Genetic Variations of Drosophila melanogaster, Carnegie Institution, Washington (1972).

7. J. Levine and D. Tracy, Structure and function of giant motor-neuron of *Drosophila melanogaster*, J. Comp. Physiol. 87:213 (1973).
8. J. Bacon and N.J. Strausfeld, Non-random resolution of insect neurons, in: Experimental Entomology Methods, T.A. Miller and N.J. Strausfeld, eds., Springer, New York (1980).
9. N.J. Strausfeld and K. Hausen, The resolution of neuronal assemblies after cobalt injection into neuropil, Proc. R. Soc. Lond. B. 199:463 (1977).
10. J. Bacon and J.S. Altman, Silver intensification of cobalt-filled neurons in intact ganglia, Brain Res., 138:359 (1977).
11. N.J. Strausfeld and M. Obermayer, Resolution of intraneuronal and transsynaptic migration of cobalt in the insect visual and central nervous system, J. Comp. Physiol. 110:1 (1976).
12. D.G. King, Anatomy of motor neurons in *Drosophila* thoracic ganglion, Neuroscience Abs. 2:904 (1976).

DISCUSSION

M. Heisenberg: How do you distinguish the protocerebrum from the deuterocerebrum?

N.J. Strausfeld: I am afraid it would take too long to explain it here, and the definition is a loose one. Basically, the deuterocerebrum is that part of the brain which contains olfactory and mechanosensory endings from the antennae.

P.A. Lawrence: I would like to ask about compartmentalization among terminals from neurons. Can the compartmental restrictions described in the leg neuromeres also be found among the wing projections?

A. Ghysen: No you cannot distinguish the borders in the central nervous system as clearly as with the cuticle.

P.A. Lawrence: Is there a rigid rule that the sensillae from different compartments follow different bundles?

A. Ghysen: Sensilla that belong to different compartments often have similar functions; on this basis we cannot say anything about compartment differences. Similarities in sensory projections need not necessarily mean absence of compartments. There is no difference in projection from anterior and posterior. They may follow the same bundle.

P.A. Lawrence: One thing we might expect from our premises of different kinds of compartments is the existence of dorsal and ventral compartments. Do they send out separate bundles in the wing?

A. Ghysen: No, they do not.

J. Palka: We did an experiment in which we made *pbx* clones in the wing so that we could not possibly confuse the bristles in the anterior and posterior compartments. We found that the projections were the same as for bithorax bristles; they go to the same region of the metathoracic ganglion. I want to add that the difference between proximal and distal companiform sensilla on the wing leads us to expect that the difference in compartments also coincides with having large and small companiform sensilla which are rather different in their cuticular morphology and potentially serve different roles. The cuticular differences should be coincident with the data we obtained.

E. Wieschaus: If the different sensilla from different compartments follow different bundles, do they have different functions? If so, it would be one of the first examples of functional differences reflected in the compartments.

A. Ghysen: I don't know the exact role of the various bristles.

D. Byers: What behavior is elicited by stimulating the anterior and posterior leg bristles?

A. Ghysen: It depends on which leg you use. On the first leg when anterior bristles are stimulated, the legs rub together; when posterior bristles are stimulated, the metathoracic leg cleans the first leg.

P.A. Lawrence: Do you mean to say that all the anterior bristles right down to the tarsus produce one effect and all the posterior bristles produce the other effect?

A. Ghysen: The efficiency of eliciting the behavior reduces as you move down the leg.

P.A. Lawrence: Are you talking of the proximal bristles or the tarsus bristles?

A. Ghysen: We stimulate at the femur and tibia.

Y. Hotta: Has there been any work on the motor system?

N.J. Strausfeld: John Coggshall (J. Comp. Neurol., 177:707 (1978)) has looked at motor projections to the longitudinal flight muscles.

USE OF NEUROTOXINS FOR BIOCHEMICAL AND GENETIC ANALYSIS OF MEMBRANE PROTEINS INVOLVED IN CELL EXCITABILITY

Linda M. Hall

Department of Genetics
Albert Einstein College of Medicine
1300 Morris Park Avenue
Bronx, New York   10461   U.S.A.

INTRODUCTION

In recent years the fruit fly *Drosophila melanogaster* has become increasingly popular for use in genetic approaches to problems in neurobiology. As a result, a variety of single gene mutations which affect nervous system function have become available to the investigator. (See other chapters in this volume and the following recent reviews for descriptions of the types of mutants which are available: Pak and Pinto, 1976; Ward, 1977; Kankel and Ferrus, 1979; Pak, 1979). Although electrophysiological studies have been used to provide clues about possible sites of action of these mutations (Ikeda et al., 1976; Siddiqi and Benzer, 1976; Jan et al., 1977; Wu et al., 1978), biochemical experiments will be necessary to identify the molecular site of the mutant defect. Mutations which affect enzymes important to nervous system function such as those involved in neurotransmitter synthesis and degradation are relatively straightforward to identify and analyze since most enzyme activities can be readily monitored in extracts. Thus, mutations affecting the enzymes choline acetyltransferase and acetylcholinesterase have been identified in *Drosophila*. (See chapter by J.C. Hall et al. in this volume).

Another class of proteins which are important to nervous system function are the structural proteins involved in ion channels and neurotransmitter receptors. These molecules play a key role in cell excitability but are difficult to monitor in disrupted cell homogenates. Ion channels, for example, are involved in mediating either voltage-induced or neurotransmitter-induced changes in cell permeabilities to small ions. While these changes can be monitored, at least in some intact preparations, by either electrophysiological recordings or ion flux studies, such assays are dependent on cell integrity and are not amenable to routine biochemical investigations which use homogenized extracts. Reconstitution of ion channels into intact vesicles or artificial lipid bilayers represents one promising approach for future analysis of these molecules but at the present time technical difficulties limit the usefulness of such approaches especially when the starting material is a crude extract such as would be obtained using *Drosophila* homogenates. Neurotransmitter receptors are similarly difficult to analyze in extracts. Although the receptor must specifically bind the neurotransmitter *in vivo*, in order to insure a rapidly responding nervous system, this binding must, by necessity, be transient in nature. So although transmitter binding represents a potential assay for receptors *in vitro*, in practice the ready reversibility of transmitter binding to its receptor imposes severe limitations on this approach.

Fortunately, in recent years neuropharmacologists have identified a variety of neurotoxins which bind specifically and with high affinity to various molecular components involved in cell excitability. For example, α-bungarotoxin, a polypeptide isolated from the venom of the snake *Bungarus multicinctus*, has been very useful for studies on the molecular characterization of the nitotinic acetylcholine receptor from a variety of vertebrate preparations (deRobertis and Schacht, 1975). More recently, saxitoxin, the molecule responsible for the toxic effects of shellfish poisoning, has been utilized as a specific molecular probe for voltage-dependent sodium channels in a variety of preparations (Ritchie, 1978). These toxins can be readily adapted to studies of receptors and ion channels in *Drosophila*. The central nervous system of *Drosophila* is a rich source of an ^{125}I-α-bungarotoxin binding component with the properties expected of a nicotinic acetylcholine receptor (Hall and Teng, 1975; Schmidt-Nielsen et al., 1977; Dudai, 1977; Dudai and Amsterdam, 1977; Dudai, 1978; Rudloff, 1978). Similarly, high affinity, saturable binding of ^{3}H-saxitoxin has recently been demonstrated in *Drosophila* extracts (Gitschier et al., 1980). In this chapter we will review how these radioactively labeled probes can be used in molecular and genetic analysis of the acetylcholine receptor and the voltage-dependent sodium channel in *Drosophila*.

DEFINITION OF TOXIN BINDING PROPERTIES

The first step in using neurotoxins as molecular probes is to develop a specific assay for the toxin binding component. Using the criteria suggested by Birdsall and Hulme (1976), the toxin probe should have the following properties. (1) The toxin binding should be saturable. (2) Binding should show localization patterns appropriate for the components being studied. (3) Binding should show appropriate pharmacological specificity and there should be a correlation between the specificity defined in binding studies and that determined electrophysiologically.

Using ^{125}I-α-bungarotoxin binding as an example, we will review how these criteria have been fulfilled using *Drosophila* extracts. In this case, the radioactive ligand is prepared by routine iodination procedures (Schmidt-Nielsen et al., 1977) and the binding to a membrane preparation is quantitated using a centrifugal assay (Schmidt-Nielsen et al., 1977) or a rapid filtration assay (Dudai, 1978). The toxin binding to a constant amount of extract can be shown to saturate with low concentrations (10 nM) of ^{125}I-α-bungarotoxin, thus fulfilling this criterion for specific binding (Schmidt-Nielsen et al., 1977). From these saturation binding experiments we estimate that there are 10 pmoles of binding activity per gram of whole flies or 88 pmoles of binding activity per gram of *Drosophila* heads. A crude extract of heads contains 0.9 nmoles toxin binding sites per gram protein (Gepner, 1979).

Differential centrifugation studies show that the α-bungarotoxin binding component is membrane associated as would be expected for a neurotransmitter receptor. Although it stays in the supernatant following low speed centrifugation (2,000xg, 10 min), the majority of the binding activity pellets with a membrane fraction following centrifugation for 20 min at 20,000xg (Schmidt-Nielsen et al., 1977). Autoradiographic localization studies of ^{125}I-α-bungarotoxin binding shows that at the light microscopic level toxin binding is confined to regions of the central nervous system of *Drosophila* known to contain synapses (Hall and Teng, 1975; Schmidt-Nielsen et al., 1977; Dudai and Amsterdam, 1977; Rudloff, 1978).

The pharmacological specificity of the α-bungarotoxin binding component is defined by comparing the ability of pretreatment with various ligands to inhibit subsequent toxin binding. As summarized in Table I, the toxin-binding sites are protected by low concentrations of unlabeled α-bungarotoxin, dihydro-β-erythroidine, nicotine, d-tubocurarine, eserine and acetylcholine. With the exception of eserine which is an anticholinesterase, these ligands are "neuromuscular nicotinic" ligands (Koelle, 1975). In contrast with these ligands, the nicotinic neuromuscular agonist decamethonium is a moderately weak inhibitor of toxin binding. The nicotinic ganglionic antagonist trimethaphan is a good inhibitor of toxin binding but other ganglionic antagonists (pempidine and mecamylamine) are among the weakest inhibitors tested. The muscarinic ligands (atropine, carbamylcholine, dexetimide, pilocarpine and DL-muscarine) are all at least an order of magnitude less effective than acetylcholine thus establishing a nicotinic specificity for this receptor.

The most convincing proof that α-bungarotoxin is binding to a functional acetylcholine receptor in the *Drosophila* central nervous system would involve a correlation of these binding studies with electrophysiological studies. Unfortunately, the small size of the

Table 1. Pharmacological Specificity of the α-Bungarotoxin Binding Component

	BINDING STUDIES				ELECTROPHYSIOLOGY
	Drosophila[a] crude extracts I_{50} (μM)	Drosophila[b] head extracts I_{50} (μM)	Drosophila[a] 1200 x purified component from heads I_{50} (μM)	Periplaneta[a] nerve cord extracts I_{50} (μM)	Periplaneta[a] 6th abdominal ganglion I_{50} (μM)
Dihydro-β-erythroidine	–	–	0.026	–	–
Nicotine	0.45	0.46	0.35	2.7	0.2
d-Tubocurarine	2.7	0.65	0.66	0.9	–
Eserine	–	5.4	1.4	–	–
Gallamine	–	–	1.6	–	–
Acetylcholine	2.8	7.9	3.0	8.3	9
Atropine	57	74	18	72	–
Carbamylcholine	–	120	23	100	400
Dexetimide	–	–	60	–	–
Decamethonium	240	250	63	170	–
Pilocarpine	–	790	150	21	800
Neostigmine	>1000	4500	540	34	–
Mecamylamine	–	–	680	–	–
DL-Muscarine	–	–	1400	–	–

[a] Gepner, 1979; [b] Detergent treated extracts (Schmidt-Nielsen et al., 1977); [c] Gepner et al., 1978 and D.B. Sattelle, personal communication.

neurons and the relative inaccessibility of the central nervous system prevents this experiment from being done in *Drosophila*. However, experiments summarized in Table I have shown that the cockroach (*Periplaneta americana*) central nervous system contains an α-bungarotoxin binding component similar to that described in *Drosophila* (Gepner et al., 1978; Gepner, 1979; D.B. Sattelle, J.I. Gepner and L.M. Hall, in preparation). The data summarized in Table I show that the order of effectiveness for ligands as defined in electrophysiological studies of the cholinergic synapse between the cercal nerve and giant fiber in the terminal abdominal ganglion in the cockroach is the same as that revealed by toxin-binding inhibition studies. In addition, D. B. Sattelle and his colleagues (personal communication) have been able to demonstrate blockade of this cholinergic synapse in *Periplaneta* at low toxin concentrations (5 x 10^{-8} M) thus providing further evidence that α-bungarotoxin binds to a functional nicotinic acetylcholine receptor in the insect central nervous system.

Similar binding studies can be done using ^{3}H-saxitoxin as a probe for the voltage-dependent sodium channel. In this case use of a rapid filtration assay is essential because although saxitoxin binds to *Drosophila* extracts with high affinity (K_D = 1.9 nM), the binding is reversible with a half-life of 18.3 s at 4°C (Gitschier et al., 1980). The toxin binding again can be demonstrated to fulfill the criterion of saturability and these experiments indicate that the number of saturable binding sites (6.4 fmol/mg head or 97 fmol binding sites/mg protein in crude head extracts) is much lower than the number of α-bungarotoxin binding sites. The larval ventral muscle fibers and their innervating nerves are readily accessible for electrophysiological recording (Jan and Jan, 1976; Wu et al., 1978) and so for the saxitoxin binding component direct comparisons of binding studies and electrophysiological measurements can be done (Gitschier et al., 1980).

In summary, the ^{125}I-α-bungarotoxin and the ^{3}H-saxitoxin binding studies provide descriptive information about their respective receptors in wild-type preparations The assays are rapid and reproducible and thus it will be possible to use these techniques to screen for potential variants affecting toxin-binding parameters, such as number of saturable binding sites, K_D, off rate and rate of thermal inactivation of the toxin binding component.

PURIFICATION STUDIES

Although the low levels of the saxitoxin-binding component in the *Drosophila* central nervous system would make purification of this component difficult,the saturation binding studies show that *Drosophila* is a rich source of the α-bungarotoxin-binding component and thus purification of this component is feasible. Neurotoxin binding can be used to monitor receptor activity throughout the pur-

Table II. Purification of an α-Bungarotoxin Binding Component from *Drosophila melanogaster*.

	Volume (ml)	pmoles 125I-α-Bgt Bound (per ml)	pmoles 125I-α-Bgt Bound (total)	mg Protein (per ml)	mg Protein (total)	Specific Activity (μmoles/ g protein)	Purification	Recovery of Binding Activity (%)
Crude head extract			2520		2851	0.0009	–	
20,000xg Pellet			1655		583	0.0028	3.2	66
Triton X-100 solubilized extract	120	4.5	540	3.3	395	0.0014	1.5	21
1st Affinity Column a) Nonadsorbed fraction			400 177		368	0.0005		
b) Carbamylcholine-eluted peak fractions	29	7.5	218	0.044	1.3	0.17	190	9
2nd Affinity Column peak fractions	(28) (16)	5.3	154 85	0.005	0.08	1.06	1200	3.4

ification thus allowing calculations of recovery and specific activity after each step. (See Table II.) In addition, the polypeptide neurotoxins can be coupled to cyanogen bromide activated Sepharose 4B (Brockes and Hall, 1975) to make an affinity column which is extremely effective for receptor purification. One potential difficulty is that the receptor binds to α-bungarotoxin with such high affinity that the receptor would be difficult to elute from the affinity column once it had bound. This problem can be circumvented by using a related snake toxin, α-cobratoxin, which binds to the receptor with similar specificity, but lower affinity. For example, when the ability of unlabeled toxins to inhibit ^{125}I-α-bungarotoxin binding is compared, a concentration of 3.6×10^{-9} M cold α-bungarotoxin inhibits binding by 50% while the amount of α-cobratoxin required for the same inhibition is an order of magnitude greater (3.3×10^{-8}M). As summarized in Table II a single passage through an α-cobratoxin column removes > 99% of the protein and allows recovery of 40% of the toxin-binding activity in the carbamylcholine eluted fractions providing greater than 100-fold purification in a single step. A second run through the affinity column removes more contaminating protein resulting in an additional 6-fold purification.

The purified receptor ^{125}I-α-bungarotoxin complex was shown to have a sedimentation coefficient of 11.5 S when compared with standards on a 5 to 20% linear sucrose gradient (Gepner, 1979). This represents a molecular weight of ∿ 300,000 daltons when estimated by the method of Martin and Ames (1961). If this receptor in *Drosophila* is similar to receptors from vertebrate neuromuscular junctions, we would expect there to be 8 to 10 μmoles toxin binding sites per g protein (Heidmann and Changeux, 1978). The specific activity of 1 μmole toxin binding sites per g protein in our *Drosophila* preparation suggests that it is ∿ 10% pure. Preliminary studies have shown that the receptor is a glycoprotein and a lectin affinity column can provide a 46-fold purification in a single step (T. Schmidt-Glenewinkel, unpublished observations). The addition of a lectin affinity column step and a Bio-Gel chromatography step (which fractionates on the basis of size differences) to the existing purification scheme should allow sufficient purification for subsequent mutant analysis.

GENETIC ANALYSIS OF NEUROTOXIN BINDING COMPONENTS

Neurotoxin-binding experiments such as those described above can provide information useful for developing mutant isolation strategies. Once binding parameters have been defined, screens can be set up to search for variants in these parameters. In addition to this brute force approach, pharmacological studies can provide information about small ligands that interact with the receptor and thus which might be useful for screening for drug-resistant variants. With respect to the α-bungarotoxin binding components, Table I shows that nicotine is one of the best inhibitors of toxin binding. Nicotine is also very effective at killing *Drosophila* when it is administered in the medium (Hall et al., 1978). Selecting for drug resistant variants provides a way to enrich for alterations in the drug receptor of interest, but since resistance can develop as the result of a variety of mechanisms, this approach must be coupled with some molecular assay to identify those variants which affect receptor structure. We have developed an isoelectric focusing procedure for the solubilized receptor $-^{125}$I-α-bungarotoxin complex which has allowed us to identify nicotine-resistant variants that also show changes in the isoelectric point (pI) of the toxin-receptor complex (Hall et al., 1978). Such changes in isoelectric point may be due either to mutations that affect a structural gene for one of the receptor subunits or to mutations that affect receptor modification. Genetic mapping studies of the nicotine-resistant HR strain have shown that the nicotine-resistance and the shift in isoelectric point exhibited by this strain both segregate with the X chromosome (L.M. Hall, B.C. Osmond and T.H. Hudson, unpublished results). Recombinational mapping studies will be undertaken to determine if they map to the same locus.

Our original isoelectric focusing procedure was very laborious since it involved slicing and manipulating tube gels. Genetic analysis would be facilitated by the use of slab gels where multiple samples could be run at once and radioactive bands could be detected by autoradiography. A difficulty in developing this procedure is that the receptor-α-bungarotoxin complex dissociates under the conditions required for slab gel runs. To overcome this problem, we have developed a procedure for crosslinking the toxin to the receptor using a photoactivatable derivative of ^{125}I-α-bungarotoxin. Monoiodo-α-bungartoxin was derivatized with methyl-4-azidobenzoimidate-HCl. This derivative was allowed to react in the dark with a 20,000xg membrane preparation from *Drosophila* and the unbound toxin was removed by centrifugation. The specifically bound toxin was then cross-linked to the receptor by illuminating the complex with ultraviolet light. This cross-links the toxin to the receptor and allows us to subject the complex to vigorous conditions including solubilization by the detergent sodium dodecyl sulfate. Although the 300,000 molecular weight receptor complex dissociates into subunits under these conditions, the toxin remains cross-linked to its binding sites. Sodium dodecyl sulfate polyacrylamide gel electrophoresis has allowed us to demonstrate that the toxin reacts primarily with a 58,000 dalton subunit (F.M. Hoffmann, unpublished results). The procedure will make is possible to screen for variants in this subunit. Since the subunit is substantially smaller (MW = 58,000) than the Triton X-100 solubilized complex (MW = 300,000) variants causing small changes in polypeptide structure should be easier to detect with this method than they would be if the entire complex were analyzed. Adaptation of this photoactivatable cross-linking procedure to other neurotoxins should make it possible to screen for electrophoretic variants using toxins that normally show rapidly reversible binding.

In looking for variants in the saxitoxin-binding component of the sodium channel, two strategies are currently being employed in our laboratory. The first is to look for temperature-induced differences in ^{3}H-saxitoxin binding parameters in the temperature-sensitive paralytic mutants (*nap^{ts}, para^{ts}* and *comatose*). Electrophysiological studies have shown these mutants have temperature-induced defects in some aspect of nerve conduction (Siddiqi and Benzer, 1977; Wu et al., 1978) and thus, these are possible candidates for sodium channel alterations. A second approach involves the construction of segmental aneuploids to screen the genome for gene dosage effects on ^{3}H-saxitoxin binding. This gene dosage approach has been successful for identifying structural genes for a variety of enzymes including acetylcholinesterase (Hall and Kankel, 1976). It remains to be seen whether it will be successful for a sodium channel component where the binding subunit is likely to be part of a multisubunit membrane-bound complex.

In summary neurotoxins that interact with specific components of excitable cells can be used to provide molecular information about these components. They also can be used in the development of genetic strategies to isolate mutants affecting these specific components. In long term studies mutant analysis coupled with neurotoxin binding sutdies will provide information about organization and regulation of genes with products involved in excitable membrane function. Identification of structural genes for receptors and ion channel proteins will open the way for isolation of temperature-sensitive mutations which will be useful in defining structure-function relationships and in defining the role of these components in development. Studies on heterozygotes for the isoelectric focusing variant in the α-bungarotoxin-binding component from *Drosophila* provides a example of how such variants can be used to deduce information concerning receptor subunit interactions *in vivo* (Hall, 1980).

ACKNOWLEDGEMENTS

The studies from the author's laboratory were supported by NIH grant NS 16204 (formerly NS 13881), NSF grant BNS 78-24594 and grant 1126A from the Council for Tobacco Research-USA, Inc. Linda M. Hall is a McKnight Scholar in Neuroscience. The following investigators have made important contributions to the work summarized in this report: J.I. Gepner, J. Gitschier, F.M. Hoffmann, T.H. Hudson, B.C. Osmond, T. Schmidt-Glenewinkel and B.K. Schmidt-Nielsen.

REFERENCES

Birdsall, N.J.M., and Hulme, E.C., 1976, Biochemical studies on muscarinic acetylcholine receptors. J. Neurochem., 27:7.

Brockes, J.P., and Hall, Z.W., 1975, Acetylcholine receptors in normal and denervated rat diaphragm muscle. I. Purification and interaction with ^{125}I-α-bungarotoxin, Biochemistry, 14:2092.

deRobertis, E., and Schacht, J., eds., 1975, "Neurochemistry of Cholinergic Receptors", Raven Press, New York.

Dudai, Y., 1977, Demonstration of an α-bungarotoxin binding nicotinic receptor in flies, FEBS Lett., 76:211.

Dudai, Y., 1978, Properties of an α-bungarotoxin-binding cholinergic nicotinic receptor from Drosophila melanogaster, Biochim. Biophys. Acta, 539:505.

Dudai, Y., and Amsterdam, A., 1977, Nicotinic receptors in the brain of Drosophila melanogaster demonstrated by autoradiography with ^{125}I-α-bungarotoxin, Brain Res., 130:551.

Gepner, J.I., Hall, L.M., and Sattelle, D.B., 1978, Insect acetylcholine receptors as a site of insecticide action, Nature, 276:188.

Gitschier, J., Strichartz, G.R., and Hall, L.M., 1980, Saxitoxin binding to sodium channels in head extracts from wild-type and tetrodotoxin-sensitive strains of Drosophila melanogaster, Biochim. Biophys. Acta, 595:291.

Hall, J.C., and Kankel, D.R., 1976, Genetics of acetylcholinesterase in Drosophila melanogaster, Genetics, 83:517.

Hall, L.M., 1980, Biochemical and genetic analysis of an α-bungarotoxin-binding receptor from Drosophila melanogaster, in: "Receptors for Neurotransmitters, Hormones, and Pheromones in Insects", D.B. Sattelle, L.M. Hall, and J.G. Hildebrand, eds., Elsevier, Amsterdam.

Hall, L.M., and Teng, N.N.H., 1975, Localization of acetylcholine receptors in Drosophila melanogaster, in: "Developmental Biology· Pattern Formation · Gene Regulation", ICN-UCLA Symposia on Molecular and Cellular Biology, Vol. 2, D. McMahon and C.F. Fox, eds., W.A. Benjamin, Inc., Menlo Park, Ca.

Hall, L.M., von Borstel, R.W., Osmond, B.C., Hoeltzli, S.D., and Hudson, T.H., 1978, Genetic variants in an acetylcholine receptor from Drosophila melanogaster, FEBS Lett., 95:243.

Heidmann, T., and Changeux, J.P., 1978, Structural and functional properties of the acetylcholine receptor protein in its purified and membrane-bound states, Ann. Rev. Biochem., 47:317.

Ikeda, K., Ozawa, S., and Hagiwara, S., 1976, Synaptic transmission reversibly conditioned by single-gene mutation in Drosophila melanogaster, Nature, 259:489.

Jan, Y.N., Jan, L.Y., and Dennis, M.J., 1977, Two mutations of synaptic transmission in Drosophila, Proc. R. Soc. Lond. Ser. B., 198:87.

Kankel, D.R., and Ferrus, A., 1979, Genetic analyses of problems in the neurobiology of Drosophila, in: "Neurogenetics: Genetic

Approaches to the Nervous System", X.O. Breakefield, ed., Elsevier, New York.

Koelle, G.B., 1975, "The Pharmacological Basis of Therapeutics", Macmillan, New York.

Martin, R.G., and Ames, B.N., 1961, A method for determining the sedimentation behavior of enzymes: Application to protein mixtures, J. Biol. Chem., 236:1372.

Pak, W.L., 1979, Study of photoreceptor function using *Drosophila* mutants, in: "Neurogenetics: Genetic Approaches to the Nervous System", X.O. Breakefield, ed., Elsevier, New York.

Pak, W.L., and Pinto, L.H., 1976, Genetic approach to study of the nervous system, Ann. Rev. Biophys. Bioeng., 5:397.

Ritchie, J.M., 1978, The sodium channel as a drug receptor, in: "Cell Membrane Receptors for Drugs and Hormones: A Multidisciplinary Approach", R.W. Straub and L. Bolis, eds., Raven Press, New York.

Rudloff, E., 1978, Acetylcholine receptors in the central nervous system of *Drosophila melanogaster*, Exp. Cell Res., 111:185.

Schmidt-Nielsen, B.K., Gepner, J.I., Teng, N.N.H., and Hall, L.M., 1977, Characterization of an α-bungarotoxin binding component from *Drosophila melanogaster*, J. Neurochem., 29:1013.

Siddiqi, O., and Benzer, S., 1976, Neurophysiological defects in temperature-sensitive paralytic mutants of *Drosophila melanogaster*, Proc. Natl. Acad. Sci. U.S.A., 73:3253.

Ward, S., 1977, Invertebrate neurogenetics, Ann. Rev. Genet., 11:415.

Wu, C.F., Ganetzky, B., Jan, L.Y., Jan, Y.N., and Benzer, S., 1978, A *Drosophila* mutant with a temperature-sensitive block in nerve conduction, Proc. Natl. Acad. Sci. U.S.A., 75:4047.

THE ACETYLCHOLINESTERASE FROM *DROSOPHILA MELANOGASTER*

S. Zingde* and K.S. Krishnan

Tata Institute of Fundamental Research
Bombay, India

* Present address: Cancer Research Institute
Bombay, India

INTRODUCTION

Acetylcholinesterase, the enzyme which hydrolyses acetylcholine, is a key component of the cholinergic system. In *Drosophila melanogaster* as in other animals the enzyme is membrane bound.[1] The enzyme from the eel *Electrophorus electricus* has been well studied.[2] Various molecular forms of the enzyme are known. All of these have a multisubunit head connected to a fibrous tail. Dudai[3] has shown that in *Drosophila melanogaster* the enzyme exists in three forms which have sedimentation coefficients of 7S, 11S and 16S. The subunit structure of acetylcholinesterase in *Drosophila melanogaster* has not yet been studied. We have labelled the enzyme in its membrane environment in the particulate fraction of the brain cells with tritiated diisopropylfluorophosphate (DFP) and examined the electrophoretic behavior of the labelled components by SDS-PAGE. In addition, Triton X-100 extracts of the particulate fraction have been fractionated on Sephadex G-200 to determine the smallest unit with which enzyme activity is associated and to confirm the polyacrylamide gel electrophoresis data.

MATERIALS AND METHODS

The wild-type strain Canton-Special (C-S) of *Drosophila melanogaster* was used. Heads of four to ten days old flies were homogen-

Abbreviations: DFP = Diisopropylfluorophosphate; SDS-PAGE = Sodium dodecylsulfate-polyacrylamide gel electrophoresis; Tris = Tris-hydroxymethylaminomethane; MW = Molecular weight.

ized in Tris-sucrose buffer. The 20,000 xg pellet of the post nuclear supernatant was used as the source of acetylcholinesterase in all subsequent experiments.

All the chemicals used were of the best grade available.

RESULTS AND DISCUSSION

Analysis of the protein pattern of membrane fragments labelled with $^3(H)$-DFP and subjected to SDS-PAGE after complete reduction with 2-mercaptoethanol shows that only one major radioactive band of molecular weight about 57,000 and a minor band of molecular weight 18,000 are obtained (Figure 1). When the membrane fragments were treated with eserine prior to labelling with (^{3}H)-DFP, the radioactivity in the 57,000 band was absent while that in the 18,000 band was still present (Figure 2). This indicated that in the particulate

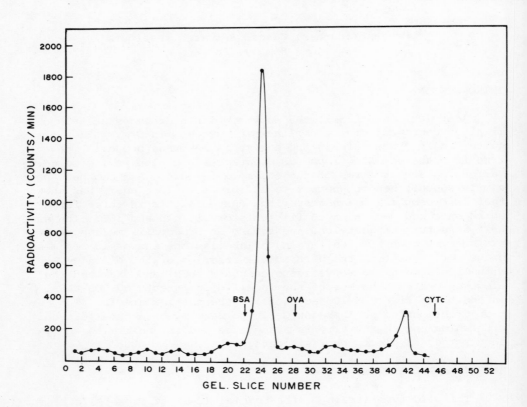

Figure 1: Radioactivity distribution on SDS-PAGE of solubilized membrane fragments labelled with ^{3}H-DFP and completely reduced with 2-mercaptoethanol. Positions of standard proteins: bovine serum albumin (BSA, MW = 68,000), ovalbumin (OVA, MW = 43,000) and cytochrome C (CYTc, MW = 11,700) are indicated.

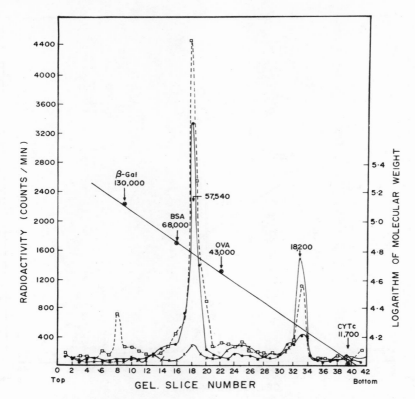

Figure 2: Radioactivity distribution on SDS-PAGE of solubilized membrane fragments labelled with (^{3}H)-DFP (□-□-□-□). (x-x-x-x-) shows the profile of a sample treated with eserine and (●-●-●-●-●) the profile of a sample treated with butyrylcholine iodide prior to ^{3}H-DFP labelling. Positions of standard proteins, β-galactosidase (β-Gal, MW = 130,000), bovine serum albumin (BSA, MW = 68,000), ovalbumin (OVA, MW = 43,000) and cytochrome C (CYTc, MW = 11,700) are plotted versus the logarithm of their molecular weights.

fraction of the *Drosophila* heads the major band that was labelled was acetylcholinesterase. When the membrane fragments were treated with butyrylcholine prior to labelling with DFP, the radioactivity in the 57,000 band was reduced only to 75% of the counts in the untreated sample, while that in the 18,000 band was reduced to 25% suggesting that the minor band is a pseudocholinesterase.

To investigate the subunit assembly of acetylcholinesterase in *Drosophila*, the membrane fragments were subjected to controlled reduction prior to SDS-PAGE. Analysis of the radioactivity distribution in the gels shows (Figure 3) that in the unreduced sample there are two radioactive bands with molecular weights of 131,000

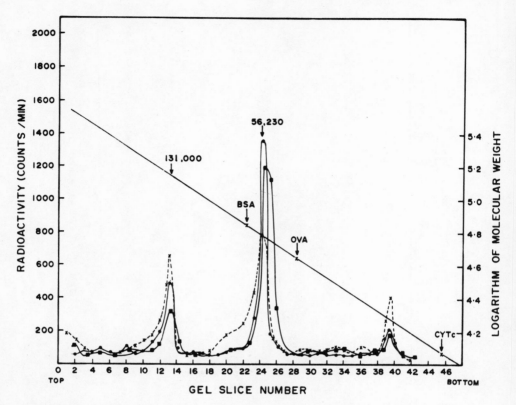

Figure 3: Radioactivity distribution on SDS-PAGE of solubilized membrane fragments labelled with (^{3}H-DFP) and solubilized without reducing agent (x---x---x). The profiles for the samples reduced with 5mM dithiothreitol (●—●—●—●—●) and 10mM dithiothreitol (□—□—□—□) prior to solubilization in SDS-containing buffer are also shown.

and 57,000 in addition to the pseudoesterase with a molecular weight of 18,000. On reduction with increasing amounts of dithiothreitol, the radioactivity in the higher molecular weight protein band is reduced and that in the 57,000 band is proportionately increased. The 18,000 band is unaffected. On complete reduction with 2-mercaptoethanol only the 57,000 band is obtained (Figure 1). The subunit of molecular weight 57,000 is therefore the monomeric form of the *Drosophila* acetylcholinesterase. The dimer of molecular weight 131,000 is composed of 2 monomers held together by disulfide bonds. No higher molecular weight forms of labelled acetylcholinesterase could be identified by SDS-PAGE.

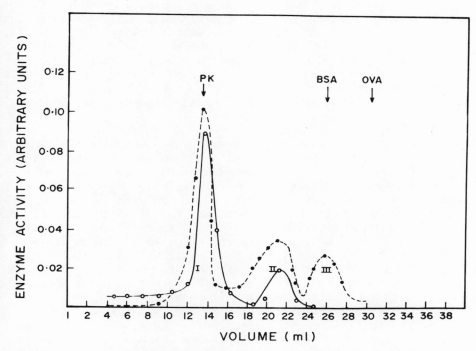

Figure 4: Elution profile on Sephadex G-200 of the Triton X-100 extract of the membrane fragments prior to (O—O—O—O) and after (●--●--●--●) reduction with dithiothreitol. The column was equilibrated and eluted with 50 mM Tris-HCl (pH 7.6), 0.1 M NaCl. Acetylcholinesterase activity was measured by the spectrophotometric method of Ellman[6]. Elution positions of pyruvate kinase (MW = 237,000), bovine serum albumin (MW = 68,000) and ovalbumin (MW = 43,000) are shown.

Studies of Rosenberry et al.[4] on the globular 11S form of the electric eel acetylcholinesterase show that in the unreduced labelled material 80% of the radioactivity is associated with the dimer, 16% with the monomer and less than 5% with the tetramer. The monomer has a molecular weight of 75,000. On complete reduction of the enzyme, the 75,000 monomer is broken down to three polypeptides of molecular weight 50,000, 27,000 and 23,000.

Steel and Smallman[5] have studied a soluble 7.4S (MW = 160,000 form of acetylcholinesterase from housefly heads and have shown that this form is a dimer which can be reduced to a monomer of molecular weight 80,000 with disulfide reducing agents. The 80,000 subunit is cleaved into three polypeptides of molecular weights 59,000, 23,000 and 20,000 on complete reduction with 2-mercaptoethanol. In *Drosophila*, however, the monomer has a molecular weight of 60,000 and complete reduction with 2-mercaptoethanol does not cleave the monomer further into smaller fragments.

To determine the smallest unit with which enzyme activity is associated, Triton X-100 extracts of the membrane fragments were chromatographed on Triton-free Sephadex G-200. Figure 4 shows the elution profile of the Triton extract. The major portion of the acetylcholinesterase activity is eluted in the void volume (molecular weight > 200,000) while there is another peak of activity which corresponds to a molecular weight 120,000. Chromatography of a partially reduced Triton extract gives three peaks of enzyme activity corresponding to molecular weight 61,000, 120,000 and greater than 200,000. This substantiates the results obtained on SDS-PAGE and suggests that the monomer has a molecular weight of about 60,000 and it is held by disulfide bonds to form a dimer of molecular weight 120,000. The activity eluted in the void volume is most likely an aggregated form of the enzyme having a molecular weight greater than 200,000. It is likely that addition of dithiothreitol to the Triton extract forms monomers and dimers which are soluble even in the absence of Triton and hence are eluted as peaks on the Triton-free Sephadex column. The smallest unit of enzyme activity in *Drosophila* is therefore the monomer form of the enzyme.

REFERENCES

1. Y. Dudai, Properties of an α-bungarotoxin binding cholinergic nicotinic receptor from *Drosophila melanogaster*, Biochim. Biophys. Acta 539:505 (1978).
2. T.L. Rosenberry, Acetylcholinesterase, in: "The Enzymes of Biological Membranes," Volume 4, A. Martonosi, ed., Plenum Publishing Corporation, New York (1976).
3. Y. Dudai, Molecular states of acetylcholinesterase from *Drosophila melanogaster*, Dros. Inform. Serv. 52:65 (1977).

4. T.L. Rosenberry, Y. T. Chen, and E. Bock, Structure of 11S acetylcholinesterase-subunit composition, Biochemistry 13: 3068 (1974).
5. R.W. Steele, and B.N. Smallman, Acetylcholinesterase of the housefly head: Affinity purification and subunit composition, Biochim. Biophys. Acta 445:147 (1976).
6. G.L. Ellman, K.D. Courtney, V. Andres, Jr., and R.M. Featherstone, A rapid spectrophotometric assay for acetylcholinesterase, Biochem. Pharmacol. 7:88 (1961).

DISCUSSION

J.C. Hall: Are the pseudocholinesterases you mention in the head, or could they come from elsewhere in the fly that might not even be in the nervous system?

S. Zingde: They are in the head since that was the only tissue used as a source of acetylcholinesterase in these studies.

ISOLATION AND CHARACTERIZATION OF MEMBRANES FROM *DROSOPHILA MELANOGASTER*

T.R. Venkatesh, S. Zingde[*] and K.S. Krishnan

Tata Institute of Fundamental Research
Bombay, India

* Present address: Cancer Research Institute
Bombay, India

This work is dedicated to the memory of Miss Devyani Desai.

As a prerequisite for examining the membranes of neurological mutants, we have undertaken to fractionate and characterize membranes derived from heads of adult and whole larvae of wild type Canton-S (C-S) *Drosophila*. Of particular interest to us are membrane fractions rich in putative brain membrane marker enzyme acetylcholinesterase.

ISOLATION OF MEMBRANE FRAGMENTS

Fly Heads

Four to ten day old adult flies were frozen in liquid nitrogen and heads were separated from the body by shaking. Detached heads were separated by sieving and suspended in 0.32M sucrose in 10 mM Tris-HCl (pH 7.6). About 10,000 flies yielded 1 g of heads. These were homogenized in 10 ml of buffered sucrose at $0°-4°C$. On homogenization, the fly head capsule breaks and the sclerotized material is pelleted in a low speed centrifugation. The supernatant consists essentially of brain tissue. The fractionation scheme is shown in Figure 1. The interfaces H1 through H6 from the sucrose gradient were collected and centrifuged at 21,000xg for 90 min and the pellets obtained were resuspended in 0.32 M sucrose. These were used

Abbreviations: Tris = Tris(hydroxymethyl)aminomethane

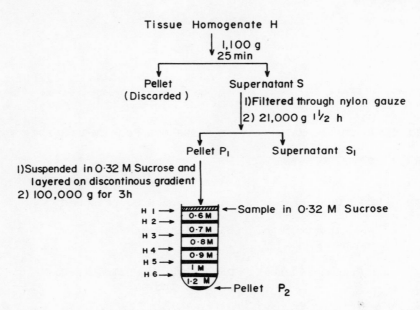

Figure 1: Scheme for isolation of membranes from fly heads.

as membrane fractions for further characterization. One gram of heads equivalent to 130 mg of protein in the homogenate yields between 1 and 2 mg membrane protein in the fractions H1 through H6.

Larvae

About 50 g of third instar larvae were washed and separated from food particles by consecutive floatation and rinsing in buffer containing sucrose. The cleaned larvae were homogenized in 500 ml of 8.5% (w/v) sucrose in 10 mM Tris-HCl (pH 7.4). The homogenate was filtered through cheese cloth. A conventional mitochondrial pellet obtained from this homogenate was resuspended in 8.5% sucrose and layered over a step gradient of 40% -25% sucrose in Tris buffer. The gradient was spun at 100,000xg for 2 hr. The interfaces were collected, pelleted and resuspended in 8.5% sucrose. This was layered over a 20%-40% linear sucrose gradient and centrifuged at 100,000xg for 2 hr. Four distinct bands were obtained after centrifugation. The fractions banding at approximate sucrose densities of 23%, 26%, 31% and 35% were labelled L1 through L4. Starting with 750 mg protein in the larval homogenate, these four fractions contained membrane protein in the range of 2 mg to 4 mg.

ENZYMATIC AND CHEMICAL CHARACTERIZATION

The profiles of enrichments for various marker enzymes in the fractions H3, H4, H5 and H6 are shown in Figure 2. Acetylcholinesterase and (Na^+-K^+)-ATPase are enriched four-fold in fractions H4 and H5, whereas Mg^{2+}-ATPase shows a two-fold enrichment in H5 alone. The activity of 5' nucleotidase is very low in fly heads. In the case of larval membranes (Figure 3) fraction L1 which bands approximately in the same sucrose density as the fraction H3 is

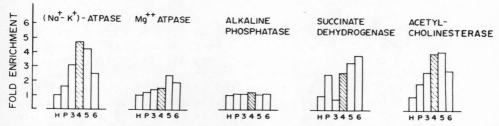

Figure 2: Enrichment profiles of marker enzymes in various fractions of the fly head. Homogenate (H), mitochondrial pellet (P), membrane fractions H3, H4, H5 and H6 (represented as 3, 4, 5, and 6) are shown. The acetylcholinesterase-rich fraction is cross hatched. Acetylcholinesterase was assayed spectrophotometrically by Ellman's[4] method. ATPase was measured by a modification of a method from Whittaker and Barker[5]. Succinate dehydrogenase was assayed as described by Arrigongi and Singer[6] and alkaline phosphatase as described by Nisman[7].

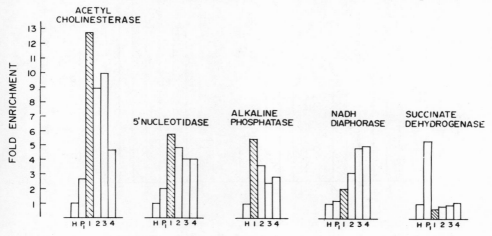

Figure 3: Enrichment profiles of marker enzymes in various fractions of the larvae. Homogenate (H), mitochondrial pellet (P), and membrane fractions L1 through L4 (represented as 1, 2, 3, and 4) are shown. The acetylcholinesterase-rich fraction is cross hatched. NADH diaphorase was assayed according to Wallach and Kamat[8] and the other enzymes as indicated in Figure 2.

enriched about 12-fold in acetylcholinesterase. Alkaline phosphatase and 5' nucleotidase are enriched in all the fractions L1 through L4, while there is a substantial reduction in the activity of succinate dehydrogenase (a mitochondrial marker enzyme). The various enzyme enrichments indicate that fly head membranes may have some mitochondrial contamination while larval membranes are likely to be contaminated by endoplasmic reticulum. Since acetylcholinesterase in *Drosophila* is known to be localized mainly in the CNS[1], the fractions H4 and H5 and L1 which are enriched in this enzyme are predominantly brain derived membranes.

Phospholipid and cholesterol contents of the various fractions from adult head and larvae are shown in Figure 4. All the membrane fractions are enriched in cholesterol and lipids. In fly heads the the relative amounts of lipids and cholesterol per mg protein is more in the lighter fractions. Preliminary electron microscopy of the various fractions combined with the difference in lipid content indicate that the separation is more on the basis of density than by size. Chromatography of the phospholipids on silica gel shows that

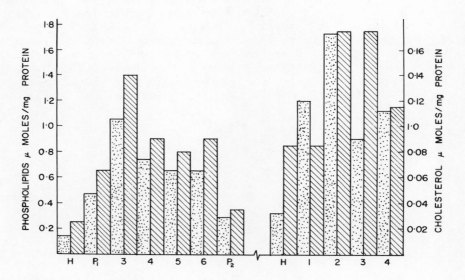

Figure 4: Phospholipid and cholesterol contents of various fractions from fly head and larvae. Cross hatched bars represent phospholipid and dotted bars cholesterol. (A) Fractions from fly heads. H, fly head homogenate, P_1 mitchondrial pellet; (3, 4, 5, 6 and P2) represent the membrane fractions. (B) Fractions from larvae. Homogenate H and membrane fractions 1, 2, 3 and 4 are shown. Phospholipid was measured by a modification of the method of Bartlett [9]. Cholesterol was estimated according to Zlatkis et al.[10].

phosphatidylcholine and phosphatidylethanolamine are the major phospholipids in the adult brain membranes. Phosphatidylserine is totally absent whereas sphingomyelin is found in trace amounts. In larval membranes the predominant lipids are phosphatidylcholine, phosphatidylethanolamine and sphingomyelin. Phosphatidyserine and cerebrosides are present in trace quantities in all the fractions L1 through L4.

Two dimensional isoelectric focusing-sodium dodecylsulfate polyacrylamide gel electrophoresis of the various fractions H3 through H6 and L1 through L4 has been performed according to the method of O'Farrell[2]. The Coomassie stained gels of the fly head fraction, Figure 5, and larval fractions, Figure 6, are shown. Over many sets of gels the patterns are reproducible. This shows that there are distinctly different proteins which are enriched in different fractions. The patterns of spots from various fractions show clear variations in intensity although the overall number of spots is about the same.

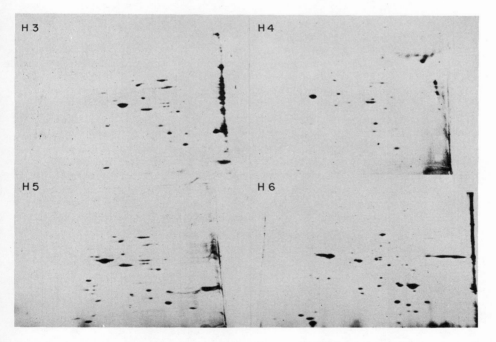

Figure 5: Protein pattern of the fly head fractions H3 through H6 as visualized by Coomassie blue staining of two dimensional polyacrylamide gels.

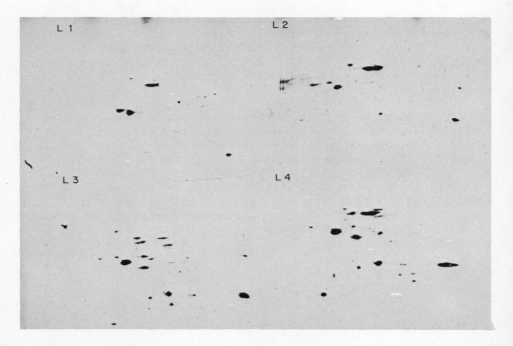

Figure 6: Protein pattern of larval fractions L1 through L4 as visualized by Coomassie blue staining of two dimensional polyacrylamide gels.

The same gels were used for visualization of glycoproteins by the method of Burridge[3] using 0.05% bovine serum albumin to reduce background. The gels were soaked in ^{125}I-concanavalin A, washed, dried and subjected to autoradiography. The autoradiographs of the fractions show many glycoprotein spots. In most cases there are more glycoprotein spots than Coomassie spots. Artists representation of the glycoprotein and Coomassie spots as well as the composite patterns for the fractions L1 and H4 which were enriched in brain derived membranes are shown in Figure 7 and Figure 8 respectively. In H4 there are 65 Concanavalin A (Con A) binding spots and 98 Coomassie stainable spots. Of these only 3 overlap, indicating that the glycoproteins are not well stained by Coomassie. In fraction L1 there are 31 Coomassie spots, 34 Con A spots and only 2 overlapping spots. In both cases a large percentage of the spots are glycoproteins. This is not due to the radioactive Con A soaking being more sensitive, since overlap of Coomassie stained and Con A bound glycoproteins is very small. Fractions H5, H6 and L2 through L4, which are not shown, have a total of about 150 spots each on two dimensional electrophoresis.

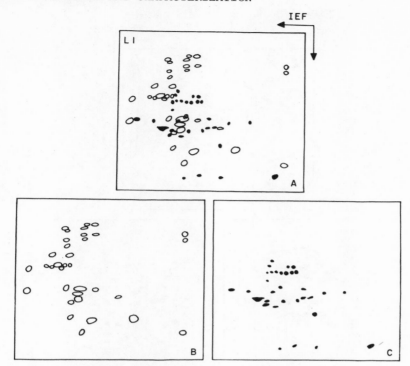

Figure 7: Artist's representation of the composite pattern (A), glycoprotein spots stained by ^{125}I-Concanavalin A (B), and Coomassie spots (C) of fraction L1 from larvae. Overlapping spots are shown by the dotted areas in the composite diagram.

Protein patterns of two dimensional gel electrophoresis followed by autoradiography of the crude membrane fractions derived from ^{35}S-labelled larvae gives about 250 spots as shown in Figure 9. This demonstrates that clearly only a small fraction of the spots have been missed by our staining procedure, and hence the two dimensional electrophoretic separation of these enriched membrane fragment proteins will be a powerful tool in the analysis of proteins of neurological mutants.

The fractionation procedure used by us is not likely to give pure membranes of specific subcellular origin. However, the results are very reproducible from one run to another with respect to the biochemical properties of the fractions. Hence the differences most likely reflect the properties of the constituents. In fraction L1 from the larvae and H4 and H5 from the adult head we have enriched surface membranes derived from the brain cells. This now makes it possible to visualize and study those components specific to nerve membrane that would otherwise be lost in the immense number of proteins present in a crude homogenate.

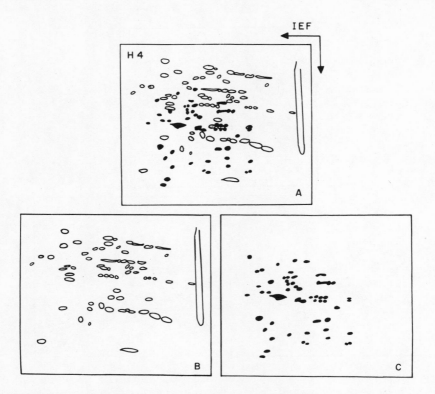

Figure 8: Artist's representation of the composite pattern (A), glycoprotein spots stained by ^{125}I-Con A (B), Coomassie spots (C) of fraction H4 from fly heads. Overlapping spots are shown by the dotted areas in the composite diagram.

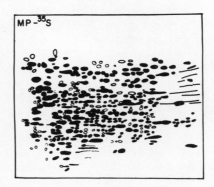

Figure 9: Artist's representation of a two-dimensional polyacrylamide gel autoradiograph of the crude membrane fraction derived from ^{35}S labelled larvae.

REFERENCES

1. J.C. Hall and D.R. Kankel, Genetics of acetylcholinesterase in *Drosophila melanogaster*, Genetics 83:517 (1976).
2. P.H. O'Farrell, High resolution two dimensional electrophoresis of proteins, J. Biol. Chem. 250:4007 (1975).
3. K. Burridge, Changes in cellular glycoproteins after transformation: identification of specific glycoproteins and antigens in sodium dodecyl sulfate, Proc. Natl. Acad. Sci. U.S.A. 73: 4457 (1976).
4. G.L. Ellman, K.D. Courtney, V. Andres, Jr., and R.M. Featherstone, A rapid spectrophotometric assay for acetylcholinesterase, Biochem. Pharmacol. 7:88 (1961).
5. V.P. Whittaker and L.A. Barker, The subcellular fractionation of brain tissue with special reference to the preparation of synaptosomes and their component organelles, in: "Methods in Neurochemistry", Vol. 2, R. Fried, ed., Marcel Dekker Inc., New York (1972).
6. O. Arrigoni and T.P. Singer, Limitations of the phenazine methosulphate assay for succinic and related dehydrogenases, Nature 193:1256 (1962).
7. B. Nisman, Techniques in demonstrating DNA-dependent protein synthesis, in: "Methods in Enzymology", Vol. XIIB, S.P. Colowick and N.O. Nathan, eds., Academic Press, New York (1968).
8. D.F.H. Wallach and V.B. Kamat, Assay of diaphorase in membrane fragments, in: "Methods in Enzymology", Vol. VIII, S.P. Colowick and N.O. Nathan, eds., Academic Press, New York (1965).
9. G.R. Bartlett, Phosphorus assay in column chromatography, J. Biol. Chem., 234:466 (1959).
10. A. Zlatkis, B. Zak, and A.J. Boyle, A new method for the direct determination of serum cholesterol, J. Lab. Clin. Med. 41:486 (1963).

PHOSPHORYLATED PROTEINS IN *DROSOPHILA* MEMBRANES

Pallaiah Thammana

Tata Institute of Fundamental Research
Bombay, India

Reversible protein modification-demodification in bacterial membranes has been shown to be an important mechanism for the adaptive behavior of bacteria in response to chemosensory stimuli.[1] It has been suggested that the protein modification mechanisms might have wider functional implications and might form the basis for an understanding of complex phenomena such as information storage and retrieval.[1,2] Phosphorylation of membrane proteins in the mammalian system is a well-documented phenomenon.[3] Greengard and coworkers have shown that phosphorylation of a set of synaptic membrane proteins, collectively known as protein I is stimulated specifically in response to cAMP and calcium.[2-4] We have explored the possibility of *in vitro* phosphorylation of proteins in membrane preparations obtained from *Drosophila* fly heads. Here we present a preliminary report of these studies.

THE METHOD OF ISOLATION OF MEMBRANES

Membranes were isolated from the fly heads of *Drosophila melanogaster* (Canton Special). Glass vials containing the flies were immersed in liquid nitrogen and the heads were dislodged from the flies by repeated freezing and shaking of the vials. The dislodged heads were recovered by sieving. Usually a head preparation of 200-400 mg (the average weight of 100 heads is 12-13 mg) was homogenized in a homogenization solution of 0.32 M sucrose in 10 mM Tris-HCl (pH 7.5) buffer containing 2 mM PMSF* to prevent proteolysis. The

---

*Abbreviations used in this text: Tris = Tris(hydroxymethyl)aminomethane; PMSF = phenylmethylsulfonylfluoride; Ac = acetate; SDS = sodium dodecylsulfate; EGTA = Ethyleneglycol-bis(β-aminoethylether) N,N'-tetracetic acid.

homogenization was done in a teflon-glass homogenizer at 4°C. The homogenate was centrifuged at 1000 x g for 15 minutes and the supernatant was centrifuged at 12,000 x g for 60 minutes. The mitochondrial pellet so obtained was the source of membranes and is referred to as membrane fraction 1 (MF1). The MF1 was stored frozen in 10 mM Tris-HCl (pH 7.5) buffer with or without 10 mM MgAc$_2$. For further fractionation, MF1 was resuspended in the homogenization solution and layered on top of a step sucrose gradient of 20-40%. The gradient was centrifuged in a Beckman SW 50.1 rotor at 33,000 rpm for 90 minutes. The interface of 20-40% sucrose containing the membrane fraction was recovered, pelleted and resuspended in a 10 mM Tris-HCl (pH 7.5) buffer with or without Mg^{2+}. This preparation is referred to as MF2. Both the membrane fractions were kept frozen in a liquid nitrogen container.

THE PHOSPHORYLATION REACTION

In vitro phosphorylation reactions were carried out with freshly thawed membrane fractions, and usually about 200 µg of membrane proteins were used in a 60 µl reaction. The reaction was carried out in 50 mM Tris-HCl (pH 7.5) buffer under the conditions indicated in the figure legends. Phosphorylation reactions were done with ^{32}P-labelled ATP (Amersham) of either low specific activity (16 C/mmole) or high specific activity (2000-3000 C/mmole). The amount of radioactivity used per reaction was 5 µC in each case. The incubation was at 23° for 20 seconds and the reaction was stopped by addition of 50 µl of SDS* stop solution. The samples were heated at 80° for 5 minutes, allowed to cool and then subjected to SDS polyacrylamide gel electrophoresis on 7.5-15% gradient gels.[6] The electrophoresis was carried out at 100V and the stained gels were dried and autoradiographed.

Figure 1 shows the results of a typical experiment of phosphorylation of proteins in a crude membrane fraction (MF1) and a fractionated membrane preparation (MF2). The experiment with the low specific activity ATP demonstrates the presence of about 14 phosphorylated bands, some of the lower molecular weight bands being very faint (lanes a-c). When phosphorylation was carried out with the high specific activity ATP, the overall phosphorylation decreased and the phosphorylation of a protein band of an apparent molecular weight of 55,000 daltons increased. In all the experiments done both with the low specific activity as well as the high specific activity ATP, no significant stimulation of phosphorylation was observed in the presence of cAMP. Presence of 1 mM 3-isobutyl-1-methyl xanthine, a phosphodiesterase inhibitor, did not lead to the stimulation of phosphorylation by cAMP. Use of as high as 1 mM cAMP was also found to be without effect. As can be seen in the lane (h) of Figure 1, the membrane fraction prepared after sucrose gradient centrifugation contains the phosphorylated proteins. The MF2 contains essentially all the phosphorylated bands with the possible exception of two bands. All the experiments shown in Figure 1 were carried out in the presence of Mg^{2+}. Virtually no phosphorylation was observed when divalent cations were omitted from the reaction.

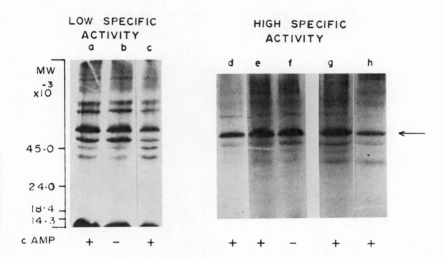

Figure 1: Effect of cAMP on protein phosphorylation in isolated membrane fractions. Membrane fractions MF1 and MF2 were incubated in the presence of 10 μM cAMP and 10 mM $MgAc_2$ as described in the text. MF1 was the membrane preparation used in lanes (a)-(g). Lane (a) is identical to lane (c) except that ethanol at a final concentration of 4% was present in lane (c). The experiment in lane (c) serves as a control for the reactions where the high specific activity ATP was employed. The high specific activity ATP obtained in 50% ethanol was used directly without prior removal of the ethanol.

The data in Figure 2 show the results of phosphorylation reactions carried out in the presence of calcium. Presence of Ca^{2+} in the incubation medium significantly stimulated the phosphorylation of a protein band termed band X (Figure 2, lane b). The band X has an apparent molecular weight of 94,000 daltons and migrates just beneath the bovine brain hexokinase marker of 98,000 daltons (Figure 3). In the presence of Ca^{2+}, a few faintly phosphorylated protein bands were also seen in addition to the band X (Figure 2). cAMP was without effect in enhancing the phosphorylation of any of the bands in the presence of Ca^{2+}. Even when the experiments were done at a lower concentration of Ca^{2+} than what is shown in Figure 2, no significant stimulation of phosphorylation by cAMP was noticed. Presence of EGTA* at a concentration sufficient to chelate the Ca^{2+}, completely reduced the Ca^{2+} mediated stimulation of phosphorylation (data not presented).

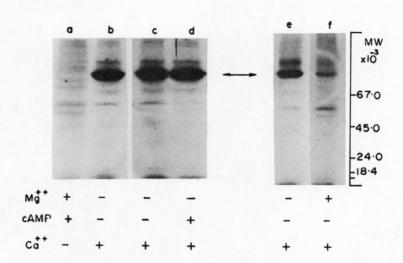

Figure 2: Effect of calcium and cAMP on protein phosphorylation. The reactions were carried out with MF1 as described in Figure 1 except 8 mM CaCl$_2$ was present where indicated. Lanes (a) and (f) contain 10 mM MgAc$_2$.

The experiment in Figure 2 lane (f) shows that the presence of Mg^{2+} counteracts the Ca^{2+} mediated enhancement of phosphorylation of band X. The decrease in phosphorylation of band X is followed by an increase in phosphorylation of a 55,000 dalton species. A further increase in Mg^{2+} concentration leads to a drastic reduction in the extent of phosphorylation of the 94,000 dalton species with a concomitant enhancement of phosphorylation of the 55,000 dalton species.

Figure 4 contains the data on the effect of variation of Ca^{2+} concentration on the phosphorylation reaction. The 94,000 dalton species is the most heavily phosphorylated protein even at 1 mM Ca^{2+} concentration. Phosphorylation of the proteins was barely detectable at 0.1 mM Ca^{2+}. It should be noted that significant phosphorylation of several bands other than the 94,000 dalton species is seen in Figure 4 (compare with Figure 2). It is also to be noted that a phosphorylated band seen above the 94,000 dalton band is missing in Figure 4. There are a few occasional differences in reaction pattern seen with different preparations. Similarly the extent of phosphorylation is qualitatively different when Figure 2 lane (a) is compared with Figure 1. The significance of these variations is not clear. These variations could result from differences in accessibility of ATP to the site of phosphorylation in different preparations.

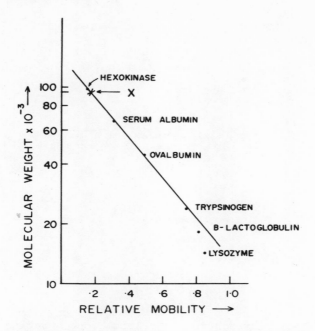

Figure 3: Determination of molecular weight of the band X by SDS polyacrylamide gel electrophoresis. The molecular weights of proteins used as standards are 14,300 (lysozyme), 18,400 (β-lactoglobulin), 24,000 (bovine trypsinogen), 45,000 (ovalbumin), 67,000 (bovine serum albumin) and 98,000 (bovine brain hexokinase).

DISCUSSION

Results presented show that there are several proteins in the membranes isolated from *Drosophila* fly heads which are phosphorylated *in vitro*. The total number was found to be close to 14 bands by one-dimensional SDS gel electrophoresis. It is not clear how many of these proteins are phosphorylated by specific kinases and if there are any multispecific kinases. Under all the experimental conditions employed there was no significant stimulation of phosphorylation in the presence of cAMP. cGMP was also without effect under the reaction conditions tested (data not shown). Ca^{2+} was found to stimulate the phosphorylation of a 94,000 dalton species, phosphorylation of which was detectable only with the high specific activity ATP. The antagonism observed in the presence of Mg^{2+} and the absence of phosphorylation in the presence of EGTA demonstrate

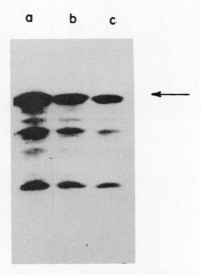

Figure 4: Effect of variation of calcium concentration on protein phosphorylation in MF1: (a) 5 mM, (b) 2 mM, and (c) 1 mM respectively.

the specificity of Ca^{2+}-mediated phosphorylation. The appearance of a 55,000 dalton phosphorylated protein band upon inclusion of Mg^{2+} in the reaction may be unrelated to the phosphorylation of a 94,000 dalton species (Figure 2). However, the data are consistent with the notion of existence of a multisubunit complex in which the extent of phosphorylation of individual components of the complex is regulated by ionic environment.

The role of the 94,000 dalton phosphorylated species is not known. A Ca^{2+}-stimulated ATPase activity responsible for calcium resorption in the muscle is known to be one of the sarcoplasmic reticulum proteins.[7,8] This enzyme of molecular weight of about 100,000 daltons has been shown to exist as a phosphorylated intermediate. Although the molecular weight of the 94,000 dalton species in our experiments is close to that of Ca^{2+} ATPase, we consider it unlikely to be a Ca^{2+} ATPase. It has been shown that there is no detectable Ca^{2+} ATPase activity in the membrane preparations made as described here (Surekha Zingde, personal communication).[5]

As mentioned earlier, the work of Greengard and his colleagues demonstrated that the phosphorylation of a set of synaptic membrane proteins of 86,000 and 80,000 daltons termed collectively as protein I, is under the regulation of Ca^{2+} and cAMP. Some of the recent experiments demonstrate that there exist multiple sites of phosphorylation of protein I which are under differential regulation by cAMP and calcium.[4] In those experiments the phosphorylation of protein I was demonstrated both in intact as well as lysed synaptosomes. The relationship between the 94,000 dalton phosphorylated species observed in our experiments and the protein I phosphorylation in the mammalian system is not clear. It remains to be seen if the 94,000 dalton species is a nervous sytem specific protein in *Drosophila*.

ACKNOWLEDGEMENT

I thank Ms. Devyani Desai for technical assistance during part of this work and Mr. Ghanshyam Swarup for a gift of purified brain hexokinase. I thank Prof. Obaid Siddiqi for his interest, encouragement, and many helpful discussions.

REFERENCES

1. M.S. Springer, M.F. Goy and J. Adler, Protein methylation in behavioral control mechanisms and in signal transduction, Nature, 280:279 (1979).
2. P. Greengard, Possible role for cyclic nucleotides and phosphorylated membrane proteins in postsynaptic actions of neurotransmitters, Nature, 260:101 (1976).
3. T. Ueda and P. Greengard, Adenosine 3':5'-monophosphate-regulated phosphoprotein system of neuronal membranes, J. Biol. Chem., 252:5155 (1977)
4. W.B. Huttner and P. Greengard, Multiple phosphorylation sites in protein I and their differential regulation by cyclic AMP and calcium, Proc. Natl. Acad. Sci., U.S.A., 76:5402 (1979).
5. T.R. Venkatesh, S. Zingde and K.S. Krishnan, Isolation and characterization of membranes from *Drosophila melanogaster*, this volume.
6. P.H. O'Farrell, High resolution two-dimensional electrophoresis of proteins, J. Biol. Chem., 250:4007 (1975).
7. A. Martonosi, R. Boland and R.A. Halpin, The biosynthesis of sarcoplasmic reticulum membranes and the mechanism of calcium transport, Cold Spring Harbor Symposia on Quantitative Biology, 32:455 (1972).
8. D.H. MacLennan, C.C. Yip, G.H. Iles and P. Seeman, Isolation of sarcoplasmic reticulum proteins, Cold Spring Harbor Symposia on Quantitative Biology, 32:469 (1972).

PHOTORECEPTOR FUNCTION

W.L. Pak, S.K. Conrad, N.E. Kremer, D.C. Larrivee[*], R.H. Schinz[**], and F. Wong[***]

Department of Biological Sciences
Purdue University
West Lafayette, Indiana 47907 U.S.A.

[*]Current Address: Department of Biology, Yale University
New Haven, Connecticut 06520
[**]Current Address: Strahlenbiologisches Institut der Universität Zürich, August Forel-Strasse 7, 8029 Zürich, Switzerland
[***]Current Address: The Marine Biomedical Institute
The University of Texas Medical Branch
Galveston, Texas 77550

For many years, we have been interested in elucidating the mechanism underlying the phototransduction process. By phototransduction we mean that process by which light signals from the environment are converted to electrical signals across the photoreceptor membrane, or, put another way, the process by which light signals modulate the permeability of the photoreceptor membrane to certain inorganic ions. The problem is of interest to many neurobiologists because the sensory transduction process is one of the basic unsolved problems in cellular neurophysiology and also because the study of phototransduction may provide insight into other neuronal excitation phenomena, all of which involve alterations in the membrane permeability to ions. In this discussion, we will forego detailed discussions of the results that have already been published (see review: Pak, 1979) and concentrate on more recent ones even though many of them are still in a preliminary form.

The phototransduction process begins with absorption of light by the visual pigment, rhodopsin. All known visual pigments, both vertebrate and invertebrate, are conjugated proteins which have as their chromophore the aldehyde of vitamin A, retinal (e.g., see

Abrahamson and Ostroy, 1967; Bridges, 1967; Wald, 1968; Morton and Pitt, 1969; Hamdorf and Rosner, 1973). In the dark-adapted state the chromophore is in the 11-*cis* form. Light isomerizes the chromophore from 11-*cis* to all-*trans* and thereby triggers the appearance of a series of intermediates that can be detected spectrophotometrically. In the case of muscoid diptera, only one pigment intermediate metarhodopsin, has been unequivocally identified (Hamdorf and Rosner, 1973; Stavenga et al., 1973; Ostroy et al., 1974; Pak and Lidington, 1974). *Drosophila* metarhodopsin is thermally stable and absorbs maximally at about 575 nm while *Drosophila* rhodopsin absorbs maximally at about 480 nm (Ostroy et al., 1974; Pak and Lidington, 1974; Harris et al., 1976).

Several lines of evidence suggest (for review and references see Pak, 1979) that photoconversion of rhodopsin to metarhodopsin, but not the reverse reaction, triggers a series of time-consuming steps (commonly referred to as the intermediate steps or process) which ultimately lead to permeability changes of the photoreceptor membranes responsible for the generation of the receptor potential. In the case of most invertebrate photoreceptors, including those of *Drosophila*, the permeability of the photoreceptor membrane increases primarily to sodium ions in response to light, leading to a depolarizing receptor potential (Kikuchi et al., 1962; Fulpius and Baumann, 1969; Millecchia and Mauro, 1969; Brown et al., 1970; Brown and Mote, 1971; Wilcox, 1980).

Available evidence is consistent with the notion that the receptor potential consists of a summation of many elementary depolarizing potential changes known as quantum bumps (Rushton, 1961; Dodge et al., 1968; Wu and Pak, 1975; Wong, 1977). Thus, a single bump corresponds to a basic, elementary unit of permeability change, resulting in an elementary unit of current flux through the ion channels. The frequency of occurrence of bumps is linearly dependent on light intensity over a wide range of intensities. As the bump frequency increases with light intensity, the individual bumps fuse to produce the receptor potential whose time course more or less parallels the time course of the stimulus.

In addition to the excitation process described above, the process controlling sensitivity (i.e. the receptor adaptation) is also thought to operate at the transduction level. Neither the mechanism of this process nor that of the excitation process has yet been elucidated.

One of the more interesting developments in the study of invertebrate photoreceptor physiology is the recent discovery of the prolonged depolarizing after-potential (PDA) (for references see Pak, 1979). The PDA is produced by a sufficiently intense and/or prolonged colored stimulus (blue for *Drosophila*) which photoconverts a sufficient net amount ($\geq$ 20%) of rhodopsin to metarhodopsin; it consists of a prolonged depolarization which persists after the stimulus offset. (The normal receptor potential which occurs during the stimulus, discussed earlier, will be referred to as the light-coincident receptor potential -LCRP). During the PDA, the photoreceptor becomes inactivated, or desensitized, to a varying degree depending on the amount of net photoconversion of rhodopsin to metarhodopsin. During a fully developed PDA, the photoreceptor is completely insensitive to further blue stimuli regardless of their intensity (Figure 1). Both the depolarization and desensitization accompanying the PDA are reversed by an orange stimulus that reconverts metarhodopsin back to rhodopsin (Figure 1). The PDA is particularly well developed in *Drosophila* and can persist for over an hour after the stimulus offset in this animal.

The PDA has been found in a wide variety of invertebrate photoreceptors (for references see Pak, 1979). A phenomenon similar to a PDA has even been found in the photoreceptors of the pineal organ of the frog (Eldred and Nolte, 1978). Moreover, the same basic membrane conductance mechanism appears to be responsible for both the light-coincident receptor potential and the PDA. For example, it has been shown that both potentials consists of bumps (Minke et al.,

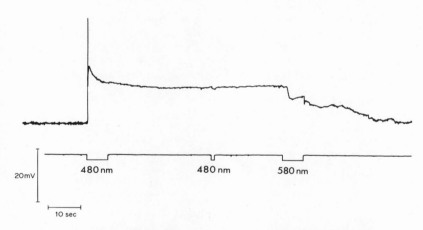

Figure 1: The prolonged depolarizing afterpotential (PDA) recorded intracellularly from an R1-6 receptor cell of a "wild-type" fly using an intense ($\sim 7 \times 10^{15}$ photons cm^{-2} sec^{-1}), blue (maximum transmission at 480 nm) stimulus. The fly had the screening pigments in its eye genetically removed using the mutation white (*w*). A blue stimulus of the same intensity delivered during the maximally developed PDA produces no further depolarizing potential; i.e. the cell is inactivated. An orange (580 nm) stimulus of comparable intensity repolarizes the cell. (Recording obtained by M. Wilcox; reproduced with permission from Elsevier North Holland, Inc. (Pak, 1979)).

1975b) and that both depend on sodium conductance increase (Brown and Cornwall, 1975; Wilcox, 1980). The physiological significance of the PDA is not yet clear. However, it may well be a form of receptor adaptation, i.e. the sensitivity controlling mechanism of the photoreceptor mentioned earlier.

The PDA process differs from the well-studied receptor adaptation process of vertebrate photoreceptors (e.g., see Dowling and Ripps, 1972; Grabowski and Pak, 1975; Engbretson and Witkovsky, 1978) mainly in its characteristic dependence on stimulus wavelength for its initiation, termination, and/or reversal. However, the peculiar color dependence may simply be a consequence of the thermostability of metarhodopsin and the relative separation of the metarhodopsin absorption peak from the rhodopsin absorption peak in animals displaying the PDA process. Thus, in these animals photoreversal of rhodopsin from metarhodopsin can readily take place during the time course of receptor dark adaptation. It may be that such photoreversal also leads to restoration of receptor sensitivity.

For the purpose of dissecting the phototransduction process, we have isolated to date over 80 independently arising mutations that affect this process. These mutations fall into the 12 complementation groups listed in Table 1.

Of these, the *norpA* and *trp* mutants have been studied in detail by us (Pak et al., 1976; Minke et al., 1975b) and the *rdgB* mutant by Harris and Stark (1977). We will give only summary statements of these results in this paper, concentrating instead on the recent results obtained from the "PDA-defective mutants" to be described later.

Table 1

Mutation (Symbol)	Name	Map Position
norpA	no receptor potential A	1– 6.5 ± 0.7
rdgB	retinal degeneration B	1–42.7 ± 0.7
trp	transient receptor potential	3–∼100
inaA through *inaF*	inactivation, but no afterpotential	5 complementation groups
ninaA through *ninaD*	neither inactivation nor afterpotential	4 complementation groups

The *norpA* mutants were the first phototransduction-defective mutants to be isolated by means of chemical mutagenesis programs (Hotta and Benzer, 1970; Pak et al., 1970; Heisenberg, 1971). There are now over 40 independent isolates of *norpA* in our laboratory. These fall into several phenotypic classes. The flies either hemizygous or homozygous for one of the more severe alleles of *norpA* are completely blind behaviorally (T. Markow, personal communication) and completely lack the photoreceptor potential (Alawi et al., 1972).

The *trp* mutant first appeared spontaneously in a highly inbred line of a wild-type strain (Cosens and Manning, 1969) and was isolated on the basis of its behavioral phenotype. (Our mutagenesis program has since yielded several *trp* alleles.) The *trp* mutants behave normally under low ambient light conditions but behave as though blind when the light level is raised (Cosens and Manning, 1969). Cosens and Manning (1969) and Cosens (1971) showed that the electroretinogram (ERG) of *trp* appears very nearly normal if a dim stimulus is used but decays to baseline during a strong, prolonged stimulus of any wavelength. Subsequently, Minke et al. (1975b) by means of intracellular recording demonstrated that the mutation, indeed, affects the photoreceptor potential.

Results obtained in our laboratory (Pak et al., 1976; Minke et al., 1975b; Pak, 1979) showed that in both *norpA* and *trp* the visual pigment, rhodopsin, has normal absorption characteristics and makes a normal phototransition to metarhodopsin. Moreover, in both cases the individual quantum bumps appear normal in amplitude and duration suggesting that the mutations do not affect the machinery of quantum bump production (i.e. the ion channels) itself. Thus, we suggested that both *norpA* and *trp* affect the intermediate steps of phototransduction. Analyses of quantum bumps suggested that the two mutations tend to alter different sets of bump parameters (rate of occurrence versus latency dispersion), leading to the conclusion that the two mutations affect different steps in the intermediate process of phototransduction. On the basis of studies with a temperature-sensitive allele of *norpA* (Pak et al., 1976), we further suggested that the effect of the *norpA* mutation is mediated through a defective polypeptide, probably the *norpA* gene product itself, and that the polypeptide is probably not the visual pigment.

The *rdgB* mutant was first described by Hotta and Benzer (1970) and studied in detail by Harris and Stark (1977). The *rdgB* gene, when defective, causes age-dependent degeneration of the photoreceptor layer. The degeneration is light-induced and tends to affect the peripheral (R1-6) photoreceptors much more severely than the central photoreceptors, R7 and R8. The most interesting finding of Harris and Stark (1977), however, is that the suppressor mutations of *rdgB* that they isolated were all alleles of *norpA*, one of them being allele-specific. That is, one particular allele of *norpA* reduces the mutant phenotype of one particular allele of *rdgB*. While suppression of a mutant phenotype can occur in a number of different ways (see Gorini and Beckwith, 1966), Harris and Stark (1977) argued that, in this particular case, suppression may indicate a direct interaction between the *norpA* and *rdgB* gene products because of allele specificity. In other words, the defect in the product of the particular *norpA* allele specifically compensates for the defect in the product of the particular *rdgB* allele. Their findings thus implicated the *rdgB* gene in the phototransduction process since the *norpA* gene had previously been shown to be involved in the process.

An important recent development in our research program has been the isolation of the "PDA-defective mutants". These mutations affect the PDA very conspicuously. While they also affect the LCRP, the effect is often not very obvious. The "PDA-defective mutants" fall into two classes (*ina* and *nina*) on the basis of their electrophysiological phenotype, as listed in Table 1 and illustrated in Figure 2. As has already been described, the PDA process in wild-

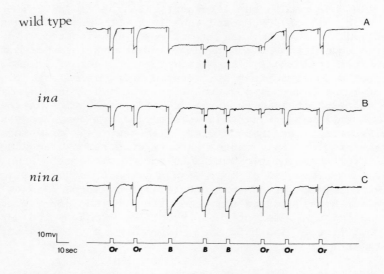

Figure 2: ERG recordings of the PDA obtained from the compound eyes of a wild-type fly (A) and the *inaA*^{P226} (B) and *ninaA*^{P228} (C) mutants. All three strains were placed on the white (*w*) background to eliminate the screening pigments in the eye. The stimuli were applied in the following sequence: two orange stimuli, three blue stimuli, and finally three orange stimuli. The illuminances of the blue and orange stimuli were approximately 5×10^{15} and 3×10^{16} photons cm^{-2} sec^{-1}, respectively. In the case of the wild-type fly

Figure 2 (cont.): (A), the initial two orange stimuli evoke only the light coincident photoreceptor potential (LCRP). The first intense blue stimulus (third stimulus) induces a fully developed afterpotential, during which further intense blue stimuli (fourth and fifth stimuli) fail to evoke responses from the R1-6 cells. These receptor cells are desensitized (or inactivated). The small responses that are observed (arrows) originate from the R7 and R8 cells (Minke et al., 1975a), in which the PDA is not induced by an intense blue stimulus. An orange stimulus (sixth stimulus) presented during the fully developed afterpotential repolarizes and reactivates the R1-6 cells, and these cells become responsive to subsequent orange stimuli (last two stimuli). In the case of the $inaA^{P226}$ mutant (B), the first intense blue stimulus (third stimulus) fails to produce the afterpotential. Yet, the R1-6 photoreceptors are inactivated, and only the R7 and R8 cells respond to the second and third blue stimuli (small responses indicated by the arrows). An intense orange stimulus reactivates the R1-6 cells, and R1-6 become responsive to orange (last two stimuli) or moderate blue stimuli once again. In the case of the $ninaA^{P228}$ mutant (C), neither the afterpotential nor inactivation is apparent. Each stimulus produces only the LCRP regardless of previous exposure to blue stimuli. (Reproduced with permission from Nature (Lo and Pak, 1978) and Elsevier North Holland, Inc. (Pak, 1979).)

type flies displays both a prolonged depolarization and desensitization (or inactivation) of the photoreceptor, both of which persist long after the stimulus is turned off (Figures 1 and 2A). In the case of the *ina* (inactivation but no afterpotential) class of mutants, however, the afterpotential (i.e. the prolonged depolarization) is either small or totally absent following an intense blue stimulus. Nevertheless, the *ina* photoreceptors are inactivated (or desensitized) by the stimulus and recover their former sensitivity only after they have been exposed to an orange stimulus (Figure 2B). In the case of the *nina* (neither inactivation nor afterpotential) class of mutants, neither the afterpotential nor desensitization of the photoreceptors is evident (Figure 2C).

From results such as the above, we have suggested that the PDA process consists of at least two distinct steps, the desensitization (or inactivation) step and the depolarization (or afterpotential) step (Pak, 1979). Our results to date are consistent with the notion that these two steps occur sequentially with the inactivation step preceding the afterpotential step.

We have determined the amount of rhodopsin contained in the R1-6 photoreceptors of some of the PDA-defective mutants by spectrophotometry of both the deep pseudopupil (Franceschini, 1972) *in vivo* and of digitonin eye extracts *in vitro*. All *ina* mutants tested had nearly normal R1-6 rhodopsin levels, while the *nina* mutants tested all showed considerably reduced levels. In the case of *inaA* and *ninaA*, the mutants we chose to represent the two classes, the R1-6 rhodopsin levels were approximately 80% and 10% of the wild-type level, respectively. The fact that the rhodopsin level is reduced in the *nina* mutants suggested that the *nina* genes might have something to do with either rhodopsin synthesis or its incorporation into the membrane. We have, therefore, decided to examine the *nina* mutants more closely in the hope of learning more about rhodopsin - the only protein unequivocally known to be involved in phototransduction so far.

One of the first problems to which the *nina* mutant were put to use was in the determination of the fractions of rhodopsin and non-rhodopsin proteins present in the rhabdomeric membrane. Since one would expect the photoreceptor membrane to contain many different kinds of proteins besides rhodopsin (e.g., ion channel proteins, triphosphonucleotide hydrolases, nucleotide cyclases, protein kinases, etc.), it is of interest to know whether the rhabdomeric membrane contains primarily rhodopsin or a substantial amount of nonrhodopsin proteins as well. Freeze-fracture electron microscopy of the compound eyes of the *nina* mutants showed that the intramembrane particles in the rhabdomeric membrane are significantly reduced in density (see Table 2). Since the intra membrane proteins are known to show up as membrane particles in freeze-fracture EM (Yu and Branton, 1976; Edwards et al., 1979), the above observations together with the observation that these mutants display reduced pigment levels suggested that a significant fraction of the protein in the rhabdomeric membrane is rhodopsin. Boschek and Hamdorf (1976) and Harris et al. (1977) had made a similar suggestion previously on

Table 2. Comparison of Relative Rhabdomeric Membrane Particle Density With Relative Rhodopsin Concentration

	Membrane Particle Density (% wild type)	Rhodopsin Concentration (% wild type)
wild-type (Oregon-R)	100	100
ninaC	70±14	35±3
ninaA	47±10	11±3
vitamin A$^-$	32± 8	< 3

the basis of their studies of vitamin A deprived *Calliphora* and *Drosophila* respectively. However, the relative amount of rhabdomeric particles remaining in the *nina* mutants or vitamin A deprived wild-type flies was consistently higher than the relative amount of rhodopsin remaining in the corresponding group of flies when these two parameters were normalized to those of wild-type for comparison (Table 2). In the case of vitamin A deprived wild-type flies, Harris et al. (1977) also showed that there appeared to be more membrane particles than could be accounted for by the visual pigment molecules that remain. The simplest interpretation of the results would be that there are both rhodopsin and nonrhodopsin proteins in the rhabdomeric membrane and that the *nina* mutations or vitamin A deprivation affect primarily the rhodopsin protein.

As had been pointed out by Harris et al. (1977) for vitamin A deprived flies, however, the rhabdomeric membrane particles that occur in excess of the rhodopsin level could be due to opsin molecules lacking the retinal chromophore. These molecules would contribute to the rhabdomeric membrane particle population but not to the spectrophotometrically detectable rhodopsin population. The above hypothesis was tested by performing SDS gel electrophoresis of digitonin eye extracts on all four classes of flies studied. Since Ostroy (1978) had previously identified the R1-6 opsin peak in SDS gels, it was possible to estimate the relative amount of R1-6 opsin (present in both the chromophore-free opsin form and the chromophore-containing rhodopsin form) from the above measurements. The results are summarized in Table 3 as fractions of the wild-type opsin content and are compared with spectrophotometrically determined

Table 3. Relative Concentration of Rhodopsin and Opsin

	Rhodopsin (% wild type)	Opsin (% wild type)
wild-type	100	100
ninaC	35±3	32± 7
ninaA	11±3	15±10 (1) < 13 (2)
vitamin A⁻	< 3	4± 8 (1) < 13 (2)

(1) and (2) refer to two different methods of estimating the opsin level (Larrivee, 1979)

rhodopsin content, also expressed as fractions of the wild-type rhodopsin level. It may be seen that the relative opsin level closely corresponds with the relative rhodopsin level for each class of flies studied (Table 3) and is substantially lower than the relative rhabdomeric membrane particle density (see Table 2). Recently, Paulsen and Schwemer (1979) have also shown that vitamin A deprivation affects the opsin concentration in the rhabdomeric membranes of the blowfly *Calliphora*. The results suggest: (1) that opsin molecules that lack the retinal chromophore apparently do not get incorporated into the rhabdomeric membrane to any significant extent and (2) that while the majority of proteins in the rhabdomeric membrane is rhodopsin, as much as 30-40% of these proteins may be something other than either opsin or rhodopsin.

The results obtained up to this point were consistent with the notion that the *nina* mutations simply mimic the effect of vitamin A deprivation. It soon became clear, however, that the effect of the mutation, at least in the case of *ninaA* (the only *nina* mutation to be studied in detail so far) is much more specific than that of vitamin A deprivation.

A major difference between *ninaA* and vitamin A deprived wild-type flies became apparent when the spectral sensitivities of their ERGs were compared. The spectral sensitivity of wild-type flies reared on normal media typically displays two sensitivity peaks, one at $\sim$ 480 nm and another in the near UV. Vitamin A deprivation depresses the overall sensitivity by about 2 log units throughout visible wavelengths (Goldsmith et al., 1964). Its effect, however is much more pronounced on the UV peak than on the 480 nm peak (Goldsmith et al., 1964; Stark et al., 1977) so that only one sensitivity peak, occurring at 480 nm, is apparent in vitamin A deprived flies (Stark et al., 1977). By contrast, the *ninaA* mutation, while also depressing the overall sensitivity, had no such selective effect on the UV peak. Thus, the shape of the *ninaA* spectral sensitivity curve was very similar to that of the wild-type curve except that the entire curve was shifted by $\sim$ 1 log unit in the direction of lower sensitivity. In order to compensate for the fact that the overall sensitivity of *ninaA* is higher than that of completely vitamin A deprived flies by about one log unit, we have examined the spectral sensitivites of partially deprived flies having overall sensitivities similar to that of *ninaA*. These flies also display a much reduced UV peak compared to either *ninaA* or wild type.

Kirschfeld et al. (1977) have explained the UV peak in the fly spectral sensitivity in terms of a UV-absorbing, photostable, sensitizing pigment of carotenoid origin. That is, the UV peak arises as a result of light energy absorbed in the UV by this carotenoid pigment and transferred to rhodopsin by some unspecified mechanism(s). Under this hypothesis, one would expect vitamin A deprivation to have a stronger effect on the UV peak than the 480 nm peak because (1) the former, unlike the latter, depends on the integrity of both the sensitizing pigment and rhodopsin and (2) vitamin A deprivation affects both these pigments. If one grants the basic validity of this hypothesis, the spectral sensitivity of *ninaA* may be explained by supposing that the mutation specifically affects rhodopsin and not, for example, the availability of carotenoids to the photoreceptor or the ability of the photoreceptor to utilize the supplied carotenoids as the retinal chromophore or the sensitizing pigment. The results thus suggest that the *ninaA* mutation might somehow affect the protein moiety of rhodopsin.

The above notion received a strong boost from freeze-fracture electron microscopy of the *ninaA* compound eye. The freeze-fracture data showed unambiguously that the observed reduction in the rhabdomeric membrane particle density in the mutant (see Table 2) is confined to the R1-6 photoreceptors (Larrivee, 1979; Schinz et al., 1980). Both the R7 and R8 rhabdomeres were found to have normal membrane particle concentrations, suggesting that the rhodopsin concentrations in the R7 and R8 cells are normal. In fact, R.S. Stephenson of our laboratory has shown recently that an apparently normal PDA can be generated from the R7 photoreceptors of *ninaA*. This may be taken as presumptive evidence that the rhodopsin concentration is normal in R7 and that the R7 photoreceptors are functionally normal as far as the PDA generation is concerned. By contrast, vitamin A deprivation has been found to affect all photoreceptors (Harris et al., 1977). Thus, the mutation *ninaA* appears to affect one specific class of opsin, i.e., the protein moiety of rhodopsin present in the R1-6 photoreceptors.

The most obvious possibility for genes that affect a specific class of protein is the structural gene for that protein. We have, therefore, carried out a gene dosage experiment to test the hypothesis that *ninaA* is the structural gene for R1-6 opsin. First, recombination and deletion mapping experiments were carried out to place *ninaA* between aristaless (2-0.01) and dumpy (2-13.0) and within the limits of the deficiency *Df(2L)S3* (see Lindsley and Grell, 1968). Using the available deficiency and translocation stocks, it was then possible to synthesize flies carrying one, two or three copies of the wild-type alleles of *ninaA*, i.e. *ninaA$^+$*. We then carried out spectrophotometric measurements on the deep pseudopupil of these flies to determine the amount of absorption changes at 578 nm when the visual pigment was photoconverted back and forth using blue and red bleaching light. The magnitude of these absorption changes was taken as a measure of rhodopsin concentration. The results are summarized in Table 4. It is readily seen that rhodopsin concentration is not dependent on dosage of *ninaA$^+$* allele. However, unlike water soluble proteins, the effect of structural gene dosage on the amount of membrane proteins is poorly understood. Thus, the above evidence alone does not seem sufficient to exclude the possibility that *ninaA* is the structural gene for R1-6 opsin

Table 4. Dependence of Rhodopsin Concentration of $ninaA^+$ Dosage

$ninaA^+$ dosage	no. flies	ΔA_{578}
1	6	.137±.013
2	4	.116±.012
3	5	.102±.004

A more crucial test for the structural gene of R1-6 opsin would be to see whether ninaA alleles produce electrophoretic variants of R1-6 opsin. This test has not yet been carried out because to date only one ninaA allele has been isolated. In the absence of the latter test, the possibility that ninaA is the structural gene for R1-6 opsin cannot yet be ruled out. If ninaA turns out to be the structural gene for R1-6 opsin, it should be possible to isolate more alleles of ninaA and to probe the functional properties of rhodopsin using these alleles. Since the functional properties of rhodopsin remain largely unknown, any technique that allows molecular alteration of opsin structure should prove immensely valuable. Even if ninaA turns out not to be the structural gene for R1-6 opsin, the ninaA mutation nonetheless would be of interest to us because of its receptor-class specificity. For example, it may turn out to be of value in elucidating any receptor-class-specific steps that may be involved in the synthesis of rhodopsin and in its incorporation into the photoreceptor membrane.

ACKNOWLEDGEMENTS

This work was supported by grants from NIH (EY00033) and NSF (BNS 77-18647). D.C. Larrivee obtained support from an NIH Institutional National Research Service Award (EY07008) to Purdue University.

REFERENCES

Abrahamson, E.W., and Ostroy, S.E., 1967, The photochemical and macromolecular aspects of vision, Prog. Biophys. Mol. Biol., 17:179.

Alawi, A.A., Jennings, V., Grossfield, J., and Pak, W.L., 1972, Phototransduction mutants of Drosophila melanogaster, in: "The Visual System: Neurophysiology, Biophysics, and Their Clinical Applications", G.B. Arden, ed., Plenum Press, New York, pp. 1-21.

Boschek, C.B., and Hamdorf, K., 1976, Rhodopsin particles in the photoreceptor membrane of an insect, Z. Naturforsch., 31c:763.

Bridges, C.D.B., 1967, Biochemistry of visual processes, in: "Comprehensive Biochemistry, Vol. 27", M. Florkin and E.H. Stotz, eds., Elsevier, Amsterdam, pp. 31-78.

Brown, H.M., and Cornwall, M.C., 1975, Ionic mechanism of a quasi-stable depolarization in barnacle photoreceptor following a red light, J. Physiol., 248:579.

Brown, H.M., Hagiwara, S., Koike, H., and Meech, R.M., 1970, Membrane properties of a barnacle photoreceptor examined by the voltage-clamp technique, J. Physiol., 208:385.

Brown, J.E., and Mote, M., 1971, Na^+ dependence of reversal potentials of light-induced current in Limulus ventral photoreceptors, Biol. Bull. 141:379.

Cosens, D., 1971, Blindness in a Drosophila mutant, J. Insect Physiol., 17:285.

Cosens, D., and Manning, A., 1969, Abnormal electroretinogram from a Drosophila mutant, Nature, 224:285.

Dodge, F.A., Jr., Knight, B.W., and Toyoda, J., 1968, Voltage noise in Limulus visual cells, Science, 160:88.

Dowling, J.E., and Ripps, H., 1972, Adaptation in skate photoreceptors, J. Gen. Physiol., 60:698.

Edwards, H.H., Mueller, T.J., and Morrison, M., 1979, Distribution of transmembrane polypeptides in freeze fracture, Science, 203:1343.

Eldred, W.D., and Nolte, J., 1978, Pineal photoreceptors: evidence for a vertebrate visual pigment with two physiologically active states, Vision Res., 18:29.

Engbretson, G.A., and Witkovsky, P., 1978, Rod sensitivity and visual pigment concentration in Xenopus, J. Gen. Physiol., 72:801.

Franceschini, N., 1972, Pupil and pseudopupil in the compound eye of Drosophila, in: "Information Processing in the Visual System of Arthropods", R. Wehner, ed., Springer-Verlag, New York, pp. 75-82.

Fulpius, B., and Baumann, F., 1969, Effects of sodium, potassium and calcium ions on slow and spike potentials in single photoreceptor cells, J. Gen. Physiol., 53:541.

Goldsmith, T.H., Barker, R.J., and Cohen, C.F., 1964, Sensitivity of visual receptors of carotenoid-depleted flies: a vitamin A deficiency in an invertebrate, Science, 146:65.

Gorini, L., and Beckwith, J.R., 1966, Suppression, Annu. Rev. Microbiol., 20:401.

Grabowski, S.R., and Pak, W.L., 1975, Intracellular recordings of rod responses during dark-adaptation, J. Physiol., 247:363.

Hamdorf, K., and Rosner, G., 1973, Adaptation und Photoregeneration im Fliegenauge, J. Comp. Physiol., 86:281.

Harris, W.A., and Stark, W.S., 1977, Hereditary retinal degeneration in Drosophila melanogaster: a mutant defect associated with the phototransduction process, J. Gen. Physiol., 69:261.

Harris, W.A., Stark, W.S., and Walker, J.A., 1976, Genetic dissection of the photoreceptor system in the compound eye of Drosophila melanogaster, J. Physiol., 256:415.

Harris, W.A., Ready, D.F., Lipson, E.D., and Hudspeth, A.J., and Stark, W.S., 1977, Vitamin A deprivation and *Drosophila* photopigments, Nature, 266:648.

Heisenberg, M., 1971, Isolation of mutants lacking the optomotor response, Dros. Inf. Serv., 46:68.

Hotta, Y., and Benzer, S., 1970, Genetic dissection of the *Drosophila* nervous system by means of mosaics, Proc. Natl. Acad. Sci. U.S.A., 67:1156.

Kikuchi, R., Naito, K., and Tanaka, I., 1962, Effects of sodium and potassium ions on the electrical activity of single cells in the lateral eye of the horseshoe crab, J. Physiol., 161:319.

Kirschfeld, K., Franceschini, N., and Minke, B., 1977, Evidence for a sensitizing pigment in fly photoreceptors, Nature, 269:386.

Larrivee, D.C., 1979, A biochemical analysis of the *Drosophila* rhabdomere and its extracellular environment, Ph.D. thesis, Purdue University.

Lindsley, D.L., and Grell, E.H., 1968, Genetic variations of *Drosophila melanogaster*, Carnegie Institute of Washington, Washington, D.C.

Lo, M.-V.C., and Pak, W.L., 1978, Desensitization of peripheral photoreceptors shown by blue-induced decrease in transmittance of *Drosophila* rhabdomeres, Nature, 273:772.

Millecchia, R., and Mauro, A., 1969, The ventral photoreceptor cells of *Limulus* II. The basic photoresponse, J. Gen. Physiol., 54:310.

Minke, B., Wu, C.-F., and Pak, W.L., 1975a, Isolation of light-induced response of central retinula cells from the electroretinogram of *Drosophila*, J. Comp. Physiol., 98:345.

Minke, B., Wu, C.-F., and Pak, W.L., 1975b, Induction of photoreceptor voltage noise in the dark in *Drosophila* mutant, Nature, 258:84.

Morton, R.A., and Pitt, G.A.J., 1969, Aspects of visual pigment research, in: "Advances in Enzymology, Vol. 32", F.F. Nord, ed., Wiley and Sons, New York, pp. 97-171.

Ostroy, S.E., 1978, The characteristics of *Drosophila* rhodopsin in wild type and *norpA* vision transduction mutants, J. Gen. Physiol., 72:717.

Ostroy, S.E., Wilson, M., and Pak, W.L., 1974, *Drosophila* rhodopsin: photochemistry, extraction and differences in the $norpA^{P12}$ phototransduction mutant, Biochem. Biophys. Res. Comm., 59:960.

Pak, W.L., 1979, Study of photoreceptor function using *Drosophila* mutants, in: "Neurogenetics: Genetic Approaches to the Nervous System",X.O. Breakefield, ed.,Elsevier North Holland, New York.

Pak, W.L., and Lidington, K.J., 1974, Fast electrical potential from a long-lived, long-wavelength photoproduct of fly visual pigment, J. Gen. Physiol., 63:740.

Pak, W.L., Grossfield, J., and Arnold, K., 1970, Mutants of the visual pathway of *Drosophila melanogaster*, Nature, 227:518.

Pak, W.L., Ostroy, S.E., Deland, M.C., and Wu, C.-F., 1976, Photoreceptor mutant of *Drosophila*: Is protein involved in intermediate steps of phototransduction?, Science, 194:956.

Paulsen, R., and Schwemer, J., 1979, Vitamin A deficiency reduces the concentration of visual pigment protein within blowfly photoreceptor membranes, Biochim. Biophys. Acta, 557:385.

Rushton, W.A.H., 1961, The intensity factor in vision, in: "Light and Life", W.D. McElroy and B. Glass, eds., Johns Hopkins Press, Baltimore, pp. 706-723.

Schinz, R.H., Lo, M.-V.C., Larrivee, D.C.,and Pak, W.L., 1980, Freeze-fracture study of insect photoreceptor membrane. II. Drosophila mutations affecting the membrane microstructure, Submitted to J. Cell Biol.

Stark, W.S., Ivanyshyn, A.M., and Greenberg, R.M., 1977, Sensitivity and photopigments of R1-6, a two peaked photoreceptor in Drosophila, Calliphora and Musca, J. Comp. Physiol., 121:289.

Stavenga, D.G., Zantema, A., and Kuiper, J.W., 1973, Rhodopsin processes and the function of the pupil mechanism in flies, in: "Biochemistry and Physiology of Visual Pigments", H. Langer, ed., Springer-Verlag, New York, pp. 175-180.

Wald, G., 1968, Molecular basis of visual excitation, Nature, 219: 800.

Wilcox, M.J., 1980, Ionic mechanism of the receptor potential in the photoreceptors of wild-type and mutant Drosophila, Ph.D. thesis, Purdue University.

Wong, F., 1977, Mechanisms of the phototransduction process in invertebrate photoreceptors, Ph.D. thesis, Rockefeller University.

Wu, C.-F., and Pak, W.L., 1975, Quantal basis of photoreceptor spectral sensitivity of Drosophila melanogaster, J. Gen. Physiol., 66:149.

Yu, J., and Branton, D., 1976, Reconstitution of intramembrane particles in erythrocyte band 3-lipid recombinants: effects of spectrin-actin association, Proc. Natl. Acad. Sci. U.S.A., 73:3891

DISCUSSION

Y. Hotta: Have you done any more freeze-fracture?

W. Pak: Yes, for most of the ina mutants, but there are no differences from wild-type.

A.S. Mukherjee: What is the effect of vitamin A on ninaA?

W. Pak: After deprivation of vitamin A, the particles still remain.

J.C. Hall: If ninaA affects rhodopsin, why do you expect a decrease in rhodopsin would suppress the effects of a no-receptor-potential mutation, in your experiment showing that the double mutant norpA; ninaA is no longer blind.

W. Pak: I wish I had not told you that. We haven't published it yet in any case. For the sake of argument, though, if *ninaA* produced rhodopsin (though this is unlikely), that protein could interact with the *norpA* gene-product in the normal fly and in the double mutant that would have a partially restored molecular interaction.

GENETIC ANALYSIS OF A COMPLEX CHEMORECEPTOR

Obaid Siddiqi and Veronica Rodrigues

Molecular Biology Unit
Tata Institute of Fundamental Research
Bombay, India

INTRODUCTION

Taste responses of Dipterans have been extensively investigated for more than three decades. V.G. Dethier (1955) used the proboscis extention test to measure the natural preferences of the flies to different chemicals. He found that blowflies accepted sugars and rejected common salt or quinine. If a mixture of an attractant and a repellent was presented, the fly's response was determined by the ratio of the two substances. Hodgson, Lettvin and Roeder (1955) first recorded the electrical responses of the Dipteran chemoreceptors. Since then a number of workers have attempted to correlate the electrophysiological and behavioral responses of a variety of flies. (See review by Hodgson, 1974.) It is evident that taste discrimination in insects, as in mammals, is based upon a small set of categoric distinctions, but the exact nature of the sensory code remains unknown (Perkel and Bullock, 1969).

In recent years several groups have isolated gustatory mutants of *Drosophila melanogaster*, which are deficient in their responses to quinine, salt or sugars (Isono and Kikuchi, 1974; Falk and Atidia, 1975; Tompkins and Sanders, 1977; Tompkins et al., 1979; Rodrigues and Siddiqi, 1978). The reasons for our interest in gustatory mutants are two-fold. A correlated study of neurophysiological and behavioral block is likely to throw light on the functional organization of the chemosensory system; secondly, and perhaps more important, working with genetic alterations of a part of *Drosophila* nervous sytem whose elements can be identified, we might hope to learn something about the manner in which genes specify the organization of complex neural networks. In this paper we present the results of behavioral and electrophysiological studies on a set of X-linked gustatory mutants of *Drosophila melanogaster*.

TASTE RECEPTORS OF *DROSOPHILA*

The gustatory receptors of *Drosophila* are sensilla trichoidea located on tarsal segments and the labellum, and sensilla basiconica on the labellum. Falk et al. (1976) have described the structure of these chemoreceptors. About 40 sensory bristles are arranged in four, somewhat irregular, rows on each half labellum. Each sensillum projects from a pit which contains 5 sensory neurons. Four of these are chemosensory cells whose dendrites run to the tip of the hair; the fifth is a mechanosensory neuron connected to the base (Figure 1). The tips of the bristles are bifid but the fine-structure of the pore and the dendritic organization remain to be analyzed.

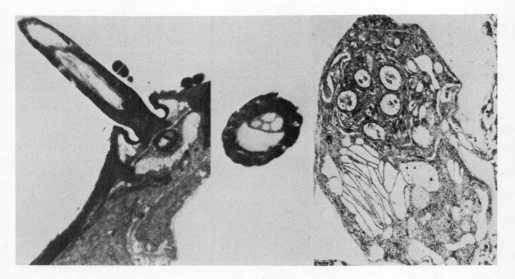

Figure 1: Labellar chemoreceptor of *Drosophila*. Left, longitudinal section through the hair X9200; center, dendrites of 4 chemosensory cells X9200; right, sensory neurons in the receptor x6700. Micrographs by R.N. Singh and Mercy Joseph.

The above description applies to a majority of the chemoreceptors, but exceptions are frequent. There are hairs with 2, 3, 4 or more than 5 neurons. (R.N. Singh and M. Joseph, unpublished work). We do not yet know whether such atypical hairs are randomly distributed or occupy characteristic positions. More than 70 percent of the labellar hairs give closely similar responses and the descriptions of the electrophysiology in the following sections refer to such "typical hairs".

A COMPLEX CHEMORECEPTOR

RESPONSES OF NORMAL FLIES

Taste responses of *Drosophila* can be examined by the proboscis extention test developed by Dethier for larger flies. If the chemosensory bristles on legs or lips are touched with a solution of sucrose, the fly extends its proboscis. By observing proboscis extention against varying concentrations of a stimulant, one can quantitatively measure the reactions of the fly to solutions of chemicals (Deak, 1976).

The response of the wild type strain Canton Special (CS) to a set of sugars is shown in Figure 2. The threshold for detection of sucrose is less than 10^{-5} M, three orders of magnitude lower than the threshold for humans. The fly is nearly equally responsive to maltose, glucose and fructose (Isono and Kikuchi, 1974) and much less so to mannose and several other sugars such as galactose, lactose and xylose (not shown in the graph). Note that the six carbon sugars which elicit a strong response include three pyranoses and a furanose.

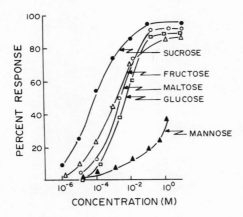

Figure 2: Proboscis extention response of normal flies. Each fly was tested ten times. The points represent averaged responses of 50 flies.

The inhibition of proboscis extention by NaCl and quinine is illustrated in Figure 3. Quinine sulfate at 1 mM and 0.5 M NaCl block the response to 10 mM sucrose. If the concentration of sugar in the stimulating solution is reduced, the curves for the inhibitory action of quinine and NaCl shift correspondingly.

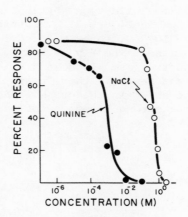

Figure 3: Inhibition of proboscis extension by quinine and NaCl against 10^{-2} M sucrose.

RECEPTOR PHYSIOLOGY

We have analyzed the electrophysiological responses of the taste sensillae. The tips of the labellar hairs were touched with a solution of the stimulant in a 30 μ wide micropipette; the same pipette served as a recording electrode. The signals arising from the 4 sensory neurons are easily distinguishable (Rodrigues and Siddiqi, 1978). Figure 4 shows the chemoreceptor responses to increasing concentrations of NaCl. The distribution of amplitudes in spikes from a single hair is given in Figure 5. At low salt concentrations we encounter, mostly, a single class of spikes 1.1 mV in amplitude. These are the W spikes. As the salt concentration is increased, the W cell is inhibited and a second cell, with a spike amplitude twice that of the first, begins to fire. We call this

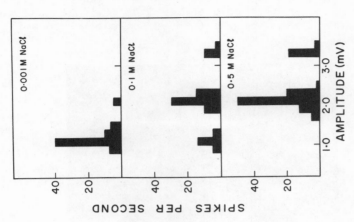

Figure 5: The distribution of spike amplitudes from W, L1 and L2.

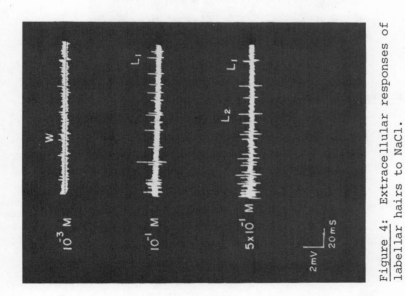

Figure 4: Extracellular responses of labellar hairs to NaCl.

cell L1. At yet higher concentrations of NaCl, a third cell, designated L2, is excited and the W cell is further inhibited. We have shown that W, L1 and L2 are also distinguishable by their adaptation characteristics (Rodrigues and Siddiqi, 1978).

The responses of normal hairs to sucrose are shown in Figures 6 and 7. Sugar inhibits the W cell and excites a single neuron which produces spikes of 1 mV. These are the S spikes. The S spikes are slightly smaller than L1.

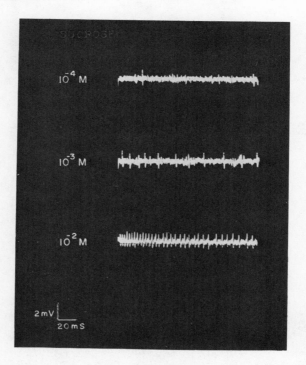

Figure 6: Extracelluar responses to sucrose.

Figure 8 summarizes the responses of labellar hairs to NaCl and sucrose. Evidently, salt concentration is measured by two neurons L1 and L2. It has been suggested that of the two L cells, one is a cation detector and the other an anion detector (Dethier, 1968; DenOtter, 1972). Responses to other sugars, glucose, maltose and fructose, are similar to sucrose, each exciting the S cell and inhibiting the W cell. Similar results have been reported earlier by Isono and Kikuchi (1974) except that they describe only one salt cell.

Unlike salt and sugar, quinine does not appear to excite individual neurons. We have not examined every hair, and it is pos-

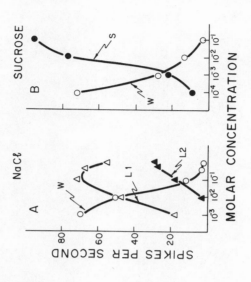

Figure 8: Effect of stimulant concentration on the firing frequency of chemosensory neurons.

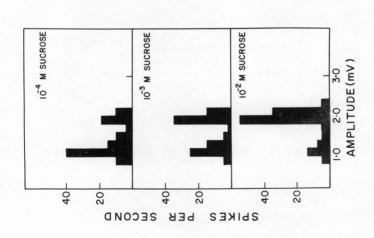

Figure 7: Amplitude distribution of sucrose response

sible that there might be specialized receptors for quinine, but as far as the majority of labellar hairs are concerned, the only effect quinine seems to have is to inhibit the firing of the other chemosensory neurons (Figure 9). Since the S cell is inhibited more strongly than the two L cells, the presence of quinine is likely to favor rejection by changing the ratio of S spikes to L spikes.

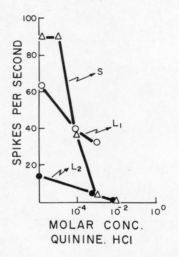

Figure 9: Effect of quinine on the firing frequency of S and L cells. The S cell was stimulated by 10^{-2} M sucrose and the L cells by 0.5 M NaCl.

A COMPLEX CHEMORECEPTOR

GUSTATORY MUTANTS

The phenotypes of six X-linked gustatory mutants are summarized in Table 1.

Table 1: Gustatory mutants of *D. melanogaster*

Strain	Quinine	NaCl	Sucrose
Canton-S	+	+	+
gustA (x1, x4)	+	+	−
gustB (x5)	+	−	+
gustC (x2)	−	−	−
†gustD (x3, x6)	−	+	+

+ normal response, − blocked
† temperature-sensitive development

Figure 10 shows the response curves for $gustB^{x5}$ and $gustC^{x2}$. Since $gustC^{x2}$ cannot taste sucrose, its response was tested in thir-

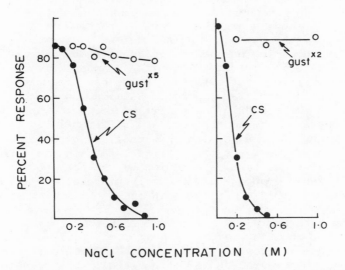

Figure 10: Proboscis extention response of NaCl-insensitive *gustB* and multiply blocked *gustC*. The response of $gustC^{x2}$ was measured against water.

sty flies, against attraction by water. The inhibitory concentration of the repellent in the control is correspondingly lower.

In order to identify the site of the genetic lesion, we tested the electrophysiological responses of the mutants belonging to the four *gust* genes. The receptor responses of the salt-insensitive mutant *gustB^{x5}*, the quinine-insensitive *gustD^{x3}* and the taste blind mutant *gustC^{x2}* seem to be indistinguishable from the wild type. The blocks produced by these genes are presumably central. As far as salt insensitivity is concerned, this is not surprising. NaCl can be detected either by L1 or L2. Mutations causing peripheral lesions in either of these detectors could not be picked in our screening which employed a high salt concentration. The behavior of quinine mutants is, however, noteworthy. If quinine is indeed perceived by the fly through a change in the ratio of S firing to L firing, the mutants whose peripheral sensitivity to this drug is unchanged might be expected to exhibit altered responses to salt or sugar. Although *gust*D mutants are not insensitive to salt or sugar, the NaCl inhibition curve for proboscis extention in these mutants needs to be carefully examined. Mutants insensitive to NaCl and quinine have also been described by Tompkins (Tompkins and Sanders, 1977; Tompkins et al., 1979). We have not yet checked our isolates against theirs for allelism.

The sugar mutant *gustA^{x1}* maps at 35 units between *v* and *f* on the X chromosome. The mutant is insensitive to pyranose sugars, sucrose, glucose and maltose but responds normally to fructose (Figure 11). The electrophysiological responses of both W and S to sucrose are blocked as are the responses to glucose and maltose. The response to the furanose sugar fructose is, however, entirely normal. The physiological defect in *gustA^{x1}* correlates perfectly with the behavioral deficit (Figure 12). Apparently, in this mutant the surface receptors for pyranose sugars have been lost while the furanose receptor sites are intact. Shimada and Isono (1978) have shown that in the fleshfly *Boettcherisca peregrina*, the labellar receptors for glucose and fructose are differentially sensitive to pronase digestion. Our results show that, in *Drosophila* too, there must be at least two types of sugar receptors on the S cell. The mutant *gustA^{x1}*, thus, carries a lesion in the most peripheral element of the chemosensory apparatus, the molecular receptors for pyranose sugars on the dendritic surface of W and S cells.

The mutants *gustD^{x3}* and *gustD^{x6}* are temperature-sensitive, the development of the taste pathway being specifically altered by the temperature of growth. Flies grown at 22°C have normal responses but, when reared at 28°C, they are unable to detect quinine.

Dr. L. Tompkins has isolated similar mutants which are sensitive to cold (Tompkins, 1979). The temperature-sensitive *gustatory* mutants have sharply defined sensitivity periods. Some are sensitive

A COMPLEX CHEMORECEPTOR   357

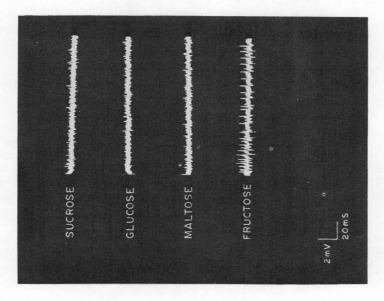

Figure 12: The response of *gustAx1* labellar receptor to sugars.

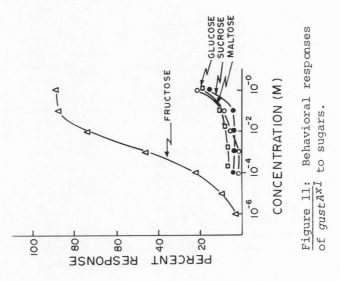

Figure 11: Behavioral responses of *gustAx1* to sugars.

during embryonic development while others are affected during late larval stages. Such mutations provide us with valuable material for investigating the development of the chemosensory pathway.

CONCLUSIONS

The gustatory system of *Drosophila* lends itself well to genetic analysis and several single-gene mutations, affecting either peripheral or central components of the taste pathway, have been obtained. Mutants blocked in relatively central portions of the nervous system provide valuable materials for investigating the genetic control of sensory projections. Of special interest from this point of view are the temperature-sensitive mutations such as *gust*D which specifically block the formation of behavioral circuits at the nonpermissive temperature.

The organization of the chemoreceptors in *Drosophila* is sufficiently simple, so that a correlated study of electrophysiological and behavioral responses of mutants is likely to throw interesting light on the sensory code.

REFERENCES

Deak, I.I., 1976, Demonstration of sensory neurones in the ectopic cuticle of spineless aristapedia (*ss^a*), a homeotic mutant of *Drosophila*, Nature, 260:252.

DenOtter, C.J., 1972, Interactions between ions and receptor membrane in insect taste cells, J. Insect Physiol., 18:398.

Dethier, V., 1955, The physiology and histology of the contact chemoreceptors of the blowfly, Quart. Rev. Biol., 30:348.

Dethier, V.G., 1968, Chemosensory input and taste discernment in blowfly, Science, 161:389.

Falk, R., and Atidia, J., 1975, Mutations affecting taste perception in *Drosophila melanogaster*, Nature, 254:325.

Falk, R., Bleiser-Avivi, N., and Atidia, J., 1976, Labellar taste organs of *Drosophila melanogaster*, J. Morphol., 150:327.

Hodgson, E.S., 1974, Chemoreception, in: "Physiology of Insecta", Vol. II, M. Rockstein, eds., Academic Press, New York, pp. 127-161.

Hodgson, E.S., Lettvin, J.Y., and Roeder, K.E., 1955, Physiology of a primary chemoreceptor unit, Science, 122:417.

Isono, K., and Kikuchi, T., 1973, A recessive autosomal mutation of a glucose receptor in labellar sensory hairs of *Drosophila melanogaster*, Jap. J. Genetics, 48:421.

Isono, K., and Kikuchi, T., 1974, Behavioral and electrical responses to sugars in *Drosophila melanogaster*, Jap. J. Genetics, 49:113.

Perkel, D.H., and Bullock, T.H., 1969, Neural coding, Neurosci. Res. Symp. Summ., 3:405.

Rodrigues, V., and Siddiqi, O., 1978, Genetic analysis of chemosensory pathway, Proc. Ind. Acad. Sci., 87B:147.
Shimada, I., and Isono, K., 1978, The specific receptor site for aliphatic carboxylate anion in the labellar sugar receptor of the fleshfly, J. Insect Physiol., 24:807.
Tompkins, L., 1979, Developmental analysis of two mutations affecting chemotactic behavior in Drosophila melanogaster, Dev. Biol., 73:174.
Tompkins, L., and Sanders, T.G., 1977, Genetic analysis of chemosensory mutants of Drosophila melanogaster, Genetics, 86 suppl. S64.
Tompkins, L., Cardosa, M.J., White, F.V., and Sanders, T.G., 1979, Isolation and analysis of chemosensory behavior mutants in Drosophila melanogaster, Proc. Natl. Acad. Sci. U.S.A., 76:884.

DISCUSSION

D. Byers: Are the hairs at the edge of the labellar lobes and sides of the proboscis chemosensory?

O. Siddiqi: Yes, but only the short ones. The long hairs outside the edge are not chemosensory.

OLFACTORY BEHAVIOR OF *DROSOPHILA MELANOGASTER*

Veronica Rodrigues

Molecular Biology Unit
Tata Institute of Fundamental Research
Bombay, India

INTRODUCTION

The olfactory system of *Drosophila* is readily amenable to genetic dissection. The response to smell involves interaction between a chemical stimulant and the receptor surface, transduction, neural excitation, transmission across synapses and integration by the central nervous system finally leading to motor activity. By an appropriate choice of mutants, attention can be focussed on any one of this series of complex processes. In this paper I will describe the properties of *olfactory* mutations on the X-chromosome and compare the behavior of mutants with normal flies.

Interest in olfactory behavior of *Drosophila* is old. In 1907, Barrows demonstrated that odor is detected by the third segment of the antenna. If one antenna is removed, the fly circles in the direction of the intact side, suggesting that the orientation to smell is achieved by comparing the stimulation of the receptors on the two sides. Begg and Hogben (1943, 1946) examined the responses of *Drosophila* to a number of chemicals and showed that mixtures of chemicals were more potent attractants than single chemicals. The first attempts to identify the genes controlling the olfactory pathway were made through a study of variants in natural populations (Becker, 1970; Fuyama, 1976, 1978). These experiments led to identification of genes on the right arm of the second chromosome which control specific odor sensitivity. Kikuchi (1973a, 1973b) isolated a mutant which is attracted by several chemicals that repel the wild type. We have described *olfactory* mutations in a number of X-linked genes which block responses to specific attractants and repellents (Rodrigues and Siddiqi, 1978). A mutant *smellblind* which is insensitive to several chemicals has been reported by Aceves-Piña and Quinn (1979).

OLFACTORY RESPONSES OF ADULT FLIES

The responses of adult flies are easily measured in the olfactometer described in Figure 1 (Rodrigues and Siddiqi, 1978). Olfactometric measurements against varying dilutions of the odorant yield a threshold and a response index (RI) which is a measure of maximal response. The responses of normal Canton Special (CS) flies to a number of stimulants are shown in Figure 1. Diluted fly food is a potent attractant; it is detected at a dilution of 10^{-4} and the response saturates at 10^{-2}. Ethyl acetate is an even stronger attractant with a threshold of detection at a dilution of 10^{-8}. The response to ethyl acetate, unlike food, falls above 10^{-3} changing rapidly into repulsion. Response reversals at higher concentrations of attractants are characteristic of several chemicals and could be due to the presence of contaminating repellents or to an intrinsic change in the response. Benzaldehyde and salicylaldehyde are strong repellents.

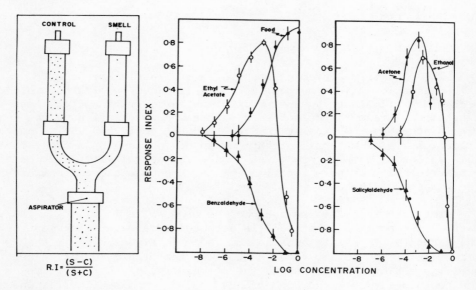

Figure 1: Olfactory responses of normal *Drosophila* adults. The olfactometer consists of a Y tube each arm of which is connected via adaptors to glass vials. One arm is connected to a smell source; the other, to control. The Response Index (R.I.) is calculated as the excess of flies on the smell side divided by the number of flies on both sides. +1 indicates total attraction; -1 total repulsion. In this and all other figures the concentration refers to the dilution factor for the odorant.

LARVAL RESPONSES

Olfactory responses of larvae can be measured on a Petri dish. A filter paper disc soaked in the odorant is placed on one side and

50 to 70 larvae are released at the center. When the dish is covered an odor gradient forms rapidly and the larvae move to one side. After 5 minutes the larvae on each half of the plate, the odor side and the control side, are counted and the difference is used to compute the response index. Larval responses to a variety of odorants are shown in Figure 2. Unlike adults, larvae are not repelled by any stimulant. Chemicals such as benzaldehyde, salicylaldehyde and pyridine, which are strong repellents for adults, attract the larvae. Ethyl acetate is a strong attractant with a detection threshold at a dilution of 10^{-6}.

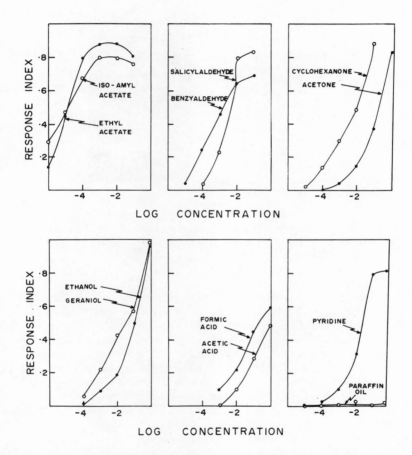

Figure 2: Olfactory responses of normal *Drosophila* larvae. 25 µl of odorant is soaked into a filter paper disc placed at one end of the dish. The larvae are released in the center. The plate is divided into two halves and larvae are counted.

R.I. = $\dfrac{\text{number on odor side} - \text{number on control side}}{\text{total}}$

In experiments with noxious stimulants such as benzaldehyde, salicylaldehyde or certain organic acids, an unusual behavior is observed. At lower concentrations of these chemicals the larvae move right up to the odorant disc; at higher concentrations the larvae go close to the disc but form a ring around it. The diameter of the ring is proportional to the concentration of the stimulant (Figure 3). It is likely that the detection of these chemicals close to the disc is not due to olfaction but due to contact chemoreception. If the filter paper is not in contact with agar but suspended above it, the ring is not formed.

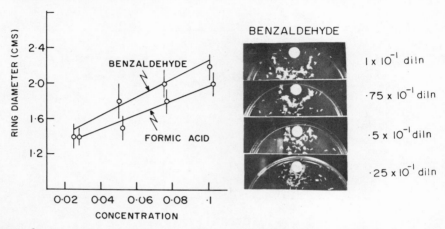

Figure 3: Larval response to high concentration of certain noxious chemicals; avoidance behavior is most probably due to detection by contact chemosensation (see text).

While the adult flies can discriminate between attractants and repellents, the larval response to odorants is restricted to attraction. The olfactory pathway must, therefore, undergo a change from larval to adult stages. Such changes in taxis are known to occur during development in Drosophila and other dipterans. Larvae of Drosophila are negatively phototactic while adults are positively phototactic (Durrwachter, 1957). Young housefly larvae prefer high humidities, but older larvae, ready to pupate, seek drier habitats (Hafez, 1950). In such cases a switch in programmed preferences must occur at a prescribed stage of development. For chemotaxis, this switch-over most probably happens during pupal development.

RECEPTOR SPECIFICITIES OF OLFACTORY STIMULI

Drosophila adults and larvae respond to a large number of volatile chemicals. One might ask how many different receptor sites on the antennal surface can independently recognize these chemicals. This question can be examined experimentally in the larval plate test by measuring the response to one chemical in a uniform background of another chemical. A background odor environment is maintained by covering the lid of the Petri dish with a filter paper soaked in an odorant; a second odorant is then used as an attractant and response curves are generated as described before. Results of such experiments are presented in Figure 4. The graphs show the responses to one chemical against an increasing background concentration of a related or unrelated chemical. It may be seen that the response to ethyl acetate is blocked by isoamylacetate (4a) but not by salicylaldehyde (4c). Similarly the response to salicylaldehyde is inhibited by benzaldehyde (4b) but not by ethylacetate (4d). We can thus infer that the two acetates compete for a common receptor (or some other element in a common path) while aldehydes and acetates are sensed through independent paths.

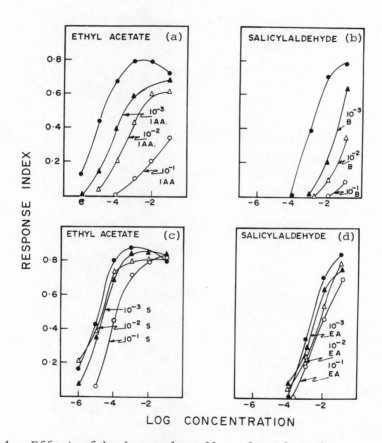

Figure 4: Effect of background smell on larval taxis.
a) test chemical is ethyl acetate; background iso-amyl acetate (IAA)
b) test chemical - salicylaldehyde; background benzaldehyde (B)
c) test chemical - ethyl acetate; background salicylaldehyde (S)
d) Reciprocal of (c).

Using the above argument, we tested a number of chemicals against each other. The results are summarized in Figure 5. Chemicals spanned by a bar interact strongly while chemicals spanned by non-overlapping bars do not interact. It may be seen that the larvae perceive major chemical groups such as alcohols, aldehydes, acids, esters and ketones as separate categories. The only exception is the interaction between ketones and acetates which appear to share common receptors. By extending experiments of this kind to a wider range of olfactory stimulants it should be possible to learn something about the categories into which odorants are classified by *Drosophila*.

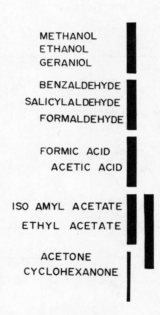

Figure 5: Chemical categories based on the competition experiment. Dark lines indicate strong competition; thin lines - weak competition.

OLFACTORY MUTANTS

We have isolated a number of X-linked olfactory mutants designated *olf*. The isolates belong to 4 different complementation groups *olfA*, *olfB*, *olfC* and *olfD* (Rodrigues and Siddiqi, 1978). The map positions of these genes are shown in Figure 6. The mutants *olfA* and *olfB* are blocked against aldehydes; *olfC* is insensitive to acetates and acetone while *olfD* has multiple defects. The response curves of these mutants are presented in Figure 7.

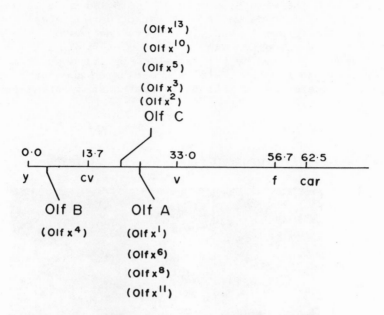

Figure 6: Map positions of olfactory genes on X chromosome. *olfD* has not yet been mapped.

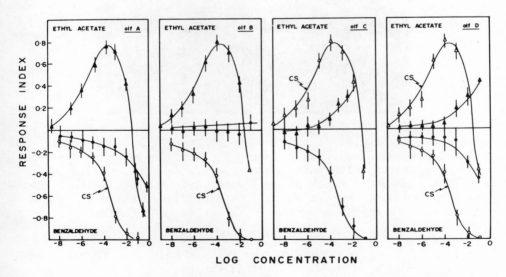

Figure 7: Olfactory responses of mutant adults to benzaldehyde and ethyl acetate.

Figure 8 shows a comparison of the larval responses of *olfB* and *olfC* with the wild type strain. *olfA* and *olfB* larvae are unable to detect benzaldehyde while *olfC* cannot detect ethyl acetate. A mutation in *olfC* simultaneously affects the response to acetate and acetone. That these two groups of chemicals share a common receptor element was suggested by the larval competition experiments. Fuyama (1978) has also reached a similar conclusion about the common odor quality of acetates and ketones through a study of naturally occurring variants.

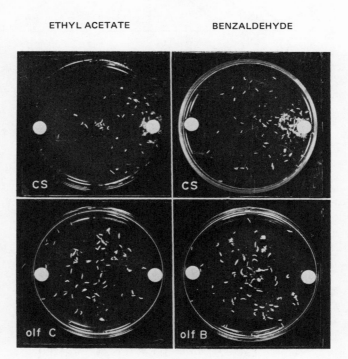

Figure 8: Larval plate test showing olfactory responses of *olfB* and *olfC*; *olfB* shows an altered response to benzaldehyde but is normal to ethyl acetate. *olfA* (not shown) shows a similar behavior. *olfC* (left) cannot detect acetates normally but response to benzaldehyde is the same as that of CS. On each plate, the disc on the right hand side contains odorant; that on left is control.

The olfactory responses of the mutants belonging to the 4 *olf* genes are summarized in the following table:

Table 1. Responses of olfactory mutants to odorants. *olfA, olfB* and *olfC* show similar defects either as larvae or adults. The *olfD* column describes the adult responses alone.

Chemical	olfA	olfB†	olfC	olfD
Benzaldehyde	−	−	+	−
Salicylaldehyde	−	−	+	−
Formaldehyde	−	−	+	−
Ethyl Acetate	+	+	−	−
Isoamyl Acetate	+	+	−	−
Methanol	+	+	+	
Ethanol	+	+	±	−
Geraniol	+	+	+	
Acetic Acid	+	+	+	
Formic Acid	+	+	+	
Acetone	+	+	−	
Cyclohexanone	+	+	±	

† Temperature-sensitive developmental mutation.

The deficiencies in the larvae parallel those of the adults; the same is true of the mutant *smellblind* described by Aceves-Piña and Quinn (1979).

CONCLUSIONS

At present, there is no direct evidence about the site of the physiological lesion in the mutants. Since these mutations specifically affect the responses to a single class of chemicals, it is likely that *olfA, olfB* and *olfC* cause peripheral defects. The olfactory organs in the larval cuticle are sloughed off during metamorphosis and replaced by different organs in the adults (Hertweck, 1931). These three genes presumably code for components used in chemoreceptors of both larvae and adults. *olfD* appears to suffer from a central lesion. Homozygous *olfD* strains exhibit severely reduced viability and the larval behavior of this mutant remains to be characterized.

REFERENCES

Aceves-Piña, E.O., and Quinn, W.G., 1979, Learning in normal and mutant *Drosophila* larvae, Science, 206:93.
Barrows, W.M., 1907, The reactions of the Pomace Fly *Drosophila ampelophila* Loew to odorous substances, J. Exp. Zool., 4:515.
Becker, H.J., 1970, The genetics of chemotaxis in *Drosophila melanogaster*. Selection for repellent insensitivity, Molec. Gen. Genet., 107:194.
Begg, M., and Hogben, L., 1943, Localization of chemoreceptivity in *Drosophila*, Nature, 152:535.
Begg, M., and Hogben, L., 1946, Chemoreceptivity of *Drosophila melanogaster*, Proc. Roy. Soc. (Lond) B, 133:1.
Durrwachter, G., 1957, Untersuchungen uber phototaxis and geotaxis einiger *Drosophila* - Mutanter nach Hufzucht in verschiedener Lichtbedingungen, Z. Tierpsychologie, 14:1.
Fuyama, Y., 1976, Behavior genetics of olfactory responses in *Drosophila*. I. Olfactometry and strain differences in *Drosophila melanogaster*, Behav. Genet., 6:407.
Fuyama, Y., 1978, Behavior genetics of olfactory responses in *Drosophila*. II. An odorant specific variant in a natural population of *Drosophila melanogaster*, Behav. Genet., 8:399.
Hafez, M., 1950, On the behavior and sensory physiology of the housefly larvae *Musca domestica*, Parasitology, 40:215.
Hertweck, H., 1931, Anatomie und variabilitat des Nervensystems und der Sinnesorgane von *Drosophila melanogaster*, Z. Wiss. Zoöl., 139:559.
Kikuchi, T., 1973a, Genetic alteration of olfactory functions in *Drosophila melanogaster*, Jap. J. Genet., 48:105.
Kikuchi, T., 1973b, Specificity and molecular features of an insect attractant in a *Drosophila* mutant, Nature, 234:36.
Rodrigues, V., and Siddiqi, O., 1978, Genetic analysis of chemosensory pathway, Proc. Indian Acad. Sci. (B), 87:147.

DISCUSSION

E. Weischaus: Where are the larval chemosensory receptors located? Are they in the maxillary regions?

V. Rodrigues: Based on ablation studies in the housefly larvae, they appear to be located in the head region; but some chemosensory receptors may be located in the body.

Y. Hotta: Can the behavior test be done on a single fly? I was wondering if you could use this for scoring mosaics.

V. Rodrigues: We are trying to develop single fly measurements. Perhaps some of the techniques described by Dr. Heisenberg would be useful.

A.S. Mukherjee: When you do the larval test, are you using a fixed time for the assay? Is it possible that if you use a longer time, you might see repulsion?

V. Rodrigues: We have followed the time course of attraction. The larvae first migrate to the odor source. After longer times, they begin to distribute randomly.

A.S. Mukherjee: How do you maintain the chemosensory mutants?

V. Rodrigues: They are maintained as homozygotes. They breed happily, eat, and mature to perfectly healthy adults.

MUTANTS OF BRAIN STRUCTURE AND FUNCTION: WHAT IS THE SIGNIFICANCE
OF THE MUSHROOM BODIES FOR BEHAVIOR?

M. Heisenberg

Institut für Genetik und Mikrobiologie der
Universität Würzburg
Federal Republic of Germany

INTRODUCTION

For approaching the enormous complexity of the insect brain one may choose first to study the sensory and motor periphery in the hope to finally work one's way up to the central processing stages of the brain or, alternatively, one may parachute in the midst of the jungle, experimentally altering the brain and try to understand the concomitant changes in behavior. While with the first approach in recent years an impressive volume of basic knowledge about neural integration of sensory data and about the generation of motor patterns has been provided, only few of these studies have really been concerned with central brain functions. Of the parachutists, on the other hand, only few signs of survival have reached the outside world (e.g., Wadepuhl and Huber, 1979).

Drosophila brain mutants have already proven to be useful in the first type of approach, particularly in the analysis of the visual system. Today I want to draw your attention to the use of brain mutants in the other kind of adventure. I will describe 4 mutants with structural defects in their mushroom bodies. With one of them we have conducted a variety of behavioral tests in order to find a functional deficit associated with the malformation. So far only "negative" evidence is accumulating. We cannot yet propose a function for the mushroom bodies, but our experiments allow us to eliminate some of the earlier hypotheses. This paper should, therefore, be considered a progress report rather than a presentation of results.

For those of you not familiar with the insect brain I will try to briefly summarize what is known about mushroom bodies in *Dro-*

sophila and what we would like to assume about them by homology to other insects. I hope this short detour into anatomy will give you an impression of the relative simplicity of this brain organelle as compared to the rest of the central neuropil. It seems that its macroscopic shape and neuronal architecture contain some important information about its function. We have the message but we do not yet understand it. This is the situation which for 130 years has led so many researchers to studying the mushroom bodies in insects.

MORPHOLOGY OF MUSHROOM BODIES

The mushroom bodies are two well-discernible structures in the median protocerebrum arranged, like most of the brain, in mirror symmetry to the sagittal midplane. (For reviews see Howse, 1975; Klemm, 1976; for *Drosophila*, see Power, 1943.) They consist each of the so-called calyx, the peduncle and generally two lobes (α, β). The peduncle can be envisaged as a huge bundle of very thin axonal fibers (Kenyon cells) extending from back to front through the whole brain on either side of the central complex, the inner-most part of the neuropil. At the front of the brain each of these Kenyon cell fibers divides sending one branch toward the midplane (β-lobe), the other one upward (α-lobe). Thus, the whole bundle consists of 3 arms (peduncle, α-lobe, β-lobe), oriented roughly parallel to the 3 main axes of the fly. The two β-lobes nearly meet in the midplane separated only by a tiny commissure of distinct fibers. The two α-lobes extend to the anterior dorsal cellular cortex. In the posterior dorsal cellular cortex the Kenyon cell fibers before entering the central neuropil have small dendritic arborizations which together with extrinsic fibers constitute the calyx.

The fine structure of mushroom bodies has been worked out in other insects. Extrinsic fibers appear to contact Kenyon cells at all levels. However, this contact is most abundant in the α- and β-lobes as well as in the calyx. Investigation of the synaptic contacts has revealed a distinct functional polarity of the mushroom body. It seems that in the calyx all contacts between extrinsic fibers and Kenyon cells have the extrinsic fibers as presynaptic elements suggesting a signal flow from the calyx to the lobes. In the lobes and to some extent also in the peduncle, however, pre- and postsynaptic extrinsic fibers are found. Also, intrinsic fibers seem to have synaptic contact among each other.

The mushroom bodies have connections to many parts of the brain. In Dipterans the most conspicuous one is that from the antennal ganglion via the olfactorio globularis tract. Fibers running diagonally all the way from the lower front of the brain to the calyces and past them into the latero-dorsal protocerebrum send side branches into the calyces. These terminate in characteristic microglomeruli by which they are thought to contact large numbers of Kenyon cells. Input from other parts of the brain to the calyces is less well documented but should not be neglected.

Extrinsic elements in the peduncle and lobes do not run in bundles. With degeneration, Golgi and recently dye injection techniques individual neurons have been followed. From such studies it has been proposed that extrinsic fibers from the α-lobe deliver a direct or indirect recurrent input to the calyx whereas in this model only extrinsic fibers from the β-lobe connect, via central complex and/or suboesophageal ganglion, to motor centers (Vowles, 1955).

While the discussion of anatomical facts and assumptions slowly leads to functional considerations, a further ultrastructural and histochemical detail should be mentioned: the lobes and the frontal part of the peduncle seem to be among the main sites where biogenic monoamines are found (Klemm, 1976).

POSSIBLE FUNCTIONS OF MUSHROOM BODIES

In contrast with the anatomical data, functional data are scarce and hardly any information comes from flies. Surgical lesions in the mushroom bodies may cause distinct behavioral changes. The operated silkworm *Cecropia* spins a two-dimensional sheath instead of a cocoon (van der Kloot and Williams, 1953). Bees and locusts display competitive motor patterns when carrying lesions in their mushroom bodies (Howse, 1974).

Electrical stimulation near the mushroom bodies can elicit elaborate sequences of courtship behavior in grasshoppers and crickets (Huber, 1960; Wadepuhl and Huber, 1979). Along these lines recently Jeffrey Hall (1979) has shown by the mosaic technique that in *Drosophila* nerve tissue in the caudal dorso-median protocerebrum just in the region of the Kenyon cells is required to be male for male courtship to occur. An elegant experiment on bees suggests the involvement of the mushroom bodies in learning (Menzel et al., 1974): cooling the α-lobe interferes with olfactory conditioning of the proboscis extension reflex. Finally, electrophysiological recording techniques have demonstrated the close relation of the mushroom bodies to the olfactory system (Maynard, 1967; Suzuki and Tateda, 1974; J. Erber, pers. comm.).

MORPHOLOGY OF MUSHROOM BODIES IN *DROSOPHILA*

In most organisms the classic model of a mushroom body as outlined above is varied in one direction or another (Hanström, 1928). In *Drosophila* (Plate I) closer anatomical inspection reveals that each mushroom body, in fact, consists of two closely associated substructures each originating in the calyx and extending through the 3 arms. One of them closely conforms to the model. It consists of Kenyon cells which run as single fibers all the way from the calyx to the frontal border of the cerebral neuropil where they split into two branches constituting much of the α- and β-lobes. This part of

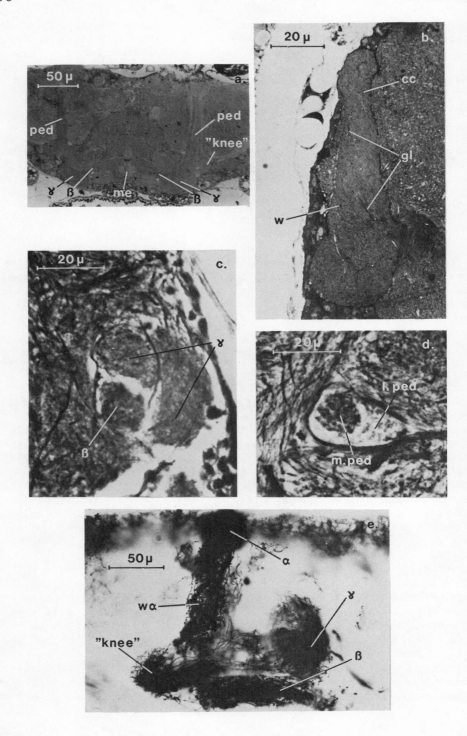

Plate I: Morphology of mushroom bodies in wild type *Drosophila melanogaster* (and *Musca domestica*).

a) Horizontal cross section of central brain at the height of pedunculi (ped.), β- and γ-lobes. Median bundle (me) separates β-lobes. 1μ, araldite, toluidine blue.
b) Sagittal section through α-lobe. Arrows point to glia (gl) lamella separating constant cross section lobe (cc) from wedge-shaped part (w). 1μ, araldite, toluidine blue.
c) Cross section through β- and γ-lobe. 7μ, paraffin, reduced silver.
d) Cross section through pedunculus at about the depth of the central body. Median pedunculus (m. ped.) appears darker than lateral one (l. ped.). 7μ, paraffin, reduced silver.
e) Frontal section through α-, β- and γ-lobe in *Musca domestica*. Many individual Kenyon cells are stained with Co^{++} followed by silver intensification. Compact bundle of fibers in β-lobe (at bottom) can be distinguished from less ordered and loosely packed fibers in γ lobe and frontal (wedge-shaped) parts of α-lobe (w-α). By courtesy of K. Hausen, Tübingen, Germany.

---

the peduncle (we will call it the *median* peduncle) and the corresponding parts of the lobes have everywhere about the same roughly circular cross section. Kenyon cell fibers are densely packed leaving little space for extrinsic fibers.

The other substructure also consists of a scaffold of Kenyon cell fibers but, in contast to the previous one, its cross section varies considerably. Starting at the calyx it soon separates from the median peduncle having at that depth only 1/3 the cross section of the latter. It soon gains in horizontal diameter forming the so-called "knee" which in horizontal sections has up to about twice the diameter of the median peduncle. The lateral peduncle (with the "knee") continues medially and is called the γ-lobe. This lobe frontally and dorsally covers the β-lobe over its entire length and has about three times the cross section of the β-lobe. From the knee the lateral peduncle also extends upwards as part of the α-lobe. It forms a wedge-shaped thickening in front of the part with constant cross section and reaches to only about half the full height.

The Kenyon cell fibers in this substructure do not grow straight. Their branching patterns have not been worked out yet. But it seems that much of the volume responsible for the variable cross section along this part of the mushroom body is due to extrinsic elements.

In summary, the mushroom bodies of *Drosophila* consist of two strikingly different substructures: one with constant cross section resembling the classical model of an insect mushroom body; the other with variable cross section and with fewer and less densely packed intrinsic fibers (M. Heisenberg and R. Böhl, unpublished).

The same two types of mushroom body subsystems exist in *Musca* as demonstrated by a cobalt impregnated mushroom body preparation of K. Hausen, Tübingen (Plate I). I mention this anatomical detail since it should remind us that mushroom bodies may not serve a uniform function.

DESCRIPTION OF MUSHROOM BODY MUTANTS

The following mutants (Fig.1, Plate II) characterize 4 different genes on the X-chromosome; numbers after the mutant name indicate position on genetic map, if known.

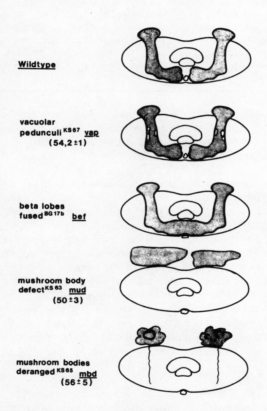

Figure 1: Schematic presentation of mushroom body morphology in wildtype and mutants (see text).

1) Vacuolar pedunculi^{KS67}, *vap* (54.2 ± 1). Vacuolar spaces are found at a certain depth along the pedunculi suggesting degeneration of extrinsic cells. Intrinsic fibers appear continuous.

MUTANTS OF BRAIN STRUCTURE AND FUNCTION

2) beta-lobes-fused^{BG17b}, *bef*. Just frontally to the elipsoid body the two β-lobes seem to be fused across the midline. This phenotype occured in a multiple mutant strain of J.A. Merriam, University of California, Los Angeles. The stock had been selected for non-phototactic behavior and was marked with *y cho sn*. The behavioral defect apparently was due to a mutant gene affecting eye and optic lobes. It seems that the gene causing mushroom body fusion is separated from genes causing other mutant phenotypes. Mushroom bodies of similar shape have been reported for termites and some lower invertebrates (Hanström, 1928). Power (1943) reported to have found one fly with fused mushroom bodies; a second such observation is described by Heisenberg et al. (1978). Mushroom bodies which are continuous across the midline are thought to represent an archaic form and Hanström (1928) adds the interesting speculation that in evolution splitting up the fused mushroom body might render its output available and subordinated to the central complex.

3) Mushroom body defect^{KS63}, *mud* (50 ± 3). This map position is derived from recombination data only. The mutation is not complemented by the deletion KA9 (12E1; 13A5), implying a map position between 45 and 48.

The antennal lobes and the tractus olfactorio globularis are grossly enlarged; at the side of the calyces hugh amounts of very thin fibers are found forming distinct lobes outside the main cerebral neuropil but connected to it by the tractus olfactorio globularis and other fibers. No mushroom bodies are found. The shape of the central complex often is distorted.

4) Mushroom bodies deranged^{KS65}, *mbd* (56 ±5). If the defect is fully developed only few fibers, typically some big ones, are found at the sites of the pedunculi. No lobes can be detected. The calyces are considerably enlarged; their appearance suggests that most of the intrinsic fibers have indeed grown but have failed to enter the cerebral neuropil, thus accumulating at the sites of the calyces. In many flies the rest of the brain seems to be little affected.

Since only the mutant *mbd* has been studied behaviorally, it will be discussed in more detail. At first it took considerable effort to "stabilize" the genetic defect. During the first half year the mutant had to be reselected several times since the expression of the defect declined so rapidly that it would have been lost in few generations. At present several somewhat more stable lines with expressivities about 70% are being maintained at the laboratory. Natural selection pressure seems at least in part to account for this fast phenotypic reversion since the trait is much more stable if the mutant is kept heterozygously balanced over an attached X-Y chromosome. This is particularly noteworthy in regard to the behavioral experiments to be described on the following page.

Plate II: Morphology of mushroom bodies in mutants.

a) Horizontal section through central brain of vap at the level of mushroom bodies. Both pedunculi show "vacuoles" at about the same depth. 1μ, araldite, toluidine blue.
b) Detail of a) a few microns below.
c,d) Two consecutive horizontal sections through central brain of bef. Note continuity of β-lobes across midline. 7μ, paraffin, reduced silver.
e,f) Enlarged antennal lobes (ant) and calyces (cyx) in the mutant mud. Frontal sections, 7μ, paraffin, reduced silver.
g) Enlarged calyces (cyx) of mutant mbd. Horizontal section about level of pedunculi. 7μ, paraffin, reduced silver.
h) Same preparation as g) but a few sections below at the level of pedunculi. Very few thin fibers seem to be left but some big ones are seen, which do not appear in wild-type. Note false pedunculi (f. ped.)

The mutant phenotype itself is variable. The defect may occur only on one side and different parts or varying amounts of normal mushroom body may be left. The α-lobe is the first to be affected. In preparations in which at first sight the α-lobe seems to be missing a very thin trace of an α-lobe hardly distinguishable from the rest of the neuropil can sometimes be detected. The next to be affected is the β-lobe. It may be somwhat thinner than normal or may be missing together with the median peduncle. Along with the reduction of the peduncle, the γ-lobe becomes less clearly visible and if the defect is fully expressed, nothing but the extrinsic fibers may be left of it. At the place of the peduncle nearly always some fibers remain, in particular some thick ones which in well-developed mushroom bodies are much less apparent. If the β-lobe (and the γ-lobe) are reduced or missing, the calyx is always bulging and has an increased volume. Even if only the α-lobe seems to be reduced, protruberances at the calyx can often be observed. The precise relation between the accumulation of fibers in the calyx and the missing fibers in the roots is not clear.

DEVELOPMENTAL OBSERVATIONS

In recent years it has been very elegantly shown that sensory nerves can find their appropriate synaptic connections even if they have to travel through foreign territory (Stocker, 1977; Ghysen, 1978; Palka et al., 1979). A similar case can be made for the Kenyon cell fibers in the mutant mbd. In about 10% of the animals (in our present stock) a huge bundle of very thin fibers grows out from the calyx along the tractus olfactorio globularis underneath the central body and, at the depth of the lipsoid body it turns dorsally to reach the site where in wildtype the β-lobe is terminating. Assuming that these are the right fibers this implies that they reach the appropriate target via the wrong route. The false peduncle, as we call it, seems to have a cross section of constant size but varying shape due to the local structure of the neuropil it invades.

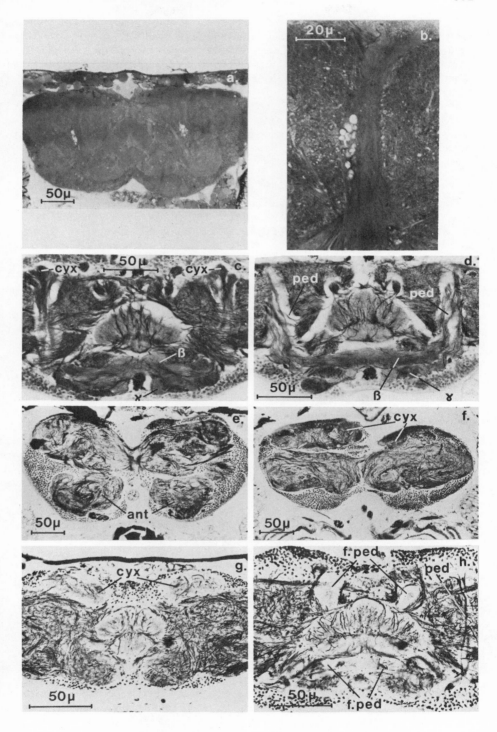

This or alternative explanations may be supported by looking at the mushroom bodies during their development. We are just beginning this study but one result may be of interest. It is known since the work of Hertweck (1931) that the 1st instar larva already possesses mushroom bodies of about the same shape and intrinsic structure as the imago. It is generally assumed that, while the brain grows during larval life and metamorphosis, more Kenyon cells are being added to the old ones. A 3rd instar larva has already larger pedunculi than what is left of the pedunculi in the adult mutant *mbd* with a fully expressed defect. This raises the question whether the *mbd* larva already has the mushroom body defect.

The preliminary answer is that nearly all *mbd* larvae in the late 3rd instar have bigger pedunculi and β-lobes than their adult siblings (G. Technau and M. Heisenberg, unpublished). Do the larval Kenyon cells retract their fibers? Do they degenerate during metamorphosis? Does the mutant reveal a developmental process of the wildtype which had been overlooked or is this diminution of the pedunculus and β-lobe during metamorphosis a specialty of the mutant? Whatever the answers may be to these questions it appears that many of the defects in the mushroom bodies of the adult flies develop during metamorphosis when much of the adult brain is already present. The false peduncle, therefore, must find its way to the site of the β-lobe through a densely packed barrier of neuropil.

BEHAVIOR

<u>Correlation with Anatomy</u>

Before describing some of the behavioral properties of the mutant *mbd*, I should briefly discuss in general terms what one would expect to find as a behavioral correlate to this anatomical defect.

It is known that people who, post mortem, were shown to have no cerebellum had lived without obvious impairment to fine motor coordination. I would like to maintain that rigorous tests nevertheless would have revealed behavioral abnormalities in these persons. Along this line I am convinced that the nervous system of an insect would not be able to fully compensate for the complete absence of mushroom bodies since the mushroom body is built in a very specific manner obviously in order to perform a specific function which other neural structures would do less well.

Another question is the extent to which mushroom body function may be suppressed in the mutant *mbd*. Neurosecretory signals from the Kenyon cells would most likely not be impaired at all. But functions requiring the appropriate output connections from the Kenyon cells should be reduced if the mutant phenotype is "fully expressed". This is defined as a state in which no lobes are distinguishable on both sides of the brain, less than 20% of the normal number of fibers is found at the site of the peduncle and no false peduncle has formed. For various reasons it seems unlikely that a significant fraction of the extrinsic fibers finding no normal target would grow into the calyx without being noticed in order to make their appropriate contacts. But even if this were the case, certain functions associated with the specific morphology of the mushroom bodies would be lost in the mutant.

In the following, the most rigorous argument can be advanced if the behavioral paradigm allows us to test individual flies. In this case if a fly shows a normal behavioral response and in subsequent histological examination is found to have the fully expressed mushroom body defect, we can conclude that normal mushroom bodies are not essential for this behavior. In experiments with groups of flies they can be separated into two groups at the end of the experiment: those flies which responded "positively" and those which did not. Subsequently the expressivity of the anatomical defect in the two groups is compared. If a behavioral defect was correlated with the mushroom body defect, the expressivity in the positively responding group would be expected to be lower than among the nonresponders.

Motor Coordination

Howse (1974) and others raised the interesting speculation that mushroom bodies might be concerned with selection and sequential organization of behavior. Flies from the mbd stock which have the fully expressed mushroom body defect as defined above can fly, jump, move their head, antennae, proboscis and show all their cleaning routines in the normal arbitrary sequence. We have never seen any motor pattern being in conflict with another one. Thus it seems that in Drosophila normal mushroom bodies are not necessary for the coordination of behavioral subroutines.

Courtship

As mentioned in the Introduction, various pieces of evidence lead to the hypothesis that mushroom bodies might be crucial in the generation of male courtship behavior. With Drosophila melanogaster this idea is particularly appealing since for this organism courtship is the longest and most complex sequence of behavioral subunits known and it has been extensively studied. However, flies with the fully expressed mushroom body defect show the whole courtship sequence in a manner indistinguishable from that of wildtype. They also spend as much time courting and are equally successful reaching copulation. Thus normal mushroom bodies seem not to be essential for the generation of male courtship in Drosophila.

Olfaction

The connection from the antennal lobe to the calyx via the tractus olfactorio globularis is very pronounced in many insects. This prompted Weiss (1974), Strausfeld (1976) and others to propose a purely chemosensory function for the mushroom body. As mentioned above, this association is also quite apparent in *Drosophila* (for instance in the mutant *mud* both the antennal lobes and the mushroom bodies are affected). We therefore have begun to compare the olfactory abilities of the mutant with those of wildtype.

Chemotaxis. The first experiment to be described was performed together with K.G. Götz, Tübingen. Hungry flies with shortened wings were placed onto an arena of fine cloth with a water ditch around. From part of the cloth a slow stream of air carrying the smell of fermenting banana was emanating. The path of the fly was automatically recorded. Götz (in preparation) had demonstrated that wildtype flies stay in the field emitting the odor and are able to sense the boundary of that field. Fig. 2a shows the path of an *mbd* fly in this experiment. The fly obviously has a strong preference for fermenting banana (as wildtype does), and it also notices the boundary between the field emitting the odor and its surround. This recording is indistinguishable from recordings of wildtype flies. In this particular fly the mushroom body defect was fully expressed.

Odor-wind orientation. In the previous experiment the orientation task may be fairly simple: the fly has to reverse its course when hitting the odor boundary. In order to do this it has to record the temporal concentration changes but not necessarily the spatial distribution. We were interested in measuring true osmotropotaxis in our flies, but as shown by Otto (1949) *Drosophila melanogaster* seems to be unable to find a source of odor if there are no local turbulences leading the fly up-wind (Fig. 2b). Thus we asked the question whether the *mbd* flies were still able to coordinate the evaluation of an odor with wind direction and with their own orientation. Fig. 2c and 2d (which document experiments designed after Flügge (1934) and Otto (1949)) again show the paths of walking flies displaying odor-wind coordination. In Fig. 2c flies immediately turn up-wind upon hitting a jet of air with the odor of fermenting pears. In Fig. 2d when the odor diffuses through a cloth in the center of the arena, the fly eventually meets the odor "down-stream" of some turbulence or draft (not monitored) and immediately finds the source. Both experiments were performed with *mbd* flies which had the fully expressed defect and are indistinguishable from the results with wildtype.

Effect of odor on phototaxis. K.F. Fischbach (unpublished) demonstrated a similar coordination between olfaction and visual behavior. He was able to show that phototaxis toward a light source is enhanced either by a repellent or upon removal of an attractant. Again this behavior is normal in *mbd* flies (Fig. 3). Since the experiment is performed with groups of 5 flies at a time correlation with anatomy is not as clear-cut.

MUTANTS OF BRAIN STRUCTURE AND FUNCTION   385

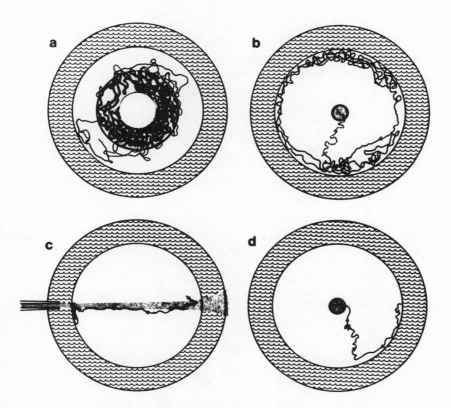

Figure 2: Olfactory orientation of wildtype and *mbd* flies.
a) Trace of an *mbd* fly walking for 200 s on a platform of fine cloth. In the shaded area odor of fermenting banana is slowly emanating from the cloth. The fly stays in the odor field. The blank area in the center is an artefact of the recording device. (Experiment was conducted in collaboration with K.G. Götz, Tübingen, Germany.)
b) Trace of a wildtype fly walking for 6 min on a platform completely enclosed in a glass container. Odor of fermenting pears is diffusing through a cloth in the center of the platform (shaded). Fly is unable to locate odor source (continuous trace) until the glass container is removed (stippled trace). Traces in b, c and d are recorded manually.
c) *mbd* fly by chance meeting a laminar jet of air carrying the odor of fermenting pears. It immediately starts walking up-wind (after Flügge, 1934).
d) Experimental conditions as in b, without glass container. *mbd* flies detect an odor source if the air is sufficiently moving. Flies with one antenna only, behave much like normal flies and can find the odor source nearly as fast (after Otto, 1949).

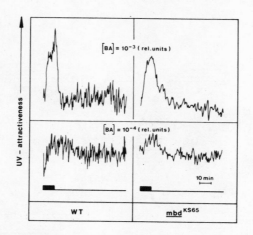

Figure 3: Attractiveness of a light source is increased by an unpleasant odor (benzaldehyde). In each experiment 5 flies are tested simultaneously. Benzaldehyde is blown into the recording chamber for 10 min (black bars) followed by clean air for 65 min. Traces are averages of 16 sweeps. Responses of wildtype and the mutant mbd are similar (K.R. Fischbach, unpublished).

Odor cues in courtship. R. Cook (unpublished) has studied courtship of mbd flies in infrared light where the males are thought to rely entirely on olfactory signals while orienting towards and following the females. mbd males with the fully developed mushroom body defect can court females in the dark very much like wildtype. Also for mdb flies, as for wildtype, courtship experience with mated females (Siegel and Hall, 1979) has the same long-lasting negative after-effect suppressing the disposition to court. Among individual mbd flies with the fully developed mushroom body defect courtship suppression is found as often as among wildtype flies. This is an example of behavioral plasticity which we have been able to test in the mutant. The result certainly does not rule out the possible role of the mushroom bodies in associative conditioning or other types of learning, but it adds a further facet to the intactness of olfactory behavior in the mutant.

"Pre-imaginal conditioning." As a further test of behavioral plasticity we repeated an experiment reported by Thorpe (1937), Manning (1967) and lately Laudien and Iken (1977). Flies were raised in two differently scented media: benzaldehyde and peppermint oil. Pupae were removed, kept on a nonscented medium until the flies hatched and were then tested for their preferences in choosing an oviposition site. If N_A is the number of eggs deposited on the food containing the scent the egg-laying fly had experienced as larva, and N_B the number of eggs deposited on the food sample carrying the other odor, we call the index of preference IP = N_A - $N_B/N_A + N_B$. Wildtype and the mutant mbd showed a strong preference for peppermint oil in this choice situation (0.03% benzaldehyde, 0.01% peppermint oil). The preference for the odor in which they had lived as larvae was rather small but significant. No difference between wildtype and mutant was found (IP = 0.13 ±0.03 for wildtype; IP = 0.11 ± 0.03 for *mbd*).

Odor concentration differences. A final test concerns the distinction of small concentration differences in a Y-maze. Flies had to choose between two slightly different concentrations of the same odor (benzaldehyde). Concentrations differing by as little as a factor of 2 could be distinguished by our wildtype. Flies from a *mbd* stock in which about 60% of the flies had the fully developed defect (and most of the others had partial defects) performed as well as wildtype (Figure 4).

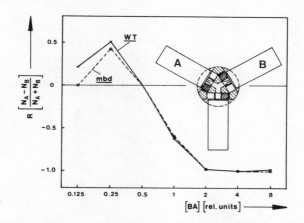

Figure 4: Flies have to choose between different odor concentrations of benzaldehyde. Reference concentration (0.5 rel. units) in all experiments corresponded to 0.0025 ml of benzaldehyde in one of the "choice-chambers" (B) of the Y-maze. Relative concentrations of benzaldehyde in the second "choice-chamber" (A) are given on the ordinate. N_A and N_B indicate number of flies in chambers A and B. The response value at the concentration BA = 0.5 rel. units is not determined.

Other behavioral tests. The avoidance of 1 M NaCl in 0.1% sucrose by hungry flies, negative geotaxis and a standard repertoire of visual responses (Heisenberg and Götz, 1975) were also tested and seemed not to differ significantly from wildtype behavior.

CONCLUSION

In view of the apparent strong selective pressure in favor of normal mushroom bodies in the *mbd* stock under our culturing conditions it is somewhat surprising that a variety of basic behavioral performances seem not to rely upon them. Obviously very important tests are still missing: associative conditioning, the ability to distinguish slightly different complex mixtures of odors or tastes and true osmotropotaxis have been omitted. So far I am still convinced that a behavioral correlate to the mushroom body defect can be found. Developmental and more refined anatomical investigations will help formulating new hypotheses. It is, however, conceivable that at some point in the exploration of a brain we will discover that our present concepts of behavior are not appropriate to the needs of a fly. The main advance of mutant analysis as compared to previous experiments of brain surgery seems to be that an abundance of flies with the same defect is available and that, therefore, different behavioral experiments can be rigorously compared.

ACKNOWLEDGEMENT

I would like to thank all the many people who participated in various aspects of this study. R. Cook, K.F. Fischbach and K. Hausen kindly provided unpublished data. K.G. Götz collaborated in an experiment on olfactory behavior. K. Böhl, R. Böhl, A. Borst, A. Moritz, W. Strube and G. Technaue enthusiastically contributed their experimental skills. K.F. Fischbach helped to improve the text; R. Wolf prepared the line drawings and plates and M. Preibler volunteered to type the manuscript. Above all I am indebted to friends and colleagues for innumerable discussions about mushroom body function. The Deutsche Forschungsgemeinschaft generously supported this work.

REFERENCES

Flügge, C., 1934, Geruchliche Raumorientierung von *Drosophila melanogaster*, Z. vergl. Physiol., 20:463.
Ghysen, A., 1978, Sensory neurons recognize defined pathways in *Drosophila* central nervous system, Nature, 274:869.
Hall, J.C., 1979, Control of male reproductive behavior by the central nervous system of *Drosophila*: dissection of a courtship pathway by genetic mosaics, Genetics, 92:437.
Hanstrom, B., 1928, Vergleichende Anatomie des Nervensystems der wirbellosen Tiere, Springer, Berlin.
Heisenberg, M., and Götz, K.G., 1975, The use of mutations for the partial degradation of vision in *Drosophila melanogaster*, J. Comp. Physiol., 98:217.
Heisenberg, M., Wonneberger, R., and Wolf, R., 1978, Optomotor-blind^{H31} - a *Drosophila* mutant of the lobula plate giant neurons, J. Comp. Physiol., 124:287.
Hertweck, H., 1931, Anatomie und Variabilität des Nervensystems und der Sinnesorgane von *Drosophila melanogaster* (Meigen), Z. wiss. Zool., 139:559.
Howse, P.E., 1974, Design and function in the insects brain, in: "Experimental Analysis of Insect Behavior", L.B. Browne, ed., Springer, Berlin, Heidelberg, New York.

Howse, P.E., 1975, Brain structure and behavior in insect, Ann. Rev. Entomol., 20:369.

Huber, F., 1955, Sitz und Bedeutung nervöser Zentren für Instinkthandlunger beim Männchen von *Gryllus campestris* L., Z. Tierpsychol., 12:12.

Huber, F., 1960, Untersuchungen über die Funktion des Zentralnervensystems und insbesondere des Gehirns bei der Fortbewegung und der Lauterzeugung der Grillen, Z. vergl. Physiol., 44:60.

Klemm, N., 1976, Histochemistry of putative transmitter substances in the insect brain, Progress in Neurobiology, 7:99.

Kloot, W.G. van der, and Williams, C.M., 1953, Cocoon construction by the Cecropia silkworm, Behavior, 5:141.

Laudien, H., and Iken, H.-H., 1977, Ökologische Prägung und Proteinbiosynthese. Versuche mit *Drosophila melanogaster* Meigen, Z. Tierpsychol., 44:113.

Manning, A., 1967, "Pre-imaginal conditioning" in *Drosophila*, Nature, 216:338.

Maynard, D.M., 1967, Organization of central ganglia, in: "Invertebrate Nervous Systems",C.A.G. Wiersma, ed., University of Chicago Press, Chicago.

Menzel, R., Erber, J., and Masuhr, T., 1974, Learning and memory in the honeybee, in: "Experimental Analysis of Insect Behavior", L.B. Browne, ed., Springer, Berlin.

Otto, E., 1949, Untersuchung zur geruchlichen Orientierung bei Insekten, Zool. Jb., 62:66.

Palka, J., Lawrence, P.A., and Hart, H.G., 1979, Neural projection patterns from homeotic tissue of *Drosophila* studied in bithorax mutants and mosaics, Devel. Biol., 69:549.

Power, M.E., 1943, The brain of *Drosophila melanogaster*, J. Morphol., 72:517.

Siegel, R.W., and Hall, J.C., 1979, Conditioned responses in courtship behavior of normal and mutant *Drosophila*, Proc. Natl. Acad. Sci. USA, 76:3430.

Stocker, R.F., 1977, Gustatory stimulation of a homeotic mutant appendage, *Antennapedia*, in *Drosophila melanogaster*, J. Comp. Physiol., 115:35.

Strausfeld, N.J., 1976, "Atlas of an Insect Brain", Springer, Berlin.

Suzuki, H., and Tateda, H., 1974, An electrophysiological study of olfactory interneurones in the brain of the honeybee, J. Insect Physiol., 20:2287.

Thorpe, W.H., 1937, Olfactory conditioning in a parasitic insect and its relation to the problem of host selection, Proc. Roy. Soc., B 124:56.

Vowles, D.M., 1955, The structure and connections of the corpora pedunculata in bees and ants, Quart. J. Microscop. Sci., 96:239.

Weiss, M.J., 1974, Neuronal connections and the function of the corpora pedunculata in the brain of the american cockroach, *Periplaneta americana* (L), J. Morphol., 142:21.

Wadepuhl, M., and Huber, F., 1979, Elicitation of singing and courtship movements by electrical stimulation of the brain of the grasshopper, Naturwissenschaften, 66:320.

DISCUSSION

J.C. Hall: In your selection procedure, isn't mating important? You have mutagenized, set up lines, and used those lines for histological screening that were successful in mating. Therefore, you may have selected against mutants which are defective in courtship. Possibly a more severe allele of mushroom-bodies-deranged would be defective in courtship. Such strains would appear to be sterile and therefore would be lost in your selection scheme.

M. Heisenberg: Yes, that is a good point.

VISUAL GUIDANCE IN *DROSOPHILA*

Karl G. Götz

Max-Planck-Institut für biologische Kybernetik
Tübingen
Federal Republic of Germany

INTRODUCTION

One of the last enquiries inspired by Theodosius Dobzhansky is entitled, "How far do flies fly?".[1] The paper refers to several field studies where a labelled strain of *Drosophila* was released and its dispersal measured by recapture of labelled flies on subsequent days. If the dispersal is simply due to random movements of the flies, then it should be analogous to the dispersal of small particles performing Brownian movements. Expected, in this case, is a normal distribution of the flies such that the increase of their mean distance from the release point is proportional to the square root of the time elapsed since the release. The expected time dependence of the dispersal seems to hold, more or less, for colonies of *D. pseudoobscura*, and the diffusion model may be considered as a reasonable first approximation of the locomotor behavior. However, the expected profile of the distribution has not been verified. Conspicuously more flies were recaptured both near the release point and at the outer periphery of the field. This discrepancy was explained by the tendency of *Drosophila* either to remain in a favorable habitat, or to cover great distances in search of such a habitat. The observation suggests that the control of locomotion can be adapted by the fly to different situations and requirements. The locomotor behavior of *D. melanogaster* has been extensively studied in laboratory experiments. Most of the results obtained so far refer to optomotor responses which enable the fly to maintain a given course and altitude over extended periods of time.

THE OPTOMOTOR CONTROL OF COURSE AND ALTITUDE

The prominent sensory cue for the control of locomotion in a freely moving fly is the displacement of the retinal images of stationary landmarks. Utilization of this cue in guidance implies that information about the rotatory and translatory components of the displacement is retrieved by an appropriate set of movement detectors in the visual system, and conveyed to appropriate sites in the motor system. The displacements perceived during free motion can be simulated in a tethered fly. The subsequent sections comment on optomotor responses to the continuous movement of striped patterns on either side of these flies. The horizontal components of this stimulus elicit yaw-torque responses during flight, or turning responses on the ground, which counteract involuntary deviations from a straight course in the corresponding mode of locomotion. The vertical components elicit covariant responses of lift and thrust which enable the fly to maintain a given level of flight.[1-5]

The prominent types of movement detectors in the optomotor control system of *Drosophila* perform direction-specific neuronal interactions between contiguous visual elements in the two compound eyes.[6,7] The hexagonal geometry of these elements delimits six preferred directions of nearest-neighbor interaction on either side in which one-way movement detection reaches its maximum sensitivity.[8] The optomotor control of legs, wings and body posture requires dense networks of directionally homologous movement detectors which cover considerable fractions, if not all, of the visual field. These networks represent the inputs of direction-specific large field units used to integrate the local signals of the elementary movement detectors, and to convey the information to the motor system.

Figure 1 illustrates the flow of information between the stimulus detectors and the response effectors in *D. melanogaster*, wild-type "Berlin", as it appears at the present state of investigation. Constituents of the responses are:

1. <u>Leg stroke modulation</u>. This effect acts on the difference of walking speed on either side. Optomotor control of turning in the walking fly is in evidence for each of the three pairs of legs.[4]
2. <u>Wing stroke modulation</u>. Difference and sum of the beat amplitudes on either side are independently controlled by horizontal and vertical movement components, respectively. Maximum modulation is roughly equivalent to a 60% change in the average thrust of the corresponding wing in still air.
3. <u>Abdominal deflection</u>. An actively-induced posture effect facilitating steering during free flight at increased airspeed.
4. <u>Hind leg deflection</u>. Same as case (3) above. With few exceptions the hind legs were deflected simultaneously and in the same direction as the abdomen.
5. <u>Hitch inhibition</u>. The term "hitch" denotes a transient reduction of wing beat amplitude which seems to occur spontaneously and independently on either side. Hitches are comparatively frequent

VISUAL GUIDANCE   393

in the absence of pattern movement. Their inhibition under visual stimulation is equivalent to an increase of the average thrust of the corresponding wing.[5]

These five constituents of the optomotor responses can be deduced from four non-covariant response types. This is to say that the optomotor control of walking speed, hind leg deflection, abdomen deflection, wing beat amplitude, and hitch inhibition requires at least four independent signal channels on either side. The simplified tentative "wiring diagram" in Figure 1 describes the control of these quantities by the corresponding minimum of eight signal channels This minimum is likely to increase when more is known, for instance about the optomotor control of important parameters such as pitch and roll during flight.[9]

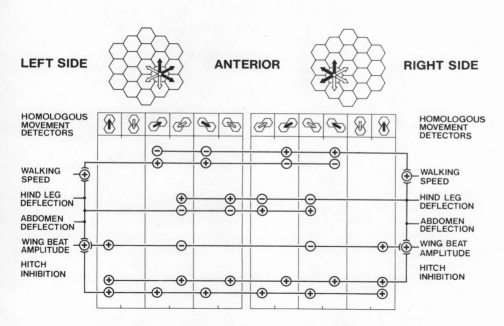

Figure 1: Tentative "wiring scheme" of the optomotor control system in *Drosophila*. Interaction between neighboring visual elements in the ganglia behind the eyes delimits six different groups of directionally homologous movement detectors. The responses of the detectors increase (+), or decrease (-), the output at particular sites of the motor system. The marginal symbols (+) refer to joint contributions of different control signals. The control of legs, wings and body posture requires, at least, four independent signal channels on either side of the fly. Recordings of muscular activity indicate further subdivisions of these channels.

The optomotor responses described so far are remarkable in many respects. They still occur if the speed of the retinal images is as slow as 1/h, or as fast as 50/s. Information processing in the optomotor control system requires several thousand of the estimated 100,000 nerve cells in *Drosophila*. Moreover, the responses can be predicted, with some accuracy, from direction of movement, speed, and structure of the visual stimulus on either side. This means that optomotor behavior, in the restricted sense, is essentially invariant to internal parameters, or the "mood" of the fly under investigation. Dissection of optomotor behavior as shown in Figure 1 invites both the classification of optomotor defective mutants,[10-14] and the identification of the targets of the descending signals within the motor system.[15,16]

OPTOMOTOR RESPONSES IN THE VICINITY OF VISUAL LANDMARKS

The optomotor control of course and altitude in *Drosophila* facilitates long range excursions in an open environment, and even the crossing of a desert[1] if this is not achieved by passive transport. Observation of the behavior of *Drosophila* in a narrow environment shows immediately the predominance of search maneuvers which seem to occur in spite of the optomotor responses.

Paradoxically, a significant fraction of the maneuvers occurs just because of these responses.[17] The following example may explain the state of optomotor instability which leads to "pseudo search" maneuvers in a narrow environment. Assume that the fly is pursuing a curved trajectory in a horizontal plane. The speed of the retinal images of the surroundings on either side can be considered as the superposition of two components. The component due to the rotatory movement of the fly is independent of the distance of the stationary landmarks, whereas the component due to the translatory movement is not. It is common experience that, in the latter case, remote landmarks (mountains) appear almost motionless, whereas landmarks in the vicinity of the observer (telegraph poles) pass by at considerable apparent speed. The three drawings on the right of Figure 2 show, from top to bottom, the increasing influence of the "translatory bias" on the actual speed of the retinal images if the mean distance between the trajectory of the fly and the surrounding landmarks is gradually diminished. The uppermost diagram refers to an open environment. The translatory bias is missing, and the rotation of the fly determines both the magnitude and the sign of the optomotor course control response. The diagram below shows how the speed of the rotatory stimulus component is modified by the superimposed front-to-back movement of a comparatively slow translatory bias. The rotation of the fly still determines the sign of the course control response. The lowermost diagram refers to a narrow environment in which the translatory bias is fast enough to reverse the direction of the rotatory stimulus component in one of the eyes. The speed of the antagonistic stimuli on either side now determines

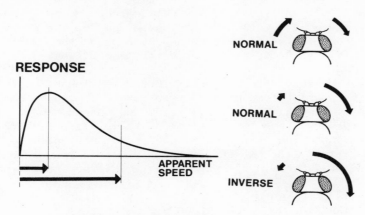

Figure 2: The collapse of course control in a narrow environment. The length of the arrows in front of the fly indicates the actual speed of the retinal images during locomotion on a curved trajectory. The translatory component of the visual stimulus is increasing with the proximity of surrounding landmarks which are moving past the fly. The diagrams show, from top to bottom, the gradual increase of the translatory bias. The concomitant inversion of the optomotor course control response gives rise to the apparently spontaneous "pseudo search" behavior.

the result of their competition for the optomotor guidance of the fly. The inverted stimulus, as the slower of the two rivals, elicits the stronger response in all situations similar to the one shown in the left of Figure 2. These situations represent states of positive optomotor feedback. By actively increasing sporadic deviations from a straight course the feedback initiates the apparently spontaneous turns of "pseudo search" in the vicinity of visual landmarks. The turning continues and the regular course control is suspended, as long as the bias induced inversion of the stimulus persists.

The tedious deduction of peculiarities in optomotor guidance was necessary in order to show that the rather complex "pseudo search" behavior of a fly is no more than the inevitable by-product of its optomotor course control response. Accordingly, "pseudo search" classifies as an inherently predictable response which is expected in one half of all possible states of optomotor stimulation. This expectation has been confirmed by "open loop" experiments with stationarily walking flies. With freely walking flies search is observed more frequently in a narrow environment than in an enlarged replica of this environment.[17] The unintended turning may be advantageous for the fly as it facilitates exploration and social contact in the vicinity of visual landmarks. It is tempting to speculate that the frequency of unintended turning be used by the fly as a cue for the distance of visual surroundings at a range that is far beyond the possible scope of binocular depth perception.

To prevent misunderstandings, it should be emphasized that *Drosophila* is capable of meaningful behavior which cannot be explained away by simple sets of ubiquitous responses. However, the distinction between these two levels of behavior is not as trivial as it may have appeared in the past. What distinguishes the "pseudo search" maneuvers in the present example from more endogenously elicited traits of behavior is predictability rather than apparent simplicity.

FIXATION AND TRACKING OF VISUAL OBJECTS

Preference of particular directions in space is not expected if the fly is surrounded by a sufficiently homogeneous distribution of visual landmarks. A cluster of landmarks, however, may acquire the character of a visual object which exerts considerable attraction on the fly. Guidance towards the object seems to be accomplished by a peculiar asymmetry of the optomotor responses which favors back-to-front movements of the fluctuating retinal image. The asymmetry thus induces turning of the fly towards the visual object, fixation of the object in the frontal visual field and also tracking of a mobile object.[18-20] Assuming random fluctuations of orientation a phenomenological theory devised by W. Reichardt and T. Poggio can be applied which gives the probability of any direction the fly may take with respect to one or more visual objects. Although these results are not always in agreement with actual histograms of the total time spent by *Drosophila* in any of these directions,[10-14,21-25] the visual guidance required for the approach of an object appears to be predictable in the restricted sense of a probabilistic description.

The formal explanation of fixation and tracking in *Drosophila* suggests that these maneuvers are determined by optomotor responses which also elicit course control and "pseudo search" in the appropriate environment. Most probably, this notion is only superficially true. The stimulus produced by a fluctuating visual object in one of the eyes seems to elicit two distinct components of optomotor guidance: a kinetic component which diverts the course of the fly in the direction of the horizontal displacement, and a flicker component which diverts the course towards the location of the stimulus.[18-20] The course control/"pseudo search" behavior of a fly receiving bilateral stimulation from many objects is mainly due to the comparatively reliable kinetic component of the response. The less reliable flicker component determines most of the fixation/tracking behavior, especially if the fly is receiving unilateral stimulation from the fluctuation of a single visual object. Comparison of the reliability of the induced behavior shows that *Drosophila* is easily held in a state of continuous course control which lasts throughout an experiment of many hours duration. However, a corresponding state of continuous fixation has not been achieved so far. The comparsion illustrates both the incongruence of the response components and the transition to partially endogenously elicited traits of behavior.

To demonstrate this transition, a flightless fly is placed on a circular disc which is surrounded by a moat to prevent its escape. The peripheral area of the disc is imaged on a scanning device which records the trace of the fly during a period of 200 s. The visual stimulus consists of two identical black objects, one on either side, which are seen against the white background of a surrounding cylinder.

The results in Figure 3 show that this situation is as perplexing to *Drosophila* as it has been to Buridan's ass in the early 14th century. The wild-type fly is running to and fro, between the objects, and may continue to do so for a very long time. The behavior entails cessation of object fixation and reversal of the course whenever the fly arrives at the moat in front of its target. Unexpectedly, the trace of the fly thus reveals a fixation-antifixation dichotomy[26] which is not inherent in the present repertoire of optomotor responses. Essentially the same dichotomy is found also in the behavior of the mutant *omb^{H31}* which has been isolated and described by M. Heisenberg and co-workers.[12] This "optomotor blind" mutant is structurally impaired in the lobula plate giant neurons of its visual system. Concomitantly, the movement-induced course control response is severely diminished. Minor defects have been found also in the fluctuation and fixation behavior of this mutant. However, the electroretinogram and the movement-induced responses of altitude control and landing seem to be intact. A more conspicuous genetic degradation of the behavior in Buridan's paradigm is found in the group of "no object fixation" mutants[25] such as *nofA^{S100}*, a mutant also isolated and described by M. Heisenberg and co-workers.[12,13] The behavior of this mutant is probably due to accelerated decay of the flicker component in a new situation. The kinetic component is not diminished, and the movement-induced course control response seems to be intact. The trace of this mutant is similar to the trace of a completely blind "sine oculis" mutant *so* which has been checked for absence of photoreceptors.

THE AVOIDANCE OF GUIDANCE

The transition between optomotorically stable orientations of a fly in an object space is superficially similar to the transition of a ball between the numbered troughs of a roulette wheel. The depth of the troughs determines the rotatory impluse required to

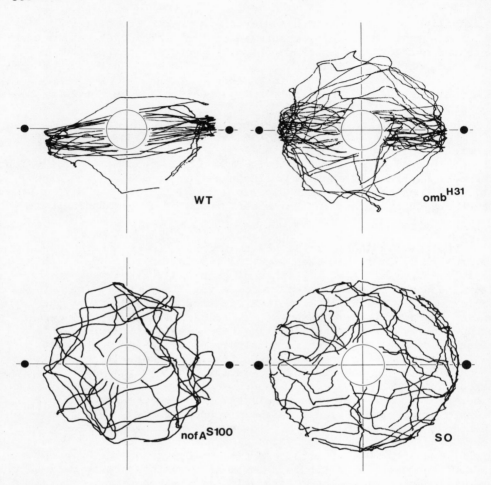

Figure 3: Choice between two identical visual objects in "Buridan's paradigm". Flies with shortened wings are allowed to walk for 200 s on an isolated disc of about 9 cm ϕ. The automatic recording of the path is blanked out in the center of the disc and also at locomotor velocities below threshold. The dots on either side of the diagrams illustrate the diameter and position of the rod-like objects outside of the disk. WT: wild-type Berlin; omb^{H31}: Heisenberg's "optomotor blind mutant"; $nofA^{S100}$: Heisenberg's "no object fixation" mutant; so: completely blind "sine oculis" mutant.

activate such transitions. The frequency of spontaneous transitions of a fly between any two stable orientations can be predicted, accordingly, from the profile of the potential troughs and from the fluctuation of the rotatory impulse.[20] Random fluctuation of these impulses is obviously not the only pattern of spontaneous activity in flies. During fixed flight, *Drosophila* also performs surprisingly uniform "body-saccades"[13] similar to those found in the freely flying hoverfly *Syritta*.[27] Direction and frequency of the saccades in *Drosophila* depend on both the visual stimulus and the mode of fixation. A fly idling in a non-fixation state often produces a train of saccades of the same polarity. Each saccade is accompanied by the directionally selective suppression of course control responses which otherwise would have inhibited the active turning.

The visual world of a fly offers a variety of meaningful cues for orientation in space. Structure and apparent speed of the surroundings are not the only aspects to which the fly responds. The spatial distribution of intensity, color, degree of polarization, and direction of polarization of the incident light provides *Drosophila* with reference systems for independent guidance in orientation.[10,11,28-30] The withdrawal from guided orientation may be equally important for the fly. As shown before, withdrawal is achieved, passively, during "pseudo search". The strategies of active withdrawal comprise the continuous "random search" due to the random fluctuation of the rotatory impulse, and the discontinuous "quantum search" due to the "body-saccades" of *Drosophila*. The regular U-turns performed by the wild-type upon arrival at one of the two objects in Buridan's paradigm are, most likely, examples of "quantum search" in the walking mode.

THE DISCRIMINATION OF OBJECTS

Unlike Buridan's ass, *Drosophila* keeps patrolling to and fro in a modified choice experiment where the two objects, or the corresponding potential troughs, represent visual stimuli of conspicuously different attraction for the fixating fly. The transient nature of the attraction by a pattern of arbitrary composition, and the recurrent conversion of fixation into anti-fixation, indicate the difficulty of assaying invariant constituents of pattern preference in *Drosophila*.

Figure 4 refers to a choice experiment where the freely walking fly is held for several hours in a state of spontaneous fluctuation around a fixed position and orientation to the patterns on either side of a tread compensator.[22,31] The cylindrical transducer on top of the compensator is sensitive to the displacements of a tiny sled of paramagnetic wire which is drawn by the fly. A sensor controlled servo system (not shown in the figure) counteracts the displacements by appropriate rotations of the ball on which the fly is walking.[4] Preference is defined, in the present experiment, as the average curvature of the path in revolutions/meter towards the preferred pattern. The data are directly obtained from the actual rotations of the ball. The broken line in the upper Figure 4 illustrates the average track of the stationary fly in a choice between differently dotted patterns of the same luminous area. The expectedly

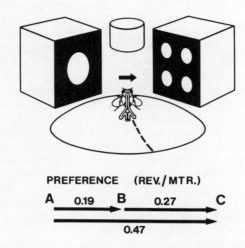

Figure 4: Choice between differently dotted patterns of the same luminous area. The fly is affixed to a sled and is kept walking for hours in nearly constant position and orientation on top of a tread compensator. The vector diagram below illustrates the seemingly linear relation of the preferences between any two of three differently dotted patterns.

small preference for the pattern on the right is invariant to the reversal of contrast, and lasts throughout the experiment.

Of particular interest is the number n of independent distinctive features accounting for the behavior of *Drosophila* in the choice experiment. This number characterizes the specificity of the discrimination between patterns, or between respective points in the n-dimensional feature space, required to describe the discrimination. The preference obtained in a choice between two patterns is functionally related with the coordinates of the two corresponding points in the feature space. It may be difficult to specify the appropriate relationship. However, in the present case of comparatively weak responses the preference should be roughly proportional to the distance between the corresponding points in an appropriately adapted feature space. At least under these conditions, the minimum number of necessary dimensions n can be derived by the following iterative procedure. The largest distance between any of the three arbitrary points A, B, and C is equal to the sum of the two smaller distances if the points are aligned in a 1-dimensional feature space. However, the largest distance is below this sum if a second dimension is required for the representation of the three points. In this case, the relations between four arbitrary points must be examined to find out whether or not a third dimension is required, and so on. The least specific discrimination is expected in a 1-dimensional feature space which resembles the ranking list derived from the results of matches between any two players in a club.

The results obtained with a series of dotted patterns of the same luminous area are summarized in the lower Figure 4. Length and direction of the arrows represent the weighted averages of preferences of 183 female flies which covered a total pathlength of 7547 m on the compensator. The arrows indicate that the average of the largest preference found in four different sets of three patterns is approximately equal to the sum of the averages of the two smaller preferences in these sets. Accordingly, the differently dotted patterns can be represented by points A, B, C,... in a 1-dimensional feature space such that the preference in a choice experiment is equivalent to the distance between the corresponding points. Within the limits of error, the result suggests that the least specific comparison is made between differently dotted patterns on either side of the visual system. The discrimination between these patterns seems to be due to the comparison of only one as yet unknown parameter which may have been derived by the fly from one or more properties of the visual objects.

The comparison of meaningless artificial patterns under the distressing conditions of the choice experiments is certainly not representative of the abilities *Drosophila* may have developed to discriminate species-, sex-, and body-specific patterns during social contact. The rich repertoire of visually guided tracking strategies found, for instance, in the hoverfly[27,32] illustrates the potentials and advantages of such a development. *Drosophila melanogaster* seems to court only on the ground. Recent investigations of the different modes of courtship tracking[33,34] have shown that the fast pursuit of the female is almost completely missing in the absence of visual input. Interestingly, the visual control of both orientation and distance in the pursuing male does not depend on the binocular perception of the female. The use of visual information is comparatively less important in the other modes of courtship tracking. The most characteristic sideward movements on a narrow arc around the female, the selection of the wing to be extended towards the female during these maneuvers, and the recognition of the female's abdomen in a tracking range of less than 1 mm body-to-body distance do not seem to require visual guidance. This may explain why the male courtship behavior maps to the posterior dorsal brain rather than to the optic lobes of *Drosophila*.[35]

Visual objects of interest to a fly rarely stand out as well circumscribed figures in the context of a natural habitat. The discrimination of figure and ground is, therefore, prerequisite to the recognition of a hidden object. One important clue for the discrimination is the apparently different speed of figure and ground in the visual field of a moving observer. The resulting spatial discontinuities of speed within the retinal images are due to the different distances of figure and ground. A neural device for the detection of discontinuous speed would disclose the contour of otherwise inconspicuous objects. Actually, this device must exist in the visual systems of both man and fly. The analysis of the neural interactions required for the spatial comparison of local speed in the visual field of the housefly is already impressively advanced.[20,36] Figure-ground discrimination has been demonstrated also in *Drosophila* and the properties of the underlying interactions seem to be similar.[13,25]

THE LINKING OF STRUCTURE AND FUNCTION

The investigation of the behavior of the visually guided fly has been accompanied by numerous attempts to identify the corresponding sites of information processing in the nervous system. A few of the more recent attempts will be mentioned here.

The understanding of relevant processes in the periphery of the visual system of *Drosophila* owes much to the genetic dissection of receptor systems and to refined electrophysiological techniques. The specific non-linear transducer processes occurring prior to movement detection can be derived, indirectly, from the optomotor responses to a particular illusion of movement which is elicited by simultaneous changes of the light flux in differently illuminated receptors of the movement detecting system in man and fly.[37]

The identification of movement-specific nervous activity has been pioneered by application of Sokoloff's method of "activity staining".[38] The visually stimulated fly is fed with tritium-labelled deoxyglucose, a substance accepted by nerve cells in need of glucose. Deoxyglucosephosphate, the product of phosphorylation, is not further metabolized and is therefore retained in the cells. This leads to enhanced accumulation of tritium in the physiologically active cells. The distribution of activity in the nervous tissue is made visible by subsequent autoradiography. To illustrate this method Figure 5 refers to the stimulation of homologous circular arrays of photoreceptors in the two eyes of *Drosophila*. The receptors on either side are equally activated by the continuous sinusoidal modulation of the local luminance on the corresponding projector screens. Simultaneous stimulation of receptors on the

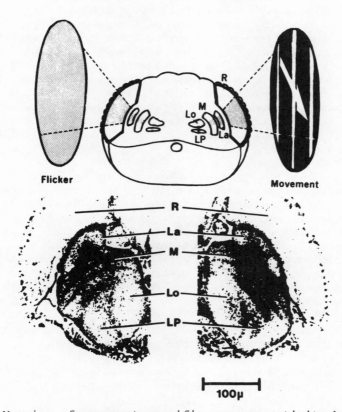

Figure 5: Mapping of movement-specific nervous activity by ^{3}H-deoxyglucose autoradiography in a horizontal section of the optic lobes of *Drosophila*. R: retina; La: lamina; M: medulla; Lo: lobula: LP: lobula plate. The schema on top shows on the left the simultaneous stimulation of a circular array of receptors by a flickering light source, and on the right the successive stimulation of a corresponding array of receptors by a moving grating. The receptors on either side are equally activated under the conditions of the experiment. The photomechanical reproduction of the autoradiogram gives a crude impression of the enhanced activity in the stimulated area of the medulla on the right which is attributed to movement-specific processing of visual information. Figure kindly provided by E. Buchner. For details see (38).

the left is achieved by the flickering of a homogeneous area whereas the successive stimulation of the receptors on the right is due to the continuous movement of a sinusoidal grating. Discrimination between these two modes of stimulation requires means for the comparison of receptor output. Different activity in the optic lobes on either side is, therefore, not expected at the level of the single receptors, but is likely to occur at the level of the movement detectors which respond specifically to successive stimulation of neighboring receptors. Comparison of the autoradiographs of these optic lobes shows a movement-specific increase of activity in a cross section of a circular area of the medulla on the right. This area is a neural projection of the successively stimulated receptor cells. Columnar elements of this area seem to process local information on movement in the corresponding retinal field. The method of activity staining thus complements the technique of genetically labelled mosaics of the acetylcholinesterase inactivating *Ace* gene which has been recently used for the localization of optomotorically relevant parts of the brain.[39]

Direct access to the neural activity of the optomotor control system in *Drosophila* is still one of the most recalcitrant technical problems. So far, movement sensitive descending channels have been identified in extracellular recordings of some of the 3578 nerve fibers in the female cervical connective.[40] More recently, a few of the direct flight control muscles have become accessible to electrophysiological probing.[15,41] Of these, the sternobasalar muscle and the anterior muscle of the first axillary are specifically activated by ventro-dorsal pattern movement in the frontal visual field.[16] Whether or not these muscles contribute to the optomotor control of altitude and pitch during flight remains to be seen. Actually, the responses of the direct muscles cannot be explained without further subdivision of the postulated descending muscular inputs in Figure 1. The subdivided inputs possibly specify position and direction of the moving stimulus in analogy to the movement-sensitive giant lobula plate neurons in the protocerebrum of the flies.[12,42-44]

To a student of visual guidance in *Drosophila* the increasing complexity of the subject may be sometimes depressing. But it would be much more so, if the several thousand neurons of the visual system were engaged in merely trivial routine.

ACKNOWLEDGEMENTS

Valuable results and competent suggestions have been contributed by my friends and colleagues E. Buchner, H. Bülthoff, R. Cook, G. Heide, M. Heisenberg, R. Hengstenberg and W. Reichardt.

REFERENCES

1. J.R. Powell and T. Dobzhansky, How far do flies fly?, Amer. Scientist 64:179 (1976).
2. K.G. Götz, Optische Untersuchungen des visuellen Systems einiger Augenmutanten der Fruchtfliege Drosophila, Kybernetik 2:77 (1964).
3. K.G. Götz, Flight control in Drosophila by visual perception of motion, Kybernetik 4:199 (1968).
4. K.G. Götz and H. Wenking, Visual control of locomotion in the walking fruitfly Drosophila, J. Comp. Physiol. 85:235 (1973).
5. K.G. Götz, B. Hengstenberg, and R. Biesinger, Optomotor control of wing beat and body posture in Drosophila, Biol. Cybern. 35:101 (1979).
6. E. Buchner, Elementary movement detectors in an insect visual system, Biol. Cybern. 24:85 (1976).
7. E. Buchner, K.G. Götz, and C. Straub, Elementary detectors for vertical movement in the visual system of Drosophila. Biol. Cybern. 31:235 (1978).
8. K.G. Götz, and E. Buchner, Evidence for one-way movement detection in the visual system of Drosophila, Biol. Cybern. 31:243 (1978).
9. J. Blondeau (in preparation).
10. M. Heisenberg and K.G. Götz, The use of mutations for the partial degradation of vision in Drosophila melanogaster, J. Comp. Physiol. 98:217 (1975).
11. M. Heisenberg and E. Buchner, The role of retinula cell types in visual behavior of Drosophila melanogaster, J. Comp. Physiol. 117:127 (1977).
12. M. Heisenberg, R. Wonneberger, and R. Wolf, Optomotor-blind $H31$ a Drosophila mutant of the lobula plate giant neurons. J. Comp. Physiol. 124:287 (1978).
13. M. Heisenberg and R. Wolf, On the fine structure of yaw torque in visual flight orientation of Drosophila melanogaster, J. Comp. Physiol. 130:113 (1979).
14. M. Heisenberg, Genetic approach to a visual system, in: "Handbook of Sensory Physiology",Vol. VII/6A Comparative Physiology and Evolution of Vision in Invertebrates, H. Autrum, ed., Berlin-Heidelberg-New York, Springer (1979).
15. G. Heide, Proprioceptorische Beeinflussung der Impulsmusterbildung im neuromotorischen System fliegender Dipteren, Verh. Dtsch. Zool. Ges. 1978, p.256 (1978).
16. G. Heide and K.G. Götz, (in preparation).
17. K.G. Götz, The optomotor equilibrium of the Drosophila navigation system, J. Comp. Physiol. 99:187 (1975).
18. W. Reichardt and T. Poggio, Visual control of orientation behavior in the fly. I. A quantitative analysis, Quart. Rev. Biophys. 9:311 (1976).
19. T. Poggio and W. Reichardt, Visual control of orientation behavior in the fly. II. Towards the underlying neural interactions, Quart. Rev. Biophys. 9:377 (1976).

20. W. Reichardt, Functional characterization of neural interactions through an analysis of behavior, in:"The Neurosciences" Fourth Study Program, F.O. Schmitt and F.G. Worden, eds., The MIT Press, Cambridge, Mass. (1979).
21. K.G. Götz, Hirnforschung am Navigationssystem der Fliegen, Naturwissenschaften 62:468 (1975).
22. K.G. Götz, Sehen, Abbilden, Erkennen - Verhaltensforschung am visuellen System der Fruchtfliege Drosophila, Verh. Schweiz. Naturf. Ges. 1975, p. 10 (1975).
23. E. Horn and R. Wehner, The mechanism of visual pattern fixation in the walking fly Drosophila melanogaster, J. Comp. Physiol. 101:39 (1975).
24. E. Horn, The mechanism of object fixation and its relation to spontaneous pattern preferences in Drosophila melanogaster, Biol. Cybern. 31:145 (1978).
25. H. Bülthoff, (in preparation).
26. H. Bülthoff, K.G. Götz, and M. Herre, (in preparation).
27. T.S. Collett and M.F. Land, Visual control of flight behavior in the hoverfly, Syritta pipiens L., J. Comp. Physiol. 99:1 (1975).
28. K.F. Fischbach, Simultaneous and successive color contrast expressed in "slow phototactic"behavior of walking Drosophila melanogaster, J. Comp. Physiol. 130:161 (1979).
29. R. Willmund, Light induced modification of phototactic behavior of Drosophila melanogaster, II. Physiological aspects, J. Comp. Physiol. 129:35 (1979).
30. B. Gebhardt, R. Wolf, R. Gademann, and M. Heisenberg, Polarization sensitivity of course control in Drosophila melanogaster, (in preparation).
31. K.G. Götz, Spontaneous preferences of visual objects in Drosophila, Drosophila Inform. Serv. 46:62 (1971).
32. T.S. Collett and M.F. Land, How hoverflies compute interception courses, J. Comp. Physiol. 125:191 (1978).
33. R. Cook, The courtship tracking of Drosophila melanogaster, Biol. Cybern. 34:91 (1979).
34. R. Cook, The extent of visual control in the courtship tracking of D. melanogaster, Biol. Cybern. (in press).
35. J.C. Hall, Control of male reproductive behavior by the central nervous system of Drosophila: Dissection of a courtship pathway by genetic mosaics, Genetics 92:437 (1979).
36. W. Reichardt and T. Poggio, Figure-ground discrimination by relative movement in the visual system of the fly, I. Experimental results, Biol. Cybern. 35:81 (1979).
37. H. Bülthoff and K.G. Götz, Analogous motion illusion in man and fly, Nature 278:636 (1979).
38. E. Büchner, S. Büchner and R. Hengstenberg, 2-Deoxy-D-glucose maps movement-specific nervous activity in the second visual ganglion of Drosophila, Science 205:687 (1979).
39. R.J. Greenspan, J.A. Finn, and J.C. Hall, Acetylcholinesterase mutants in Drosophila and their effects on the structure and

and function of the central nervous system, J. Comp. Neurol. (in press).
40. R. Hengstenberg, The effect of pattern movement on the impulse activity of the cervical connective of Drosophila melanogaster, Z. Naturforsch. 28c:593 (1973).
41. A.W. Ewing, The neuromuscular basis of courtship song in Drosophila: the role of the direct and axillary wing muscles, J. Comp. Physiol. 130:87 (1979).
42. K. Hausen, Functional characterization and anatomical identification of motion sensitive neurons in the lobula plate of the blowfly Calliphora erythrocephala, Z. Naturforsch. 31c: 629 (1976).
43. R. Hengstenberg, Spike response of "non-spiking" visual interneurone, Nature 270:338 (1977).
44. H.E. Eckert and L.G. Bishop, Anatomical and physiological properties of the vertical cells in the third optic ganglion of Phaenicia sericata, J. Comp. Physiol. 126:57 (1978).

EFFECTS OF A CLOCK MUTATION ON THE SUBJECTIVE DAY -- IMPLICATIONS FOR A MEMBRANE MODEL OF THE *DROSOPHILA* CIRCADIAN CLOCK

Ronald Konopka and Dominic Orr

Division of Biology
California Institute of Technology
Pasadena, California 91125 U.S.A.

INTRODUCTION

The fly brain, like the human brain, possesses an oscillatory system which enables the brain to form a model of the 24 hour cycle in the external environment. This oscillator is termed circadian, since it continues to oscillate in the absence of environmental cues with a period of about a day.[1] The molecular mechanism of the circadian oscillator in any organism is as yet unknown. Since the fruit fly is an excellent organism for the study of the effects of mutations on nervous system function, chemically induced mutations affecting circadian rhythmicity were isolated.[2,3] These mutations behave as alleles of a single genetic locus, *per*, which has been mapped by means of deletions to the 3B1-2 region of the X chromosome.[2-4] They affect the periodicity of the eclosion rhythm and the adult locomotor activity rhythm in a similar manner: the short-period allele, *per^s*, shortens the period from 24 hours in wild-type to 19 hours in the mutant; the long-period allele, *per^l*, lengthens the period to 29 hours; and an arrhythmic allele, *per^o*, produces aperiodic eclosion and activity.

The *per* locus is contained within the zeste-white region (bands 3A3-3C2), where the number of salivary chromosome bands apparently corresponds to the number of complementation groups formed by lethal mutations induced in this region.[5] The *per* locus is not, however, allelic to any of these lethal mutations.[4] Furthermore, genetic deletions which overlap only in the 3B1-2 region produce an aperiodic phenotype.[4] These results are consistent with the proposition that the *per* locus is a nonessential locus, the deletion of which results in aperiodicity but not lethality.

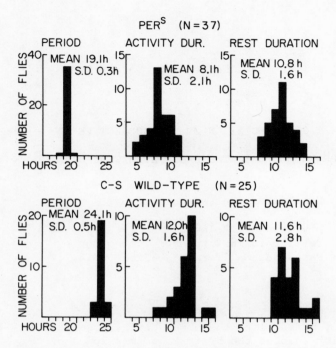

Figure 1: The period and duration of activity and rest in locomotor activity rhythms of individual per^S and wild-type flies.

In order to determine how the subjective day and night in the circadian cycle are affected by the per^S mutation, we measured the duration of activity and rest in the locomotor activity cycle of adult per^S and wild-type flies. These results suggest that this mutation is affecting the duration of the molecular processes corresponding to the subjective day, while the duration of the subjective night is unaffected. This suggestion is consistent with previous results from tests of per^S on the duration of the light-insensitive part of the light response curve for the eclosion rhythm.[2,6]

METHODS

Flies were grown on cornmeal medium in constant light. The method for measuring eclosion and activity rhythms has been described previously.[2,3] The activity and rest durations in the cir-

Table 1. Parameters of the Light Response Curves of the per^s and Wild-type Eclosion Oscillators.[2,6]

Genotype	Maximum Delays and Advances	Duration of Light-Insensitive Portion	Duration of Light-Sensitive Portion
Wild-type	about 0.15 cycle	12 hr	12 hr
per^s	0.5 cycle	7 hr	12 hr

cadian activity cycle of wild-type and per^s flies were measured in the following manner. Digital recordings of the hourly activity of individual flies were made in constant infrared light. The activity onset for each cycle was determined to be the time at which 20% of the total activity of that cycle had occurred. Since activity onsets are not as sharp as offsets, this procedure helped minimize differences due to variability in small amounts of activity early in the subjective day. The activity offset for each cycle was determined as the time at which 95% of the activity of that cycle had occurred. Duration of activity for each cycle was measured as the difference between time of onset and time of offset. Duration of rest was measured as the difference between time of offset and time of the following onset. The period of each activity rhythm was measured as the slope of the regression line fit to the activity offsets. Each run lasted from 4 to 12 cycles at a temperature of 22°C.

RESULTS AND DISCUSSION

Differences Between Wild-type and per^s Activity Cycles

Figure 1 shows the distribution of average periods of the activity cycles for 37 per^s and 25 wild-type flies, as well as the distribution of the average durations of the active and rest portions of the cycles. The duration of the active part of the per^s cycle is significantly shorter than that of wild-type ($p < 0.001$, t-test), while the durations of the rest part of the per^s and wild-type cycles are not significantly different ($p > 0.5$). Thus, the shortening of the activity cycle by the per^s mutation can be entirely accounted for by a shortening of the subjective day, during which the fly is active.

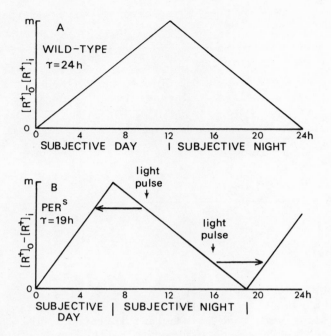

Figure 2: The establishment and dissipation of a hypothetical ion gradient during the circadian cycle of wild-type and per^S flies.

Differences Between Wild-type and per^S Eclosion Oscillators

Circadian oscillators can be reset by external stimuli. Depending on the time within the circadian cycle it is administered, a single light or temperature pulse may advance the oscillator and the rhythm being assayed relative to an unstimulated control, or delay them, or have no net effect on them. A plot showing the net advances and delays produced by stimuli given throughout one cycle is known to as a response curve. These response curves usually have a stimulus-sensitive portion, when the stimulus produces delays and advances, as well as a stimulus-insensitive portion when the stimulus has no net effect. To determine the effect of the per^S mutation on the light response curve of the eclosion oscillator, the net advances and delays produced by light pulses of white fluorescent light, 40 minutes in duration and 300 foot-candles in intensity, were determined for the wild-type and per^S strains.[2] The

Table 2. Comparison of Advances and Delays in the Eclosion Rhythm Induced by Light Pulses Administered to the per^S Mutant During the Subjective Night with Those Predicted by the Model

Hours Into Subjective Night	Advances and Delays (Hours)	
	Observed	Predicted
2	-3	-3
4	-6	-6
6	-9	-9
8	+6	+6
10	+3	+3
12	0	0

results indicated that the mutation had two effects (Table 1). First, the per^S mutation increased the ability of the oscillator to be reset by these light pulses; that is, the amplitude of the response curve of the mutant was increased relative to that of wild-type. In addition, the duration of the light-insensitive (subjective day) portion of the response curve was decreased, while the duration of the light-sensitive portion (subjective night) remained the same as in wild-type (Table 1). Again, the decrease in the length of the short-period cycle can be entirely accounted for by the decrease in duration of the subjective day.

These results lead to a simple model in which two different molecular processes are involved in the basic oscillator, one during the subjective day and one during the subjective night. The per^S gene product is involved in the subjective day process, while the subjective night process involves molecules coded for by another genetic locus or loci.

A Membrane Model for the *Drosophila* Clock

We propose that the basic oscillation consists in the establishment and depletion of an ion gradient across a membrane. During the subjective day, an ion pump establishes the gradient, while during the subjective night the gradient is depleted through open photosensitive ion channels. The ion pump is thought to require metabolic energy, and thus may have ATPase activity. We envision the entire process as follows. At the start of the subjective day, there is no gradient. The absence of the gradient (in darkness) or the beginning of the light period (in a light-dark cycle) turns on the ion pump and initiates the formation of the gradient (Figure 2). At the end of the subjective day, the gradient reaches a maximum value, m. The attaining of this value m turns the pump off and allows the photosensitive ion channels to open in darkness. If the oscillator is in a light-dark cycle and the light period is prolonged, the light maintains the channels in the closed state and the gradient at its maximum value. Once the dark period begins, the channels open and allow the gradient to be depleted. When the gradient reaches a zero value, the pump is again activated, and the process repeats itself. Resetting during the subjective night occurs in the following way. A maximal light stimulus causes all the channels to close, and the pump to be activated. The value of the gradient is at first the same as during the light pulse, but the pump now gradually increases the gradient until it again reaches its maximum value, at which time the subjective night begins again. Thus a light pulse early in the subjective night will produce a net delay, while a pulse late in the subjective night will produce a net advance (Figure 2, lower). We do not intend this model for the basic oscillator to account for complex transients in light resetting, which are probably due to interactions of a "slave", or driven oscillator, with the driver.[7] Therefore, for purposes of this discussion, we consider a large delay as equivalent to a small advance, and a small delay equivalent to a large advance. This model for resetting is similar in some respects to that proposed by Pavlidis for the strong resetting of the *Drosophila pseudoobscura* eclosion oscillator.[8] The predicted advances and delays are in good agreement with actual values observed for the per^S mutant (Table 2), which also exhibits strong resetting (maximum advances and delays of ½ cycle). The model does not, however, successfully predict weak resetting behavior (maximum advances and delays of about 0.15 cycle) as is shown by wild-type *D. melanogaster* (Table 1). Perhaps weak resetting is also due to an interaction between the basic driving oscillator with a driven system, and therefore does not solely reflect the mechanism of the driver.

Our model is similar to the membrane models proposed by Njus et al.[9] and Sweeney[10] in that all three models propose active and passive membrane processes as the mechanism of the oscillator. All three models also implicate ions and ion gradients in the mechanism. The model of Njus et al., like the *Drosophila* model presented here, includes photosensitive ion gates. In their model, light opens the gates, whereas in our model light closes the gates. The model of Njus et al. postulates a requirement for lateral diffusion of proteins in the membrane, wherease our model has no such requirement. A major difference between our model and those of Njus et al. and Sweeney is the phase of the active and passive processes. In our model, one process coincides with the subjective day and the other with the subjective night. In the other two models, the active process occurs from the midpoint of the subjective day to the midpoint of the subjective night. In our model, the per^S mutation increases the activity of the ion pump, resulting in a shorter subjective day, since the maximum value of the gradient is attained more rapidly. This is in agreement with genetic evidence suggesting that increased activity of the *per* locus product results in a shortening of the period of the oscillator.[11]

With the model we have presented, it is possible to obtain short-period and long-period mutants in several ways. Short-period mutants may be obtained by an increase in pump activity (as we have suggested for per^S), by decreasing the value of the gradient needed to begin the subjective night, or by increasing the number of channels so that the gradient is depleted more rapidly. Long-period mutants may be obtained by decreasing pump activity, raising the maximum value of the gradient, or decreasing the number of channels. Additional genetic loci affecting periodicity might code for different subunits of the pump or channel proteins.

The main feature of our model is to suggest separate molecular processes for subjective day and subjective night. Analysis of light response curves and activity cycles of other *per* alleles and clock mutations at other loci should yield additional insight into the function of gene products within the circadian cycle.

ACKNOWLEDGEMENTS

We thank David Hodge and Michael Walsh for help with the design and construction of the activity monitors. This research was supported by grants from the USPHS and the Whitehall and Pew Foundations.

REFERENCES

1. F. Halberg, Circadian (about twenty-four-hour) rhythms in experimental medicine, Proc. Roy. Soc. Med. 56: 253 (1963).
2. R. Konopka, Circadian clock mutants of *Drosophila melanogaster*, Ph.D. thesis, California Institute of Technology (1972).
3. R.J. Konopka and S. Benzer, Clock mutants of *Drosophila melanogaster*, Proc. Natl. Acad. Sci. USA 68: 2112 (1971).
4. M.W. Young and B.H. Judd, Nonessential sequences, genes and the polytene chromosome bands of *Drosophila melanogaster*, Genetics 88: 723 (1978).
5. B.H. Judd, M.W. Shen and T.C. Kaufman, The anatomy and function of a segment of the X chromosome of *Drosophila melanogaster*, Genetics 71: 139 (1972).
6. R.J. Konopka, Genetic dissection of the *Drosophila* circadian system, Fed. Proc. 38: 2602 (1979).
7. C.S. Pittendrigh, V.G. Bruce and P. Kaus, On the significance of transients in daily rhythms, Proc. Natl. Acad. Sci. USA 44: 965 (1958).

8. T. Pavlidis, A mathematical model for the light affected system in the *Drosophila* eclosion rhythm, Bull. Math. Biophysics 29: 291 (1967)
9. D. Njus, F.M. Sulzman and J.W. Hastings, Membrane model for the circadian clock, Nature 248: 116 (1974).
10. B.M. Sweeney, A physiological model for circadian rhythms derived from the Acetabularia rhythm paradoxes, Int. J. Chronobiol. 2: 25 (1974).
11. R.F. Smith and R.J. Konopka, Genetic dosage effects at the *per* locus, Manuscript in preparation.

APPARENT ABSENCE OF A SEPARATE B-OSCILLATOR IN PHASING THE CIRCADIAN RHYTHM OF ECLOSION IN *DROSOPHILA PSEUDOOBSCURA*

M.K. Chandrashekaran
School of Biological Sciences
Madurai University
Madurai, India

ABSTRACT

A well known 2-oscillator model proposed by Pittendrigh and Bruce (1957) postulates that (1) a light sensitive and temperature compensated A-oscillator (master) and (2) a temperature sensitive and light refractory B-oscillator (slave) underlie, govern and gate the circadian rhythm in the eclosion of *Drosophila pseudoobscura*. The principal merit of this coupled oscillator model is that it seemed to "explain" the phenomenon of "transients" and at least one of its tenets has found experimental proof (Chandrashekaran, 1967). There has not been any proof for nor any evidence against the existence of a separate B-oscillator. Experiments carried out by me to evoke the responses of the B-oscillator to the exclusion of the A-oscillator in *Drosophila pseudoobscura* have failed. Temperature pulses and light pulses (regardless of the lighting conditions of the cultures and experiments) always shifted or reset what has been considered the A-oscillator leaving no evidence or clue for the existence or involvement of a separate B-oscillator.

Much before genetic mutants (Konopka and Benzer, 1971) of the *Drosophila* clock were reported or transplantation of circadian oscillators in *Drosophila* were made (Handler and Konopka, 1979), Pittendrigh and Bruce (1957) proposed an explicit formal model to account for the response features of the circadian rhythm in the eclosion of *Drosophila pseudoobscura* flies. The model envisaged a coupled oscillator set up with one of the oscillators: the A-oscillator being light sensitive and the master/pacemaker; and the B-oscillator presumed to be insensitive to light but sensitive to temperature and the slave to the A-oscillator. The A-oscillator was temperature-compensated (thus qualifying for chronometry) and instantaneously phase shifted by light perturbations (Chandrashekaran, 1967) and B-oscillator gradually caught up and re-established phase with the A-oscillator after a few cycles. The efforts of the B-oscillator to regain phase with the A-oscillator were reflected in the "transients" that follow light and temperature perturbations. The principal merit of this coupled oscillator model was that it seemed to "explain" the phenomenon of transients rather picturesquely even though other models could do the same (Bunning and Zimmer, 1962) implicating only a single oscillator. One of the important tenets of the coupled oscillator model, the instantaneous resettability of the light sensitive basic oscillation could even be experimentally proved (Chandrashekaran, 1967). More recent work on *Drosophila* (Engelmann and Mack, 1978) and humans (Wever, 1979) implicate a multiplicity of oscillators governing a heirarchy of circadian functions. The Pittendrigh and Bruce (1957) model and its specific postulate of a separate temperature sensitive, light refractory, slave B-oscillator have, however, lingered on with neither proof for nor evidence against the latter. I have experimentally reexamined the hypothesis and report here my failure to find any evidence or even cues for the existence of a separate B-oscillator.

METHODS

Large cultures of *Drosophila pseudoobscura* (strain PU 301 collected by Pittendrigh in the Sierra Nevada in 1947) were raised in "safe" red light (610 nm, 50% transmission at 605 nm) at 20 ± 0.5°C in plastic troughs, 22 cm in diameter and 10 cm deep. A circadian rhythmicity was set in motion in the otherwise arrhythmic dark-reared pupal populations by exposing them in darkness to high temperature pulses (30°C) or to low temperature pulses (10°C) for periods of 12 hours. The beginning of the temperature treatment, which was non-recurrent, was arbitrarily taken as "0" hour of the experiment. The lower and higher temperature levels of 10°C and 30°C chosen in these experiments for the rhythm-inducing events did not appear to have any detrimental effects as was evident from the 85-90% eclosion.

The pattern of eclosion was measured using a light beam and photocell device designed by Engelmann and described in Maier (1973). The hourly rate of eclosion could be recorded over periods of 7-8 days after temperature and light treatment on the 20 channels of an Esterline Angus Event Recorder. Median values of the peaks of eclosion were calculated from the fly counts made directly from the charts. In all calculations of phase shifts only the medians of eclosion peaks occurring on days 4, 5, 6 and 7 are taken into consideration thus precluding the "transient" peaks of eclosion from days 1 to 3.

EXPERIMENTS AND RESULTS

Temperature Evoked Rhythms and Light Perturbations

The process of eclosion in *Drosophila pseudoobscura* flies is totally arrhythmic when cultures of pupae are raised in constant light (LL) or constant darkness (DD) from egg-stage onwards. A single non-recurring light or temperature perturbation will induce a rhythmicity in the eclosion activity of the flies which then emerge in "gates of time" of 5-8h. Rhythms initiated by light pulses and temperature pulses have different time courses but about the same period of about 24h. In other words they have different phase angles (Aschoff, 1965) relative to their Zeitgeber events.

Two basic assumptions are made in this work, neither of which is proven but both of them find provision in the Pittendrigh and Bruce (1957) model. Assumption (1): if there indeed exists a temperature-sensitive light refractory B-oscillator, then a rhythm induced by a temperature perturbation in cultures that had never seen light (and hence were arrhythmic until the temperature shock) ought to be the overt expression of this B-oscillator. The model makes no allowance for temperature to act directly on the master A-oscillator and explicitly rules out feedback from B to A. Feedback is strictly one-way from A to B.

Assumption (2) is the elegant suggestion of Pittendrigh and Minis (1964) that phase response curves represent "the time course and wave form" of the basic oscillation(s) responding to perturbations.

In the present series of experiments I investigated the response of rhythms induced by high and low temperature pulses of 12h duration to short light perturbations. The phase response curve arising out of such information was then compared with the standard phase response curve (Pittendrigh and Minis, 1964) which is made on rhythms induced by light. The spectral properties, intensity, duration and similar features for the light perturbations employed in evoking both categories of phase response curves were identical. Even the phases along the cycles that were perturbed were the same. The only difference lay in the manner in which the rhythms were induced in otherwise arrhythmic cultures. Figure 1 presents the phase response curves for 15 min white (fluorescent) light of 1000 lx pulses obtained on i) temperature induced rhythms (solid curve) and on ii) LL/DD-transfer induced rhythms (broken line curve). There is hardly any difference in the "wave form" of the 2 phase response curves even though the "time course" is dislodged by about 4h on the horizontal axis. This discrepancy in the "time course" is traceable to the difference in the phase angles of temperature-induced and light-induced rhythms alluded to earlier. It must be pointed out here that the time course of the eclosion peaks can be 180° out of phase relative to each other (Engelmann, 1966; Chandrashekaran and Loher,

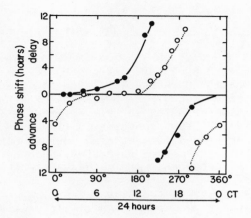

Figure 1: Response of the different phases of the *Drosophila pseudoobscura* eclosion rhythm to white fluorescent light pulses of 1000 lx intensity and 15 min duration. The solid line phase response curve to light pulses was obtained on primary dark-raised arrhythmic populations in which a single high temperature (30°C) pulse of 12 hours duration, scanning 0°-180° of day 1 of the experiment, induced the rhythms before the light pulses were administered. The broken line phase response curve to light pulses was obtained on populations made rhythmic by a singular transfer from constant light to darkness at hour "0" of the experiments on day 1 (12 CT on the circadian time scale of Pittendrigh and Minis, 1964). The values are averages of phase shifts measurable on days 4, 5, 6 and 7 after light and/or temperature treatment. The line partitioning the delays from advances indicates the phase position of untreated control populations. 0 CT indicates beginning of subjective day phase and 12 CT half a cycle later indicates beginning of subjective night. Thus 0°-180° of the duration of the rhythm inducing temperature pulse and of the figure correspond to 0 CT - 12 CT.

1969). Similarly rhythms induced by "warm" pulses and rhythms induced by "cold" pulses in DD-raised cultures are 180° out of phase (Chandrashekaran, unpublished). The close similarity between the 2 phase response curves represented in Figure 1 was not anticipated on the assumption of a separate temperature sensitive light refractory B-oscillator. Indications are that the A-oscillator is responding in both classes of rhythms.

THE CIRCADIAN RHYTHM OF ECLOSION

Temperature Evoked Rhythms and Temperature Perturbations

It could be argued that it is still the B-oscillation that is being set in motion by the temperature pulse, but B being light refractory, as adduced in the original model, it is the light sensitive A-oscillator that is responding to the light pulses (hence the similarity of the phase responses). Therefore, the following series of experiments were performed. In this series the rhythms were i) evoked by temperature and subsequently ii) phase shifted by temperature. The pupae had at no time seen light until eclosion. If there is a B-oscillator, then it must respond to temperature. The phase response curve arising out of this information was again compared with a phase response curve which was made on light induced rhythms but with temperature perturbations. Figure 2 compares temperature pulse phase response curves made on (1) temperature evoked rhythms (solid curve) and on (2) LL/DD-transfer induced rhythms. Once again there are no differences to be seen in "the time course and wave form" of the 2 phase response curves. Both phase response curves in Figure 2 (and also those in Figure 1) are of the "strong" (Winfree, 1970) type and coincide within a few hours in phase at which "delay" responses switch to "advance" responses.

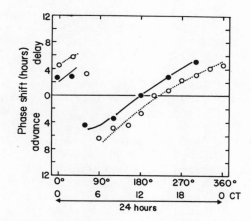

Figure 2: Response of the different phases of the *Drosophila* eclosion rhythm to low temperature (10°C) pulses of 12 hours duration scanning a whole cycle 0°-180° each time. The solid line phase response curve was obtained on populations in which a rhythm was induced by singular 12 hour high temperature (30°C) pulses. The broken line phase response curve was obtained on populations whose rhythm was set in motion by a non-recurrent light/dark transition. Other details as in Figure 1.

DISCUSSION

The most forthright conclusion to arrive at is that there is only one oscillator responding to the several light and temperature treatments. This oscillator is the A-oscillator of the Pittendrigh and Bruce (1957) model. There appears to be no evidence in the results of my experiments for any separate B-oscillator with the qualities attributed to it in the coupled-oscillator model. In fact the results of the experiments of Engelmann et al. (1973, 1974) on the influence of temperature pulses on the *Kalanchoe* petal rhythm showed many similarities to results from light pulse treatments. The leaf movement rhythms of the plants *Phaseolus* (Moser, 1962) and *Solanum nigrum* (Junker and Mayer, 1974) also respond nearly identically to temperature as they do to light perturbations. All these authors imply the same oscillator is involved, and Engelmann et al. (1974) even assume that temperature and light perturbations enter the system at the same point in the oscillation.

Further evidence that makes a separate identity and existence of a B-oscillator unlikely are the series of papers by Winfree (1970, 1973a, and 1973b) in which he characterized a "singularity" point along the light sensitive oscillation (T^+) at which a critical but extremely delicate amount of monochromatic light energy (S^+) could send the oscillation into a "phase-less state". The point of singularity lies 6.8 h into the subjective night of the *Drosophila* rhythm and exposing it to weak blue light of 10 µW cm for 50 sec results in severe derangement of the rhythm. If under such circumstances, a B-oscillator could still exist and be functional, then the feedback of A to B must be nearly absolute and presumably without further attenuation of the perturbation. Another important tenet of the Pittendrigh (1964) model and one of the 2 assumptions of this paper that the phase response curve represents "the time course and wave form" of the basic oscillation has also become unlikely ever since Winfree (1970) showed that even for *Drosophila pseudoobscura* there are 2 types (type 0 or "strong" and type 1 or "weak") of phase response curves even for light stimuli. The strength of stimulation apparently decides the "time course and wave form" of phase response curves. Even the "sign" of the stimulus, i.e. whether it is "positive" (as in light pulses or high temperature pulses) or "negative" (darkness pulses and low temperature pulses) emphatically determines the "time course and wave form" of phase response curves both in the *Drosophila* (Chandrashekaran, 1974), and mammalian circadian systems (Subbaraj and Chandrashekaran, 1978). There certainly are temperature-dependent processes, as suggested by Hamm et al. (1975) for the *Drosophila* system, causing a slowing down or speeding up of light signals arriving at the oscillator, or specifically influencing the phase angle features of the rhythm relative to the Zeitgeber (Pittendrigh et al., 1973). But these processes have been recognized by the authors themselves as being non-oscillatory in nature and origin. In the absence of more compelling evidence for the existence of a temperature sensitive light refractory, slave B-oscillator in "gating" the *Drosophila* rhythm it might be proper to conclude that there is none.

ACKNOWLEDGEMENTS

This work is dedicated to Professor P.K. Menon on the occasion of his seventieth birthday. The experiments reported in this paper were carried out in the laboratory of Professor W. Engelmann at the Institut für Biologie I of the University of Tübingen. I am grateful to Professor Engelmann for placing at my disposal excellent laboratory facilities and his sage advice during my many years stay with him. Professors E. Bünning, A. Johnsson and A.T. Winfree had suggested critical improvements to the manuscript in its various stages of completion for which I am grateful to them. This work was supported by grants under the Schwerpunktsprogramm of the Deutsche Forschungsgemeinschaft to Professor W. Engelman (En42/14).

REFERENCES

Aschoff, J., 1965, The phase angle difference in circadian periodicity, in: "Circadian Clocks", J. Aschoff, ed., North Holland Publishing Co., Amsterdam, pp. 95-112.

Bünning, E., and Zimmer, R., 1962, Zur Deutung der Phasenverschiebungen und "transients" nach exogener Störung endogener Rhythmen, Planta (Berl.), 59:1.

Chandrashekaran, M.K., 1967, Studies on phase shifts in endogenous rhythms. I. Effects of light pulses on the eclosion rhythms in *Drosophila pseudoobscura*, Z. vergl. Physiol., 56:154.

Chandrashekaran, M.K., 1974, Phase shifts in the *Drosophila pseudoobscura* circadian rhythm evoked by temperature pulses of varying durations, J. Interdiscipl. Cycle Res., 5:371.

Chandrashekaran, M.K., and Loher, W., 1969, The relationship between the intensity of light pulses and the extent of phase shifts of the circadian rhythm in the eclosion rate of *Drosophila pseudoobscura*, J. Exp. Zool., 172:147.

Engelmann, W., 1966, Effect of light and dark pulses on the emergence rhythm of *Drosophila pseudoobscura*, Experientia, 22:606.

Engelmann, W., and Mack, J., 1978, Different oscillators control the circadian rhythm of eclosion and activity in *Drosophila*, J. Comp. Physiol., 127:229.

Engelmann, W., Karlsson, H.G., and Johnsson, A., 1973, Phase shifts in the *Kalanchoe* petal rhythm, caused by light pulses of different duration. A theoretical and experimental study, Int. J. Chronobiol., 1:147.

Engelmann, W., Eger, I., Johnsson, A., and Karlsson, H.G., 1974, Effect of temperature pulses on petal rhythm of *Kalanchoe*: an experimental and theoretical study, Int. J. Chronobiol., 2:347.

Handler, A.M., and Konopka, R.J., 1979, Transplantation of a circadian pacemaker in *Drosophila*, Nature, 279:236.

Hamm, U., Chandrashekaran, M.K.,and Engelmann, W., 1975, Temperature sensitive events between photoreceptor and circadian clock?, Z. Naturforsch., 30c:240.

Junker, G., and Mayer, W., 1974,Die Bedeutung der Epidermis für licht und temperaturinduzierte Phasenverschiebungen circadianer Laubblattbewegungen, Planta (Berl.), 121:27.

Konopka, R.J., and Benzer, S., 1971, Clock mutants of *Drosophila melanogaster*, Proc. Nat. Acad. Sci. U.S.A., 68:2112.

Maier, R.W., 1973, Phase shifting of the circadian rhythm of eclosion in *Drosophila pseudoobscura*, J. Interdiscipl. Cycle Res., 4: 125.

Moser, I., 1962, Phasenverschiebungen der endogenen Tagesrhythmik bei *Phaseolus* durch Temperatur und Lichtintensitätsänderungen, Planta (Berl.), 58:199.

Pittendrigh, C.S., and Bruce, V.G., 1957, An oscillator model for biological clocks, in: "Rhythmic and Synthetic Processes in Growth", D. Rudnick, ed., University Press, Princeton, pp. 75-109.

Pittendrigh, C.S., Caldarola, P.C., and Cosbey, E.S., 1973, A differential effect of heavy water on temperature-dependent and temperature-compensated aspects of the circadian system of *Drosophila pseudoobscura*, Proc. Nat. Acad. Sci. U.S.A., 70:2037.

Pittendrigh, C.S., and Minis, D.H., 1964, The entrainment of circadian oscillations by light and their role as photoperiodic clocks, Amer. Naturalist, 98:261.

Subbaraj, R., and Chandrashekaran, M.K., 1978, Pulses of darkness shift the phase of a circadian rhythm in an insectivorous bat, J. Comp. Physiol., 127:239.

Wever,R., 1979, "The Circadian System of Man", Springer-Verlag, New York, 256p.

Winfree, A.T., 1970, The temporal morphology of a biological clock, in: "Lectures on Mathematics in the Life Sciences, Vol. 2", M. Gerstenhaber, ed., American Mathematical Society, Providence.

Winfree, A.T., 1973, Resetting the amplitude of *Drosophila's* circadian chronometer, J. Comp. Physiol., 85:105.

Winfree, A.T., 1973b, Suppressing *Drosophila's* circadian rhythm with dim light, Science, 183:970.

HIGHER BEHAVIOR IN *DROSOPHILA* ANALYZED WITH MUTATIONS THAT DISRUPT THE STRUCTURE AND FUNCTION OF THE NERVOUS SYSTEM

Jeffrey C. Hall[1], Laurie Tompkins[1], C.P. Kyriacou[1], Richard W. Siegel[2], Florian von Schilcher[3], and Ralph J. Greenspan[4]

[1]Department of Biology, Brandeis University, Waltham, MA., U.S.A.
[2]Department of Biology, University of California at Los Angeles, Los Angeles, CA., U.S.A.
[3]Zoologisches Institut der Universität München, München, Federal Republic of Germany
[4]Department of Physiology, University of California School of Medicine, San Francisco, CA., U.S.A.

INTRODUCTION

Higher behavior in lower organisms seems an ideal candidate for genetic analysis. At first glance, this is because fixed action patterns such as courtship rituals seem almost certain to be programmed by the action of genes that control the development and function of their nervous systems.

There is a long history of the genetic approach to behavioral study in *Drosophila*, as it has been applied to investigations of mutants affecting the external morphology (reviewed by Manning, 1965; Grossfield, 1975; Ehrman, 1978); and as it relates to multigenic or interspecific studies, with their implications on the evolutionary significance of higher behaviors (e.g. Manning, 1965, 1975; Ewing and Manning, 1967).

Genetic dissection of reproductive behavior is important from another point of view that has perhaps come to the fore only recently; namely, the desirability of using genetic variants to make informative experimental perturbations of the <u>intact</u> organisms whose complex behavior is to be observed. Thus, in *Drosophila*, single gene mutants and genetic mosaics have been used to analyze courtship, permitting the perturbed flies to move about freely -- except with respect to

the aim of the specific genetic "treatment". The flies have been
able to assume normal postures and usage of their appendages, because
the genetic variants used have been "endogenous" disruptions, without
the non-specific debilitations of the nervous system that can result
from surgical intervention, drug treatment, or tethering of the
courting flies with recording electrodes.

This article summarizes several of the genetic experiments we
have carried out on *Drosophila* reproduction, which have mostly involved specific genetic variants of neurobiological interest (see
Hall and Greenspan, 1979, for a review). We have therefore tried to
make extensions of the earlier work, moving beyond the use of cuticular mutants and complicated multigenic differences. Some of the
mutants under current investigation appear directly to disrupt courtship, without major (or any) effects on other behaviors. These mutations may therefore define loci that are in the *Drosophila* genome
specifically for the control of the fly's nervous system as it functions in reproduction. Other mutants we have used were found first
on behavioral criteria ostensibly unrelated to courtship, or on neurochemical and neurobiological criteria known not to be limited to the
nervous system's control of reproduction. But these "non-courtship"
mutants have been extremely interesting in their effects on mating
rituals. From the discovery of these effects, we have realized that
it would have been difficult both technically and conceptually to
demonstrate a connection between, say, optomotor responses, olfaction
or acetylcholine metabolism and courtship, unless the relevant mutants had been available. Moreover, some of the connections -- apparently rather direct ones -- between courtship and other higher
behaviors in this organism have proved surprising. Again, mutations
such as those disrupting learning or circadian rhythmicity and, as
we describe in this report, courtship behavior as well, have been
crucial in analyzing the relationships among these higher neural
functions.

VISUAL RESPONSES IN COURTSHIP

For a male and female to see one another would seem to be important in triggering courtship and allowing it to continue. This
is obviously true for *Drosophila* species that will not mate in
the dark (e.g. *D. subobscura*, or *D. auraria*, see Grossfield, 1966,
1972). Males and females of *D. melanogaster*, though, will mate in
the dark, suggesting that vision might not be relevant to reproductive behavior in this species. However, we have learned that courtship vigor and copulation success (monitored during relatively
short observation periods) decline by at least a factor of two when
genetically blind males of *D. melanogaster* are put with normal females
(Siegel and Hall, 1979). "Vigor" here is a simple courtship index,
i.e. the fraction of an observation period spent by the male in
courtship of a female who is in the same chamber (cf. Hall, 1978a;
Jallon and Hotta, 1979). The males behaving abnormally expressed

mutant alleles of the no-receptor-potential (*norpA*, X-chromosomal) gene, or of the glass-eye (*gl*, 3rd chromosomal) gene. Such mutants have no light-triggered electrical activity in their visual system (e.g. Pak et al., 1969).

The use of the mutants here was a convenience, permitting easy observation of the blind males in normal light; thus, components of courtship could be recorded with more detail than a mere score for mating success. We also noted, more subjectively, that the mutant males were often misdirected in their attempts to orient toward or follow females. This abnormal behavior is in no way surprising; but we realized that it is important to dissect further the role of visual behavior in courtship. Perhaps it would be sufficient for a male of this species merely to see the female. Or perhaps he must in addition detect her movement. The latter is suggested by the fact that immobilized females of this species, in the presence of wild-type males, are courted with at least a two-fold decrement and are mated with a much greater reduction, in comparison to the good mating success observed with mobile females (Hall, 1978a; Seigel and Hall 1979).

Poor or absent courtship by males of etherized females has been found in other species (e.g. Streisinger, 1948); and this is the kind of experiment we first performed to quantify the effects of female immobility on *D. melanogaster* courtship. Parenthetically, it should be noted that prolonged exposure (30 min. or longer) of a male to an etherized female causes his courtship substantially to decline, almost certainly due to the ether per se (!). Thus, we have augmented the anesthesia tests by putting normal males in the presence of females expressing shibire-temperature-sensitive mutants (*shi^{ts}*, Grigliatti et al., 1973). Such females become immobile at 27°C, and the effects of their movement on the males are similar to those seen with etherized females. In addition, we have been able to take a freely moving pair of flies, and selectively immobilize the female to show that the male's courtship declines rather abruptly when the female becomes paralyzed (L. Tompkins, J.C. Hall, and R.W. Seigel, in preparation). In the "opposite" experiment males were found to court females much more vigorously, after the latter had recovered from the temperature-induced paralysis.

Movement of the females in courtship situations may require her performance of fixed action patterns per se in some species (such as *D. subobscura*), i.e. postures and movements of appendages that rival those of the male in apparent complexity (e.g. Brown, 1965). These are species for which, indeed, light is necessary for mating (see above); and, also as expected, immobilization of the female in these cases blocks mating and male courtship actions (e.g. Spieth, 1966). *D. melanogaster* males, though, may only have to detect general movements of their con-specific females; and the males seem to do this solely by visual cues. We have found this to be so by testing the effects on male courtship of an optomotor-blind mutation (omb^{H31}), isolated by Heisenberg et al., (1978) on criteria of generaly defective turning responses (and then found to be missing key neurons in one of the optic lobes). The use of this mutant in courtship tests was more informative than the experiments involving immobile females, because the mutant males should have a highly selective impairment in movement detection via the visual system; mechanoreception of female movement, for example, should be left normal. We found that the omb^{H31} males show a 20-fold drop in the number of quarter-turns they make per min., compared to the wild-type male's actual optomotor response to a female. And the mutant males have a courtship index and copulation success which is dramatically depressed, and to the same degree as the decrement associated with totally blind mutants (Tompkins, J.C. Hall, and R.W. Siegel, in preparation). We conclude that visually triggered detection, by *D. melanogaster* males, of optomotor stimuli from females is an important component of the courtship ritual. Therefore, both sight and the female's active participation in courtship of this species seem more important than previously assumed from the absence of an absolute requirement for light.

OLFACTORY RESPONSES IN COURTSHIP

Olfaction has been suggested to be important in male-female interactions in *D. melanogaster*, in part from the results of olfactometric experiments (e.g. Averhoff and Richardson, 1974); and from the fact that female "sex appeal" has a focus on or in the abdomen, a common source of pheromones in insects. That the abdomen is important comes from studies of male-female sex mosaics. These gynandromorphs are courted by normal males essentially only if the mosaics have abdomens that are all or part female (reviewed by Hall, 1978b; Hall and Greenspan, 1979).

Following the above leads, we have distilled and extracted volatile compounds from males and females of *D. melanogaster* (Tompkins et al., 1980). The aggregate substances from females stimulate dramatic increases in courtship behavior, as performed by two males (and recorded with the usual courtship index). Two such flies will court one another hardly at all in the presence of: no other stimulus, the solvents used to extract and dilute the putative pheromones, or materials from mature males. The volatile compounds from females act at a distance (i.e. do not have to be present in the courtship chamber with the two males), but only when they are within 7-8 mm of the males.

These volatile substances may be involved in species recognition, which would appear to be important in the courtship of members of this genus. This conclusion comes from the fact that the volatile compounds from *D. melanogaster* females do stimulate two *D. simulans* males to court each other but do not stimulate two *D. hydei* males to do so (Tompkins et al., 1980). The former kind of flies are closely related to and will mate with *D. melanogaster*, but the latter species is a much more distant relative.

Additional force is given to the idea that olfactory information is important in *D. melanogaster* courtship from tests of mutant males that appear not to be able to smell. These are from a smellblind (*smb*) strain isolated by Aceves-Piña and Quinn (1979) on criteria of inability to recognize synthetic volatile compounds. We learned that these males (expressing a genetic variant or variants on the X chromosome) do not respond to the pheromones extracted from females of their species, even at extremely high concentrations of these substances (Figure 1). Note, too, that there is an optimum and rela-

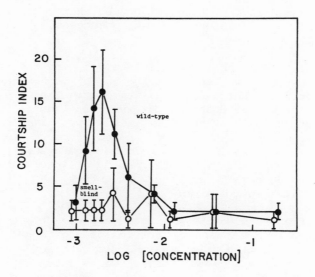

Figure 1: Courtship between two males stimulated by volatile compounds from *D. melanogaster* females. The putative pheromones were distilled and extracted as in Hedin et al. (1972), diluted in ethanol, then dispersed through small pieces of filter paper that were placed in courtship chambers (Tompkins et al., 1980). The concentration of pheromone that stimulates the highest courtship index (plotted ± SEM) -- involving the pair of males in each test -- corresponds to ca. 0.02 female, based on the number of such flies used as the source of the volatile compounds during the distillation. Filled circles, wild-type males; open circles, *smb* mutant males.

tively low concentration of these materials, with respect to the maximum courtship stimulation induced between two males (Figure 1). This finding may parallel the fact that "too-high" concentrations of pheromones -- studied in a variety of insects -- lead to a decrement in their behaviorally stimulating effects (e.g. Rust, 1976).

When *smb* males were put with females in the usual tests, the courtship of the mutants was dramatically subnormal, in a fashion that was quantitatively and qualitatively similar to the behavior of visually deprived males. For these olfactory-deficient males, courtship indices were down by at least two-fold, and the mutant's orientation toward females appeared to be poorly executed. Since a visual defect "removes" approximately half the normal courtship, and an olfactory deficit does the same, we imagined that doubly mutant males would be behaviorally sterile. They were, in fact, at least 10-fold depressed in courtship indices (thus barely above the "background" value recorded between two males lacking any other stimuli). Yet, *smb*; *gl* males were able to copulate with females, when given several days to do so in food vials (Tompkins et al., 1980). Thus, *smb* may still allow a bit of olfactory stimulation to trickle in (*gl* does utterly eliminate visual activity). Or, perhaps it is the case that other sensory modalities, such as contact chemoreception or mechanoreception, can be sufficient for mating success, albeit at very low levels.

In spite of our inability completely to interpret the effects of *smb* on courtship, we feel that the use of this mutational tool is very important. Whereas visual deprivation and its relationship to courtship can be assessed by turning off lights, or perhaps by painting over the eyes of the fruit flies, putative roles for olfactory input cannot so easily be studied without mutants; because it is difficult or impossible to eliminate olfactory sensations in *Drosophila* through surgery or other kinds of intervention. The fly has several locations for olfactory receptors, which are most easily "blocked" by a genetic lesion.

In addition to studying abnormal responses to courtship odors, we have looked into the genetic control of pheromonal production in *Drosophila*. A starting point was the discovery of a dramatic stimulating effect that young adult males exert on the courtship performed by older males. The latter are 2 days or more post-eclosion, while the former must be at most one-day old. These young adults stimulate almost as much courtship as do virgin females, though such males do not themselves court other females or males (Jallon and Hotta, 1979; Tompkins et al., 1980). We thought that the young males could be producing sex-stimulating pheromones; and this was shown to be the case, through the usual distillation and extraction procedures: materials from young males had strong effects in the bioassays (cf. Figure 1), but substances from their older brothers did not.

There is a mutation that causes males of any age to be rather similar in their behavior to that of young wild-type males. This is the 3rd chromosomal fruitless (*fru*) mutation of Gill (1963). Hall (1978a) showed that the mutant males do not mate with females, but they both court other males and stimulate other males to court them. The stimulation can be "passive" in that immoblized *fru* males, or even abdomens cut away from them, stimulate abnormally high-level courtship (Hall, 1978a). It was thus suggested that *fru* males generate volatile compounds, not made by mature wild-type males, that are sex-stimulating. This was shown to be the case, first in bioassays performed with materials extracted from mutant males; but there was no effect of substances from their brothers heterozygous for this recessive mutation (Tompkins et al., 1980). Moreover, the compounds from the mutant males generated a gas chromatogram that was different from that produced by normal males (Figure 2). Note also that wild-type males yield a chromatogram that is different from the wild-type female profile (confirming Hedin et al., 1972), which is in turn the same as the profile from *fru* females (this mutation has no apparent effect when homozygous in females).

Preliminary chromatographic work on the pheromones from young wild-type males suggests that the compounds from these flies are similar to those from the mutant males, especially with respect to the prominent *fru*-specific peak at ca. 11 min retention time (cf. Figure 2). Thus, *fru* may block the male's normal, post-developmental "turn-off" of pheromone production, which occurs concurrently with the maturation of his ability to court and mate with females (e.g. Jallon and Hotta, 1979). This idea suggests that the too-high level of sex-stimulating pheromones associated with the mature *fru* males not only stimulates these self-same males to court other males and stimulates other males to court them, but also somehow blocks the mutants' responses to females. This possibility is now being tested, and it will include post-eclosion maturation in the continual presence of materials from *fru* males or young fru^+ males. This is because we have already learned that a wild-type male that is put simultaneously with a female and with *fru* substances -- but was not previously exposed to *fru* pheromones -- are still able to mate with the females.

The foregoing discussion makes it obvious that *Drosophila* males do not always <u>fail</u> to stimulate courtship behavior. In addition females do not always <u>stimulate</u> high-level courtship, and in these situations olfactory communication is very much at issue. It has long been known that *D. melanogaster* females, once they have mated, will tend to "inhibit" further copulations for at least 4-7 days (e.g. Manning, 1967; Burnet et al., 1973). Recently, we have learned that this inhibition does not require active "rejection" behaviors (cf. Cook, 1975), because mated females that have been etherized or that express paralytic shi^{ts} mutations are still less stimulating than are immobilized virgin females (Siegal and Hall, 1979). Further, a male need not "see" anything associated with a mated female in order to be relatively depressed in his behavior, because norpA or gl males show a significant decrement in courtship indices in the presence of mated vs. virgin females (Siegel and Hall, 1979).

We therefore suggested that a female generates "new" volatile

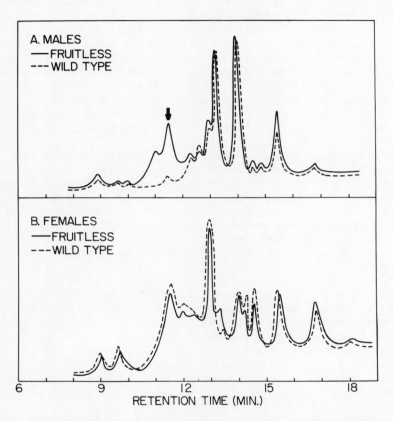

Figure 2: Gas chromatograms of volatile compounds from *D. melanogaster*. These materials were distilled and extracted as in Hedin et al. (1972), using about 10^4 3-5 day old adults for each of the four genotypes shown. The ordinates represent the amounts of material in arbitrary units, and the abscissas show the retention times on the column in min., re a 100-300°C temperature gradient, increasing at 12°C per min. (See Tompkins et al., 1980 for details of chromatography). The arrow designates the peak of material that is much more prevalent in the *fru* than the wild-type male extract.

compounds with adverse effects after she has copulated. This has been borne out by distillation and extraction work, followed by bioassay experiments using two males and materials from mated females vs. the more stimulating ones from their virgin sisters (L. Tompkins and J.C. Hall, in preparation). We also know that olfactory stimuli are relevant to these phenomena from tests of *smb* males: these mutants cannot discriminate between virgin and mated females, when put with these two kinds of females in successive tests (R.W. Siegel and D. Gailey, in preparation); whereas *smb*+ males show at least 5-fold less courtship in the second test with the mated flies (Siegel and Hall, 1979).

LEARNING AND MEMORY IN COURTSHIP

Male-female interactions in a lower form such as *Drosophila* are assumed to involve thoroughly innate neural mechanisms that produce entirely fixed action patterns. We have just discussed the fact that certain male-female interactions can be very different in different circumstances. And we learned in the experiments on mated females that there is a strong after-effect, exerted on a male that had previously been paired with a mated female (Siegel and Hall, 1979). Such males, if left with these non-stimulating and unresponsive females for as little as 20 min have their courtship behavior directed at a virgin female depressed in a subsequent test. This after-effect lasts for 2-3 hours (Siegel and Hall, 1979). But we imagined that no real associative learning was involved -- merely a debilitation that could have come from "noxious" odors or clogged olfactory receptors; these could cause males to receive relatively small degrees of stimulation from the subsequent virgin female.

These dreary explanations have been undermined by a genetic demonstration that authentic conditioned behavior is involved in the interactions of males with mated females. First, we showed that males expressing the memory mutation, amnesiac (*amn*), are different from wild-type in the degree of after-effect induced by exposure to mated females. The mutation was isolated on memory criteria unrelated to courtship (Quinn et al., 1979); here, the mutants were found to "forget" cues associated with electrical shocks much more rapidly than do wild-type *Drosophila*. In mating experiments, as well, the mutants are defective, because *amn* males have an abnormally brief after-effect after having been with these unreceptive flies. That is, *amn* allows males to court virgin females normally within 15-30 min. after their previous "training" (Siegel and Hall, 1979).

The mutant males "sense" mated females normally, because the former are dramatically depressed in courtship responses when directly in the presence of such females. If debilitation or faulty olfaction per se were the primary results of such experience, then *amn*, like wild-type males, should show a lengthy after-effect.

One interpretation of these findings on the relationships of learning to courtship would be the notion that males -- while they are being trained -- associate an adverse odor with the presence of a female. Subsequently, then, they tend not to court females which "should" be attractive, because such objects were previously coupled with a negative stimulus. That odors associated with the mated females are the main cues is suggested by some recent findings, showing that males can easily be trained by exposure to mated females that had been immobilized by *shi^{ts}* mutations. Therefore, actions such as rejection behaviors that can be performed by mated females are not necessary for training. Wild-type males in the dark can be reasonably well trained by exposure to mated females (Table 1), suggesting that visual input received by males from mated females is not of paramount importance. However, genetically blind males, which can discriminate between virgin and mated females (Siegel and Hall, 1979), were not trainable. This is simply puzzling and is being investigated further. Flies with deficits in olfactory processing cannot be trained either, but this is as expected from the fact that these mutant males cannot tell mated from virgin females in the first place (see previous section). These results are noted in Table 1, which also summarizes findings on primary and experience-dependent responses of males to females that have been derived from tests of normal males and those with defects in the reception or processing of sensory information.

Are learning and memory in courtship situations of adaptive value? A positive answer can certainly not be entertained currently. In fact, effects of previous experience on the behavior of males with subsequent females can only be shown in a clear-cut fashion if these virgins (in the second tests) are immobile. If the virgins are free to move about, not as much of an affect-effect can be revealed (Siegel and Hall, 1979). But we generally sense that even a slight drop in mating potential as recorded in laboratory tests would, in a natural situation, have devastating consequences for the probability of a successful copulation. This is not proved by any means, but it is quite possible that a male that has been previously exposed to a mated female, or is blind, or cannot smell would have no chance of mating in stringent wild conditions, in contrast to the reasonable fertility of such males that we have recorded in less natural situations (Table 1).

AUDITORY COMMUNICATION IN COURTSHIP

Drosophila males have long been known to produce a "courtship song", when they are orienting toward or following females and performing their wing display (reviewed by Bennet-Clark and Ewing, 1970). We have wondered if the normal performance of courtship song is entirely "pre-programmed" during embryonic, larval, and pupal development -- or, could it be that storage of auditory information by young adult males is relevant to their eventual ability to sing?

Table 1: Responses of normal and mutant males to virgin and mated females. A summary of the effects of putting males with females of different reproductive history, and of introducing various kinds of mutations into the males, is presented by showing qualitatively what happens when: (1) males are with virgin females; (2) they are asked to discriminate between virgin and mated females in successive tests; (3) they are tested for an after-effect -- i.e. relatively poor subsequent courtship -- caused by previous experience with mated females ("learning"); and (4) they are tested for the duration of this after-effect ("memory"). These results are described, either for the first time in this report; or come from Siegel and Hall (1979); Tompkins, et al., (1980); + means a strong response or effect; ± means a definite response, but one that is measurably reduced; and - means an absence of response or effect.

	DOES THE MALE:			
	(1)	(2)	(3)	(4)
Male Genotype	Mate with a virgin female?	Discriminate between a virgin & a mated female?	"Learn" that he was with a mated female?	"Remember" for a fairly long time that he was with a mated female?
wild-type	+	+	+	+
wild-type (in dark)	±	?	±	?
no-receptor-potential or glass-eye	±	±	-	-
amnesiac	+	+	+	-
smell-blind	+	-	-	-

While perhaps it is outlandish to suggest that there are experience-dependent effects on *Drosophila* courtship song, (1) such effects are well-documented in other organisms (e.g. Konishi and Nottebohm, 1969; Marler and Peters, 1977); (2) these phenomena in flies would be another possible reason for the animal to have evolved learning mechanisms; (3) there is ample opportunity for young males to hear songs from mature males, because the former stimulate so much courtship through their unexpected production of pheromones (see above);

and finally (4) courtship songs of *Drosophila* males that had been raised in total isolation have never been recorded to ask if they are completely normal. To show that a *Drosophila* male's wing vibrations during courtship are "completely normal" requires, in fact, a demonstration of several auditory components -- and more than were previously realized to be present. The well-known pulses of tone, coming every 30 msec or so in the song of *D. melanogaster* males, are not the only feature of this sonic output. For instance, there is also "sine song" (e.g. Schilcher, 1976) a relatively low amplitude, low frequency hum, that usually blends into the pulses (which sound like rapid clicks) during a given bout of singing .

We have recently investigated with genetic techniques the neural control of pulse song and sine song in this species, to augment the more standard neurophysiological data that are emerging from studies of this fruit fly (Ewing, 1977, 1979). We analyzed the songs of gynandromorphs that were monitored for male-like behavior directed at normal females. Earlier, it had been shown that the brain of such mosaics must be part male, in order that these gynandromorphs exhibit any male-like behavior (Hall, 1977). Further work using markers for the male and females genotypes of internal cells in these mosaics -- unlike earlier work in which only cuticular tissues were marked (e.g. Hotta and Benzer, 1976) -- showed that nerve cell perikarya in the dorsal, posterior brain are crucial. If such cells were male on one side of the brain or the other (or both), the mosaic would orient and perform wing display toward a female (Hall, 1979). Many gynandromorphs that performed wing display produced no courtship song or only aberrant sounds that did not contain any regular pulses of tone (Schilcher and Hall, 1979). The focus for male-specific control of normal song was found almost certainly to be in the ventral thoracic nervous system, in that haplo-X neurons on one side of the thorax <u>or</u> the other were found to be sufficient to allow a normal *D. melanogaster* song to be produced by either the wing corresponding to the male neurons or the opposite wing (Schilcher and Hall, 1979).

We are looking further into the neural control of courtship song in mosaic experiments that have an additional genetic component: a courtship mutant, cacophony (*cac*), with its altered sonic patterns. Males hemizygous for this X chromosomal mutation -- induced by Schilcher (e.g. 1977) -- generate definite pulses of tone, but they are louder and have inter-pulse-intervals (ipi's) that are decidedly longer than normal, i.e. ca. 45 msec instead of the wild-type value of ca. 35 msec. Sine song and wing-beat frequency are normal in *cac* males, implying that the "pulse generator" has been rather selectively affected (Schilcher, 1977). Does the *cac* affect the nervous system in a specific way or at least a specific place? This is now being answered with mosaics that will express *cac* in haplo-X tissues, marked with the usual enzyme mutation - i.e. acid phosphatase-minus -- that we have been using to score nerve cells, (e.g.

Schilcher and Hall, 1979; cf. Kankel and Hall, 1976). The non-*cac* tissues in these mosaics are diplo-X, which would seriously confuse the interpretability of data from cases that we are trying to score as normal or mutant singers. Thus, we are circumventing problems of sexual dimorphism by "transforming" the diplo-X, *cac/+* tissues in each mosaic into phenotypically male ones, through the use of the transformer (*tra*) mutation (Sturtevant, 1945). The use of this tool will re-appear later. Here, we should note that *XX;tra/tra* "males" are apparently normal in all male behaviors, including courtship song and copulation (cf. Hall, 1979; Schilcher and Hall, 1979), though they are sterile after a copulation (cf. Marsh and Wieschaus, 1978).

We suggest, then, that *cac* may have a focus in the thorax, which is where cells in *cac+* flies must be male in order that a normally patterned song be produced. The mosaic analysis of the transformed *cac* mosaics is possible because we have already shown that *cac* is a recessive mutation (in *cac/+*; *tra/tra* flies); thus only the *cac/O* tissues in the mosaics will lead to abnormal behavior, i.e. if such cells include the focus. We must also entertain the possibility of a *cac* focus elsewhere than in the thorax, because the mutation may not be defective solely with respect to its courtship song. This is because the mutant males are defective courters, and when their wings are removed they are still less successful at achieving copulation than are wingless wild-type males (Schilcher, 1977). Thus, *cac* may have a focus for aberrant courtship in more than one part of the nervous system.

We have in fact recently made findings suggesting that male song, as other courtship behavior, is under the control of more than one neural ganglion in the fly -- i.e. the brain as well as the thoracic ganglia. These results have to do with still another feature of the courtship song, previously undetected, but a new "complexity" that is hopefully of much neurogenetic interest. We have found that the ipi's -- intervals of silence between pulses of tone -- are not constant (e.g. "34 msec", as has often been reported for *D. melanogaster*). Rather, the ipi's fluctuate rhythmically and sinusoidally, with an amplitude of ca. 3-5 msec and a period length of ca. 55-60 sec in 20 tested wild-type males (C.P. Kyriacou and J.C. Hall, in preparation). These data were collected by recording song during several minutes of courtship, and plotting the rhythmic fluctuations re series of 10-sec intervals (cf. Figure 3). There is no question that the ipi's are different in different time periods, because the standard errors within the various bouts of recorded song are small (see Figure 3). And the period lengths show little variation from male to male. Moreover, this newly discovered aspect of the courtship song has been found in males from several wild-type strains of *D. melanogaster*, and in males expressing a variety of cuticular mutants. Moreover, *cac* males have normal patterns of fluctuations, regarding period length and amplitude per se, though the fluctuations occur in the range of 45 msec (see above).

From the records of many males, we have concluded that an "oscillator" controlling variation in the song's ipi "runs" even when a male is not singing. Such breaks occur routinely, as a male may interrupt his courtship several times in an observation period (during which a "vigorous" courtship index means that the male is following a female in 6-8 out of 10 min.). When the male resumes his wing display and singing, he usually returns with ipi's that are "in phase", i.e. that one would predict if the mechanism of an underlying sinusiodal variation kept operating during bouts of silence.

Mechanisms controlling variation in ipi are not non-specific features of the neural programming for wing vibrations that are monotonously the same in all *Drosophila*. This is concluded, in part from our records of *D. simulans* males. These males in general have ipi's that are longer than those of *D. melanogaster* (e.g. Bennet-Clark and Ewing, 1970). In the former species as well as the latter, ipi's fluctuate, but the components of the oscillating behavior are different, and this may be as much of a species "signature" as are other aspects of auditory behavior. In *D. simulans* males, the amplitude of ipi change is dramatic, ca. 10 msec; and the period length is different in this sibling species, i.e. approximately 35 sec instead of ca. 1 min as in *D. melanogaster* (Kyriacou and Hall, in preparation). Hence, ipi fluctuations are under genetic (at least species-specific) control. But even within *D. melanogaster* there are striking alterations in these rhythmic behaviors that can be effected by changing the genotype, as we discuss below.

CIRCADIAN RHYTHMICITY RELATED TO TIMING PHENOMENA IN REPRODUCTIVE BEHAVIOR

Drosophila courtship behavior may be optimal at particular portions of the diurnal cycle (e.g. Hardeland, 1972). But most of the behavioral data on circadian rhythmicity in this genus comes from investigations of eclosion rhythms and those related to general locomotor activity (reviewed by Saunders, 1976). Genetic analysis has been introduced into this kind of work, through the isolation of variants with altered circadian rhythmicity, the most striking of which are three mutations of the *per* gene of *D. melanogaster* (Konopka and Benzer, 1971). One allele (per^o) abolishes eclosion and activity rhythms; another (per^s) shortens period length of the diurnal rhythms from ca. 23 to ca. 19 hours, and a third allele (per^ℓ) leads to longer periods of ca. 29 hours.

We wondered if the timing mechanisms that would appear to be under the control of this *per* gene might also be involved in the rhythmic oscillations in courtship song that occur over much shorter time frames. This possibility was strikingly realized, because per^s males had ipi oscillations with period lengths of ca. 40-45 (N = 15 males tested); per^ℓ males had ca. 80-90 sec. period lengths (N = 12) and the arrythmic per^o males (N = 12) showed different ipi's during different courtship bouts, but without any regular, let alone sinusoidal pattern of fluctuation (Figure 3).

Additional features of the fluctuating characteristics of courtship song, influenced by *per* mutants, are as follows: (1) Raising these males (per^+ or *per*-mutant) to adulthood in constant light, and storing them after eclosion under such conditions, does not alter the nature of ipi variations; thus, these rhythmic behaviors need not be entrained by light signals, as the diurnal ones must be (cf. Konopka and Benzer, 1971); (2) The period length of ipi oscillation is invariant over a temperature range of 20°C; thus this fluctuating behavior is "temperature-compensated", as are most circadian rhythms; (3) Heterozygotes involving various *per* alleles suggest that per^o and per^s are semi-dominant alleles, each of which act in the direction of shortening period lengths. These mutations are X-chromosomal; therefore the $per^o/+$ and $per^s/+$ flies had to be turned into males by the action of the transformer mutation in order that such heterozygous flies could be tested (recall that *tra* by itself does not derange the periodicity of courtship song).

The results just described allow us to develop the following mnemonic for thinking about the effects of these mutants on timed behavior in courtship (see Figure 4): The clock mechanism per se is not directly influenced by action of the *per* gene, but the length of an abstractly considered "pendulum" is affected by this gene. If this component of the timing mechanism is of normal length (as in per^+ males), the clock ticks at a normal rate, and periodicity is normal. If it is longer (in per^ℓ), the clock ticks more slowly and there are longer periods. If it is shorter, there is faster ticking and too-short periods (per^s), or no periods at all when the length of such a hypothetical entity reaches zero (per^o). This admittedly very formal model predicts that per^s and per^o act in the same direction; so -- while these two mutants are different when in hemizygous condition -- they might have similar effects when heterozygous. Indeed the mean period length for ipi's in $per^o/+$ (transformed) males is only ca. 40 sec. (N = 14), and for $per^s/+$ males is reduced to ca. 43 sec. (N = 12 (Kyriacou and Hall, in preparation).

The discovery of a connection between diurnal rhythms and timing phenomena in reproductive behavior suggests that the courtship song (for example) may in part be under the control of the brain (cf. thoracic control of song by male genotype, Schilcher and Hall, 1979). This idea is suggested by experiments showing that *per* alleles affect control by the brain of diurnal rhythms (Konopka, 1972; Handler and Konopka, 1979). However, to show directly that *per* mutations also affect courtship song via brain defects will require mosaic experiments. Indeed, we are now constructing flies that will be, for instance, part $per^\ell/+$ and part $per^\ell/0$, to ask which tissues will lead to longer-than-normal periodicity of ipi's when they are haplo-*X*. We will also test these self-same mosaics for diurnal rhythms of locomotor activity, to map the control of these two

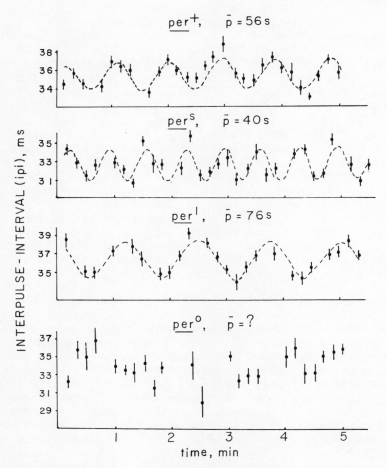

Figure 3: Variations in inter-pulse-interval for courtship songs of *Drosophila* males. *D. melanogaster* males that were wild-type (per^+) or expressed a mutant allele of the *per* gene (see text) had the wing vibrations they directed at females recorded. Mean intervals between pulses of tone (ipi's) were calculated for successive 10-sec. fractions of the 5-6 min. observation periods. Each point represents such a mean ipi, (in milliseconds = ms) together with the standard error. Points do not appear for all time-frames monitored, because males often broke off their courtship during some of the 10-sec. intervals. A non-linear regression was fitted to the points by the method of least squares, and the best fits to the data are shown as dashed lines. For per^+, per^S, and per^ℓ males, sinusoidal functions (i.e. the dashed lines) gave highly significant reductions in the total variation ($p < .01$ in each case). However, the data from the per^o males gave a very poor fit to both sinusoidal and other functions. The period ($\bar{p}$) in seconds for each of the sine waves was computed from the regression equation.

GENETIC ANALYSIS OF HIGHER BEHAVIOR   441

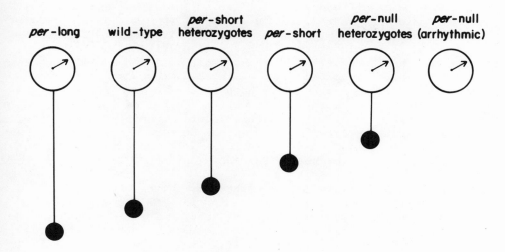

Figure 4: Formal view of effects of the *per* gene on oscillating behavior in courtship song. An underlying "clock" mechanism, controlled by unknown genetic factors and symbolized by these clock faces, is assumed to be the same in all 6 genotypes. But the length of the pendulum for this clock is under the control of *per*, such that the long allele lengthens this component, and the short and null (arrhythmic) alleles shorten it -- with *per*-null causing the most severe decrease. Thus, these hypothetical pendulum lengths are drawn proportional to the duration of periods for inter-pulse-intervals of courtship song in these 6 kinds of males (see text for the actual values).

oscillating phenomena directly with respect to each other (in collaboration with R.J. Konopka).

In order routinely to sing with any sort of normal pulses, these gynandromorphs whose rhythms are analyzed should be thoroughly male. Thus, they are being constructed so as to express the *tra* mutation. The mosaics will also carry the genetic variants for marking the mutant (i.e. $per^{\ell}/0$) genotype of internal cells and tissues (cf. Kankel and Hall, 1976).

Some mosaic work has already been initiated on the oscillating features of coutship song, and these experiments bear on the possible adaptive significance of oscillating ipi's: Do variations influence female receptivity, as the pulses of tone per se are assumed to (e.g. Schilcher, 1976)? To ask this question, we would like to "play" songs to females through the use of electronic devices (cf. Bennet-Clark and Ewing, 1969). These artificial songs will be generated with and without fluctuations in the intervals between tone pulses. However, we have already been able to create flies that produce these pulses, with and without the normal oscillations. These genotypes involve mutationally altered neurotransmitter metabolism, and such experiments are discussed in the next section.

MUTANTS OF ACETYLCHOLINE METABOLISM AND HIGHER BEHAVIOR

We have isolated single gene mutations in the genes -- both of which are on the 3rd chromosome -- coding for choline acetyltransferase (CAT) and acetylcholinesterase (AChE). These enzymes synthesize and degrade, respectively, the neurotransmitter acetylcholine (Hall and Kankel, 1976; Hall et al., 1979). The mutation eliminating either of the enzymes is lethal in embryogenesis, which obviously does not permit behavioral studies to be done. However, we have been able to generate mosaics that express *Cat* (CAT-minus) or *Ace* (AChE-minus) mutations in only a portion of the nervous system (e.g. Greenspan et al., 1980). Many of these mosaics survive to adulthood and have exhibited informative defects in the structure and function of the CNS, including abnormalities of courtship.

Most of the mosaic work has been done with *Ace* mutations. If such a mosaic, with a mutant clone in the CNS, survives, the neuropile that is adjacent to AChE-minus tissue in the cortex surrounding a given ganglion is defective in morphology: reduced in volume, altered in the appearance of staining for AChE, possibly disorganized in the arrays of axons, or even degenerative. The implications of these structural changes in the CNS need not be detailed here (cf. Greenspan et al., 1980). What is interesting in the current context is that many of the *Ace* mosaics are able to carry out complex behaviors, though possibly with rather "subtle" defects. For instance, *Ace* mosaics with mutant clones in the optic system (including portions of the brain proper that process visual information) usually have defects in optomotor behavior and associated abnormalities in visual physiology (Greenspan et al., 1980). These defects are not in the realm of a simple collapse of all aspects of posture, movement, or light-triggered visual responses. The implication is that cholinergic function is selectively impaired in the mutant clones, leading to specific defects in visual processing and responses instead of gross overall sensory and motor abnormalities.

These mosaics happened to be male in the AChE-minus clones and female in Ace^+ tissues, which comprised the majority of each mosaic (the 3rd chromosomal Ace^+ allele had been translocated to the X-chromosome, so that mosaics losing this Ace^+ allele would have also lost an X-chromosome from those nuclei; see Hall et al., 1979). We wondered if any of these gynandromorphs could court, since their male tissues were neurochemically and structurally defective (Greenspan et al., 1980); and since haplo-X tissues in the brain are required for male-specific behavior (see above). It turned out that, indeed, several Ace mosaics could court, though rather weakly so; all mosaics responding positively to females had male, Ace^- tissue in the dorsal, posterior protocerebrum, found earlier to be the "male courtship focus" (e.g. Hall, 1979).

Since defective Ace^- tissues permit processing of information necessary for motor output involved in courtship, we thought that AChE-minus clones in the system might have rather subtle and thus informative defects in courtship song. For these experiments, we generated the usual kinds of haplo-X (Ace^-), diplo-X (Ace^+) mosaics; but all parts of each mosaic were transformed into male tissues, so that abnormalities of song would relate only to the Ace^- genotype (Kyriacou and Hall, in preparation). We have found that these transformed male mosaics usually court rather vigorously, unless there is quite extensive mutant tissue in the brain (N = 2 such non-courting individuals). Songs of the courting mosaics were always normal if there was no internal mosaicism (N = 18) or if Ace^- clones were limited to a small portion of the anterior brain, optic lobes or abdominal ganglion (N = 4). But we have found one striking case of defective courtship song, correlated with mutant tissue unilaterally present in the thoracic nervous system (Figure 5). For this mosaic, there were normal pulses of tone from either wing, but the wing corresponding to the mutant clone showed very sporadic, non-sinusoidal oscillation in ipi, whereas the opposite wing had the usual pattern (Figure 5). In controls (non-mosaic, tra males) each wing shows sinusoidal oscillations. And in fact there appears to be a "master" control over both wings because a wing that is vibrating during a given moment of courtship -- whether it is left or right, and whether or not it will be the wing used in the next bout -- is "told" to be on the correct part of the sine wave.

In these transformed Ace mosaics, then, we have induced a selective alteration in one feature of courtship, and we suggest that the oscillations are in part under thoracic control. The clone in the thoracic ganglia in this mosaic was too extensive to allow a fine "focusing" of the specific part of the thoracic nervous system that is involved here. But further mosaics with this particular kind of behavioral defect may yield a better localization of the key cells. And it is useful to reiterate that these behavioral defects are quite selective and sharply defined -- not involving, say, a complete failure of wing display or sonic output. This is in spite of the apparent morphological "damage" induced by this defect in acetylcholine metabolism, which can be seen on the mutant side of the thoracic nervous system shown in Figure 5 (cf. Greenspan et al., 1980).

We have not yet addressed neurochemical interpretations of the behavioral and neural abnormalities in these *Ace* mosaics. We do not know, for instance, if the morphological changes occur (1) because of over stimulation of post-synaptic neurons, that might result from an absence of hydrolysis of the neurotransmitter in the AChE-minus tissues; or (2) an eventual failure of the post-synaptic cells to respond, which could result from desensitization in the fact of chronically accentuated levels of acetylcholine. Too much stimulation of post-synaptic cells by acetylcholine and other neurotransmitters (or their agonists) can lead to morphological defects in these targets, especially with respect to vertebrate neuromuscular junctions (e.g. Laskowski et al., 1975; Leonard and Salpeter, 1979), and possibly in more central tissues as well (e.g. Herndon and Coyle, 1977). But too little stimulation at cholinergic synapses (possibly resulting from desensitization, see above) might also lead to abnormalities of target cells, especially with respect to the cell differentiation and pattern formation of such cells in central ganglia (e.g. Black and Geen, 1974; Freeman, 1977; and Hildebrand et al., 1979).

To investigate further the neurochemical causes of *Ace*-induced defects in the CNS (from overstimulation or desensitization?), we have studied mutants of choline acetyltransferase (CAT), with putatively lower-than-normal levels of acetylcholine. Some of this work has involved both mosaics and tests of higher behavioral functions, and has provided a way to begin to ask if "too much" acetylcholine could have the same eventual effect as none.

Two of the four *Cat* mutants are temperature-sensitive (ts), permitting development to adulthood at low temperatures (18°-22°), but leading to complete and very early death when exposed to 29° or above (Greenspan, 1980). Just as is the case with mosaics, these conditional mutants allow the developmentally lethal effects of the mutations to be bypassed, in order that adult behaviors can be analyzed (cf. other work on temperature-sensitive *Ace* mutants, Hall et al., 1980). These *Cat*^{ts} mutants, when exposed to heat as adults, pass out. Such paralysis takes a long time (Figure 6) but it can be induced at temperatures that have no apparent effect on *Cat*$^+$ flies. The paralysis is reversible (as long as the flies are not left at high temperature for too lengthy a period of time, Greenspan, 1980). And, during the course of the gradual paralysis, both the levels of the CAT enzyme itself and acetylcholine fall dramatically (Greenspan, 1980). Courtship behavior of heat-treated *Cat*^{ts} males has also been monitored, and it is "turned-off" well before there is any decrement in the general ability of these flies to move (Figure 6).

In *Cat* mosaics -- which expressed ts mutations only in male tissues, female portions of these mosaics remaining *Cat*$^+$ -- defects in general locomotor ability are even more dramatically separable from turn-offs of sexual behavior. That is, one such mosaic had its courtship ability eliminated by heat treatment, but it remained

GENETIC ANALYSIS OF HIGHER BEHAVIOR 445

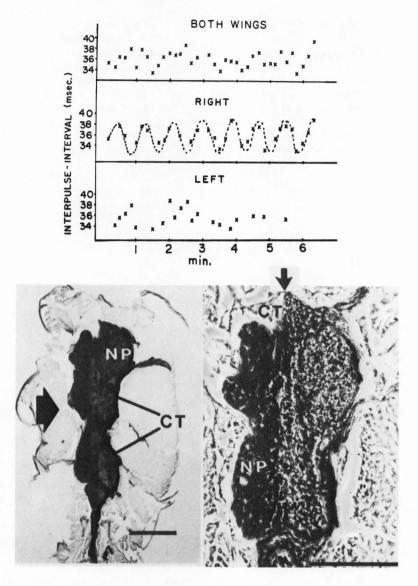

Figure 5: Courtship song and internal tissue marking of an acetylcholinesterase mosaic. This diplo-X, haplo-X mosaic expressed an Ace mutation in the XO tissues present throughout most of one side of the thoracic nervous system (large arrow, re brightfield micrograph of this 10 μm section); but there was no mosaicism elsewhere in the brain. All tissues in the mosaic were male, due to expression of the transformer mutation. A "dividing line" between mutant Ace⁻ (unstained) tissue and normal Ace⁺ tissue (darkly stained for

(continued on page 446)

Figure 5 (cont.): AChE) can be seen in the anterior portions of the cellular cortex (CT) of the thoracic nervous system: see small arrow in the phase contrast micrograph (right-hand picture, which comes from a slightly different plane of sectioning through this mosaic). The axonal and synaptic neuropile (NP) on the mutant side shows morphological abnormalities characteristic of mosaicism for *Ace* expression (see text). Behaviorally, this male courted females and displayed both wings during its various courtship bouts. While pulses of courtship song were recorded during display and vibration of either wing, the interpulse-intervals did not fluctuate sinusoidally in the song produced by the wing on the same side of the thorax as the Ace^- tissue. Yet in the song coming from the other wing, there was a close fit of ipi fluctuations to a sine-wave function (see legend to Figure 3). The scale bars = 100 µm.

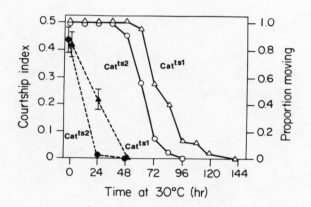

Figure 6: Heat-sensitive paralysis and courtship deficits in Cat^{ts} mutants. Flies expressing either of two temperature-sensitive choline acetyltransferase mutants were raised at 18°C, a permissive temperature for either mutant. Each kind of mutant was heterozygous for a Cat^{ts} allele and a deletion of the 3rd chromosomal *Cat* locus. Approximately 50 adults of each genotype were tested. The adults were subjected to 30°C after eclosion and either their general postures and movements were monitored; or, for some of the males, their courtship behavior was recorded during successive days (see text for definition of the "courtship index"). Wild-type males, raised at 18°C then subjected to 30°C, show neither paralysis nor decrements in courtship during the time-period shown on the abscissa. Filled symbols, courtship indices (plotted ± SEM if such values were larger than the data points); open symbols, paralysis data. After Greenspan (1980).

otherwise normal for many days (Figure 7). The control gynandromorphs (with no CAT-minus tissue) continued their reasonably high-level courtship in repeated tests (cf. Hall, 1979). The other mosaic studied here had both paralysis and failure to court induced by high-temperature treatment (Figure 7). This mosaic may have had a very extensive clone of Cat^{ts}-mutant tissues, which would then have been available for turn-off of normal acetylcholine synthesis throughout much of the nervous system. This possiblity is difficult to investigate, because a histochemical stain for CAT activity in the *Drosophila* CNS is not available. We realized, though, that gynandromorphs such as those described in Figure 7 provided an ideal way to recognize internal mosaicism for this enzymatic function, using a behavioral marker. This is because any *Cat* mosaic that courts must have male, and hence *Cat*-minus tissue, at least in the dorsal, posterior brain (cf. Hall, 1979; Greenspan et al., 1980). Thus, one can allow such mosaics to develop with a chronic absence of *Cat* functioning in their male tissues -- even using the non-ts *Cat* mutations (Greenspan, 1980) -- then make histological observations of the dorsal brain of the gynandromorphs which had courted females. Some such mosaics may be able to court in spite of their neurochemical defect, by analogy to the perhaps surprising courtship abilities of certain of the *Ace* mosaics discussed above. In fact, one question we wish to ask of the *Cat* mosaics is: is a decrement in acetylcholine synthesis of similar consequences as a failure to degrade this neurotransmitter? It might well be that Cat^- clones in these mosaics will have morphological abnormalities in the CNS that are similar to those seen in the *Ace* mosaics with the "opposite" enzymatic defect.

CONCLUSIONS

We believe that the genetic studies on courtship described here show that much information on what is necessary for higher behavior in *Drosophila* can be obtained from the analysis of neurobiological mutants. For instance, certain components of animal communication seemed as if they "should" be necessary for triggering sexual behavior in these fruit flies. But the mutations that disrupt the sensory functions, or the processing of sensory information related especially to vision and olfaction, allowed real experimental tests of the roles played by sight and smell in mediating courtship behavior. And, whereas decrements in vision would seem to be easily inducible by manipulation of the environment, we know of no way other than the genetic tool we had available for perturbing olfaction to ask what happens to his courtship performance when a male cannot sense the female's pheromone.

Once we turned our consideration to the more intrinsic factors necessary for courtship, we saw that the tool of sex mosaics was a

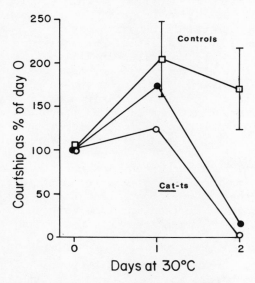

Figure 7: Courtship behavior of gynandromorphs carrying Cat^{ts} mutations. Mosaics were generated as described in Hall et al., (1979), using an X chromosome onto which had been translocated the normal allele of the Cat gene. The third chromosomes carried either a Cat^{ts} mutation in hemizygous condition (cases symbolized with O and ●), or a Cat^{ts} mutation heterozygous with Cat^+ (control mosaics symbolized with ☐). All of these gynandromorphs had been raised at 18° then shifted to 30°C after a preliminary courtship text on the first day. Subsequent courtship indices (see Figure 6 and text) were recorded by testing the mosaics individually with females on later days; these values, for a given day, are plotted as a proportion of the courtship index from the initial tests of each gynandromorph. The indices for the control mosaics (N = 3) are plotted ± SEM. The mosaic designated with ● ceased its general movement by day 4, but the other experimental mosaic and the controls remained vigorous in their activity, throughout and beyond the time period shown on the abscissa. After Greenspan (1980).

powerful one for gaining information on the control of reproductive behavior that is assumed to be effected by the brain; but this is not easy to study experimentally without the use of gynandromorphs. One might use brain ablation experiments in attempts to knock out courtship abilities of adult flies. But these techniques often do not allow fine enough lesions to be made or detected; and they say nothing about the sex-specific control of the many particular subcomponents of the courtship sequence; nor do they bear on the development of the CNS which must be at least partially under the control of the fly's sex chromosome constitution.

The analysis of mosaics by techniques of internal cell marking also allowed us to study connections between specific aspects of the fly's neurochemistry and particular features of courtship behavior. Again, one assumes the acetylcholine-mediated transmission is important in controlling many aspects of behavior; but it is not easy to induce the relevant neurochemical defects and still meaningfully test the flies for their ability or inability to perform complex action patterns. Thus, the mutants of acetylcholine metabolism have made their expected contribution in the genetic dissection of courtship behavior. These studies of sexual behavior, in turn, have made unexpected contributions to neurobiological investigations of this transmitter substance. In our work on genetically manipulated acetylcholine levels, we have been surprised to learn of the anatomical defect in the CNS that is induced by acetylcholinesterase mutations; and we are more surprised that the choline acetyltransferase mosaics, tested in courtship, provide a potential way to learn more about why the deficits in the degradative enzyme induce structural defects in the nervous system.

We have analyzed genetic variants other than those that were expected to be thoroughly relevant to courtship behavior. These additional mutants might not have been expected to have anything to do with reproduction: such as those with defects in learning and memory and mutations affecting circadian rhythmicity. We are intrigued by the finding that centrally mediated conditioning may play an important role in several aspects of male-female and even male-male interactions; and perhaps these possibilities that we have raised run counter to some older ideas that have been too well fixed. As for mechanisms controlling diurnal oscillations and their relationship to rhythmic fluctuations occurring, during courtship, over much shorter time frames -- we suggest the exciting possibility of a very basic and broadly important timing mechanism that is controlled in part by the *per* gene. Parenthetically, it is interesting to recall that dominance relationships among *per* alleles could only be studied with respect to male courtship by transforming *XX* flies into males (Figure 4). The *tra* gene was also very useful for turning gynandromorphs into mosaic males; without this tool, certain of the tests involving mosaicism and courtship song could not have been done (e.g. Figure 5). Another variant not directly related to the control of courtship, but a valuable tool in our experiments on the role of female behavior in reproduction, was the temperature-sensitive paralytic mutant *shi^{ts}* which allowed us to avoid some of the artifacts involved in the use of anesthesia to immobilize females.

In general, we are impressed by the fact that mutants used "merely" as tools, or to disrupt important higher functions, provide a very powerful way to establish connections among the mechanisms underlying different behaviors -- including those related specifically to reproduction. Ultimately, these mutants should allow us to find out what neural mechanisms comprise the biological entities which provide the actual connections. The neural relationships among the behaviors controlled by these genes will be possible to establish when experiments on mutants such as amnesiac and *per* are merged with the mosaic technology. That technique can now tell us what part of the internally marked nervous system is influenced by a gene that affects courtship and mating or any other behavior.

Finally, we tip our hats to the investigators who have been isolating all of these remarkable mutants in *Drosophila*. And we sit somewhat in awe of the mutant genes themselves; because not only have these higher-behavioral mutants provided unique tools, the very existence of these genetic variants was the key factor which suggested that one might actually try to show how courtship behavior is intimately related to other complex neurobiological and behavioral phenomena.

ACKNOWLEDGEMENTS

The work was supported by grants from the U.S. Public Health Service, (GM 21473 and NS 12346), by a Research Career Development Award from the U.S.P.H.S. (to J.C.H., #GM 00297); a NATO Fellowship (C.P.K.); Deutsche Forschungsgemeinschaft (F.v.S.); and by both a U.S.P.H.S. Training Grant plus a Brandeis University Goldwyn Fellowship (R.J.G.). We thank Don Gailey, Kristin White, Anne Gross, Gary Karpen, James Finn and Margaret Stewart for assisting in these experiments.

REFERENCES

Aceves-Piña, E.O., and Quinn, W.G., 1979, Learning in normal and mutant *Drosophila* larvae, Science, 206:93.
Averhoff, W.W., and Richardson, R.H., 1974, Pheromonal control of mating patterns in *Drosophila melanogaster*, Behav. Genet., 4: 207.

Bennet-Clark, H.C., and Ewing, A.W., 1969, Pulse interval as a critical parameter in the courtship song of Drosophila melanogaster, Anim. Behav., 17:755.

Bennet-Clark, H.C., and Ewing, A.W., 1970, The love song of the fruit fly, Sci. Amer., 223 (No. 1):84.

Black, I.B., and Geen, S.C., 1974, Inhibition of the biochemical and morphological maturation of adrenergic neurons by nicotinic receptor blockade, J. Neurochem., 22:301.

Brown, R.G.D., 1965, Courtship behaviour in the Drosophila obscura group, Part II. Comparative studies, Behaviour, 25:281.

Burnet, B., Connolly, K., Kearney, M., and Cook, R., 1973, Effects of male paragonial gland secretion on sexual receptivity and courtship behaviour of female Drosophila melanogaster, J. Insect Physiol., 19:2421.

Cook, R., 1975, Courtship of Drosophila melanogaster: rejection without extrusion, Behaviour, 52:155.

Ehrman, L., 1978, Sexual behavior, in: "Genetics and Biology of Drosophila, Vol. 2b," M. Ashburner and T.R.F. Wright, eds., Academic Press, London, p. 127.

Ewing, A.W., 1977, Neuromuscular basis of courtship song in Drosophila: the role of the indirect flight muscles, J. Comp. Physiol.A, 119:249.

Ewing, A.W., 1979, The neuromuscular basis of courtship song in Drosophila: the role of the direct and auxillary wing muscles, J. Comp. Physiol.A, 130:87.

Ewing, A.W., and Manning. A., 1967, The evolution and genetics of insect behavior, Ann. Rev. Entomol., 12:471.

Freeman, J.A., 1977, Possible regulatory function of acetylcholine receptor in maintenance of retinotectal synapses, Nature, 269:218.

Gill, K.S., 1963, A mutation causing abnormal courtship and mating behavior in males of Drosophila melanogaster (Abstr.), Amer. Nat., 3:507.

Greenspan, R.J., 1980, Mutations of choline acetyltransferase and associated neural defects in Drosophila melanogaster, J. Comp. Physiol. A, in press.

Greenspan, R.J., Finn, J.A., Jr., and Hall, J.C., 1980, Acetylcholinesterase mutants in Drosophila and their effects on the structure and function of the central nervous system, J. Comp. Neurol., in press.

Grigliatti, T., Hall, L., Rosenbluth, R., and Suzuki, D.T., 1973, Temperature-sensitive mutations in Drosophila melanogaster, XIV. A selection of immobile adults, Mol. Gen. Genet., 120:107.

Grossfield, J., 1966, The influence of light on the mating behavior of Drosophila, Drosophila Stud. Genet., (Univ. Texas Publ. 6615) 3:147.

Grossfield, J., 1972, The use of behavioral mutants in biological control, Behav. Genet., 2:311.

Grossfield, J., 1975, Behavioral mutants of Drosophila, in: "Handbook of Genetics, Vol. 3," R.C. King, ed., Plenum, New York, p. 679.

Hall, J.C., 1977, Portions of the central nervous system controlling reproductive behavior in Drosophila melanogaster, Behav. Genet. 7:291.

Hall, J.C., 1978a, Courtship among males due to a male-sterile mutation in Drosophila melanogaster, Behav. Genet., 8:125.

Hall, J.C., 1978b, Behavioral analysis in Drosophila mosaics, in: "Genetic Mosaics and Cell Differentiation," W.J. Gehring, ed., Springer-Verlag, Berlin, p. 259.

Hall, J.C., 1979, Control of male reproductive behavior by the central nervous system of Drosophila: dissection of a courtship pathway by genetic mosaics, Genetics, 92:437.

Hall, J.C., and Kankel, D.R., 1976, Genetics of acetylcholinesterase in Drosophila melanogaster, Genetics, 83:516.

Hall, J.C., and Greenspan, R.G., 1979, Genetic analysis of Drosophila neurobiology, Ann. Rev. Genet., 13:127.

Hall, J.C., Greenspan, R.J., and Kankel, D.R., 1979, Neural defects induced by genetic manipulation of acetylcholine metabolism in Drosophila, in: "Society for Neuroscience Symposia, Vol. IV," J.A. Ferendelli, ed., Society for Neuroscience, Bethesda, Md., p. 1.

Hall, J.C., Alahiotis, S., Strumpf, D.A., and White, K., 1980, Behavioral and biochemical defects in temperature-sensitive acetylcholinesterase mutants of Drosophila melanogaster, submitted to Genetics.

Handler, A.M., and Konopka, R.J., 1979, Transplantation of a circadian pacemarker in Drosophila, Nature, 279:236.

Hardeland, R., 1972, Species differences in the diurnal rhythmicity of courtship behavior within the melanogaster group of the genus Drosophila, Anim. Behav., 20:170.

Hedin, P.A., Niemeyer, C.S., Gueldner, R.C., and Thompson, A.C., 1972, A gas chromatographic survey of the volatile fractions of twenty species of insects from eight orders, J. Insect Physiol., 18:555.

Heisenberg, M., Wonneberger, R., and Wolf, R., 1978, optomotor-blind[H31] -- a Drosophila mutant of the lobula plate giant neurons, J. Comp. Physiol. A, 98:217.

Herndon, R.M., and Coyle, J.T., 1977, Selective destruction of neurons by a transmitter agonist, Science, 198:71.

Hildebrand, J.G., Hall, L.M., and Osmond, B.C., 1979, Distribution of ^{125}I-α-bungarotoxin-binding sites in normal and deafferented antennal lobes of Manduca sexta, Proc. Natl. Acad. Sci. U.S.A., 76:499.

Hotta, Y., and Benzer, S., 1976, Courtship in Drosophila mosaics: sex specific foci for sequential action patterns, Proc. Natl. Acad. Sci. U.S.A., 73:4154.

Jallon, J.-M., and Hotta, Y., 1979, Genetic and behavioral studies of female sex appeal in Drosophila, Behav. Genet., 9:256.

Kankel, D.R., and Hall, J.C., 1976, Fate mapping of nervous system and other internal tissues in genetic mosaics of Drosophila melanogaster, Devel. Biol., 48:1.

Konishi, M., and Nottebohm, F., 1969, Experimental studies in the ontogeny of avian vocalizations, in: "Bird Vocalizations; Their Relation to Current Problems in Biology and Psychology," R.A. Hinde, ed., Cambridge University Press, Cambridge, p. 29.

Konopka, R.J., 1972, "Circadian Clock Mutants of *Drosophila melanogaster*," Ph.D. Thesis, California Institute of Technology, Pasadena, Ca.

Konopka, R.J., and Benzer, S., 1971, Clock mutants of *Drosophila melanogaster*, Proc. Natl. Acad. Sci. U.S.A., 68:2112.

Laskowski, M.B., Olson, W.H., and Dettbarn, W.D., 1975, Ultrastructural changes at the motor end plate produced by an irreversible cholinesterase inhibitor, Exp. Neurol., 47:290.

Leonard, J.P., and Salpeter, M.M., 1979, Agonist-induced myopathy at the neuromuscular junction is mediated by calcium, J. Cell Biol., 82:811.

Manning, A., 1965, Drosophila and the evolution of behavior, Viewpoints Biol., 4:125.

Manning, A., 1967, The control of sexual receptivity in female Drosophila, Anim. Behav., 15:239.

Manning, A., 1975, Behaviour genetics and the study of behavioural evolution, in: "Function and Evolution in Behaviour," G. Baerends, C. Beer, and A. Manning, eds., Clarendon Press, Oxford, p. 71.

Marler, P., and Peters, S., 1977, Selective vocal learning in a sparrow, Science, 198:519.

Marsh, J.L., and Wieschaus, E., 1978, Is sex determination in germ line and soma controlled by separate genetic mechanisms?, Nature, 272:249.

Quinn, W.G., Sziber, P.P., and Booker, R., 1979, The *Drosophila* memory mutant *amnesiac*, Nature, 277:212.

Pak, W.L., Grossfield, J., and White, N.V., 1969, Nonphototactic mutants in a study of vision of *Drosophila*, Nature, 222:351.

Rust, M.K., 1976, Quantitative analysis of male responses released by female sex pheromones in *Periplaneta americana*, Anim. Behav. 24:681.

Saunders, D.S., 1976, "Insect Clocks," Pergamon Press, New York.

Schilcher, F.v., 1976, The function of pulse song and sine song in the courtship of *Drosophila melanogaster*, Anim. Behav., 24:622.

Schilcher, F.v., 1977, A mutation which changes courtship song in *Drosophila melanogaster*, Behav. Genet., 7:251.

Schilcher, F.v., and Hall, J.C., 1979, Neural topography of courtship song in sex mosaics of *Drosophila melanogaster*, J. Comp. Physiol. A, 129:85.

Siegel, R.W., and Hall, J.C., 1979, Conditioned responses in courtship behavior of normal and mutant *Drosophila*, Proc. Natl. Acad. Sci. U.S.A., 76:3430.

Spieth, H.T., 1966, Drosophilid mating behavior: the behavior of decapitated females, Anim. Behav. 14:226.

Streisinger, G., 1948, Experiments on sexual isolation in Drosophila. IX. Behavior of males with etherized females, Evolution, 2:187.

Sturtevant, A.H., 1945, A gene in *Drosophila melanogaster* that transforms females into males, Genetics, 30:297.

Tompkins, L., Hall, J.C., and Hall, L.M., 1980, Courtship-stimulating volatile compounds from normal and mutant Drosophila, <u>J. Insect Physiol.</u>, in press.

DISCUSSION

<u>V. Nanjundiah</u>: Can you say something about the ganglion transplants that affect rhythm?

<u>J.C. Hall</u>: Konopka's group has transplanted per^s ganglia into the abdomen of arrythmic adults and in some cases they were able to induce short-period oscillations in the host.

<u>V. Nanjundiah</u>: Did they look at the interpulse interval?

<u>J.C. Hall</u>: No, rhythmic aspects of song have not been examined in any transplantation experiments. We will eventually ask if it is the case that clock mutants affect the brain or the thoracic nervous system in the control of courtship song. The male with the acetylcholinesterase clone suggests that the thoracic ganglion is involved in oscillating interpulse intervals and recordings of songs from gynandromorphs implicate thoracic control. But the period mutants could affect song through a primary focus in the brain, which is their apparent focus for circadian defects.

<u>J.A. Campos-Ortega</u>: In the picture of the acetylcholinesterase mosaic fly, the left side was said to be acetylcholinesterase-negative, yet the adjacent neuropile appeared to be more darkly stained. How do you explain this?

<u>J.C. Hall</u>: The dark staining in the neuropile on the mutant side (which is by the way one of the characteristic defects in morphology with *Ace*-minus clones) comes from wild-type axons containing normal esterase which cross over from the normal side.

<u>P.A. Lawrence</u>: When you calculate an abnormally low Courtship Index, how do you eliminate non-specific effect of the mutants you use on other aspects of behavior?

<u>J.C. Hall</u>: The general motor ability of these generally appears to be normal;they may, of course, be sick on some subtle way. One method of trying to eliminate non-specific behavioral problems, due to variations in something like genetic background, is to test more than one mutant allele of a given gene and see if they all have the same effect. Secondly, we can make the mutant isogenic with wild-type by back-crossing it for several generations. Even though it can never be ruled out that a mutation does not have some pleiotropic, non-specific, or unknown effects, the use of mutants is critical in some

of these behavioral experiments. For instance, to measure the role of vision in courtship, you can turn off the lights, but to test the importance of olfaction you cannot remove all smell receptors with forceps. So you need a mutant such as smellblind that is probably blocked in all olfactory responses, perhaps because of a central defect.

A. Shearn: Is the pheromone in young males the same as that in the mutant fruitless?

J.C. Hall: Chromatographically, these materials are similar but not identical. In the bioassay, the pheromones from these two kinds of males are also similar.

Y. Hotta: Jallon and I have also been working on courtship behavior. By putting a male and female in vials separated by a glass partition through which the flies can see each other, we can eliminate the olfactory cue. We can still see courtship activity. If a blind male is within 3 mm of a female, we find courtship activity is induced, probably by the female's pheromone.

J.C. Hall: We also found that extracts from female flies contain a very short-range pheromone. There is no pheromone from females that acts over a distance of greater than one cm to trigger male courtship.

ATTRACTANTS IN THE COURTSHIP BEHAVIOR OF *DROSOPHILA MELANOGASTER*

Renée Venard

Molecular Biology Unit
Tata Institute of Fundamental Research
Bombay, India

Chemical communication among insect individuals is specially important for sexual communication.[1,2] One or both sexes produce sex-pheromone(s) which induces a specific behavioral response in the mate. This behavior may include: locomotor stimulation, attraction, arrest of movement. Several chemicals are known to be responsible for this phenomenon.

Since Sturtevant's early studies on *Drosophila*,[3] many workers have described the details of the sequential behavioral pattern displayed by both sexes during courtship.[4,5] Wing vibrations at right angles to the body axis is the most conspicuous among male courtship signals. Jallon and Hotta have defined the female sex-appeal as a stimulus or a set of stimuli which induce wing vibration in the male courters.[6] However, the physical nature of female sex-appeal was not clearly demonstrated although they have observed that a blind male will start vibrating its wings at a distance of 3 mm from a female. The fact that a female washed with hexane cannot induce this behavior, suggests the involvement of a female specific stimulant compound.

In the first part of this report, I shall present a study done in collaboration with Jallon showing evidence for the induction of male wing vibration by a female-specific odor. In the second part the analysis of wing vibration behavior for some olfactory mutants[7] will be discussed.

To try to study the physical nature of sex-appeal, we allowed 100 flies of either sex to run around for a fixed time (1 hour) in a small vial (1 ml) which was then washed with 0.5 ml hexane. 20 µl of this hexane extract was poured into a watch-glass, the hexane evaporated rapidly and the watch-glass turned over and placed on a glass-plate to form a courtship chamber. The ceiling of this chamber is 3.5 cm^2 and the volume is 0.7 cm^3.

The male courters, which were introduced through a hole in the glass-plate, were 4 days old and were isolated from females within 12 hours after eclosion. The behavior of courter males alone or in pairs inside different courtship chambers was observed for ten minutes and the cumulative time of wing vibration was measured (ΣT). Our results are summarized in Table 1. In experiment A we observed the behavior of male pairs inside a blank chamber. We observed a short vibration time only in a few cases and never for long time periods. Our observations are in agreement with Hall's published data.[8] In experiment B we found that when males where introduced individually into a feminine chamber, the male exhibited only a very short vibration time. In contrast, in experiment C when male testers were introduced in pairs into a feminine chamber, after a certain latency time, they start to show bouts of wing vibration toward each other. The male toward whom the courtship is directed, shows immediately a very clear rejection response: kicking with his legs, flickering his wings. In a quantitative measurement of ΣT, we considered only courting vibrations. We have studied 117 male couples and have determined an average cumulative vibration time of 36 seconds. As shown in D when a male held for ten minutes in a feminine chamber was introduced into a clean chamber with a naive male, this naive male showed characteristic courtship toward the other. Finally, in E an experiment analogous to C was performed with a male pair introduced inside a masculine chamber. No significant vibration could

Table 1. Induction of male wing vibration by female-specific stimulant. The different experimental conditions are described in the text. For each experiment, the latency time and the cumulative vibration time (ΣT) have been measured during the 600 second observation period and the individual values have been averaged.

Experiment Chamber	Number of males/chamber	Number of experiments	Average latency time (s)	$\overline{\Sigma T}$ (s)
A - blank	2	43	15 ± 7	2
B - feminine	1	54	-	1
C - feminine	2	117	240 ± 40	36 ± 8
D - blank	2	26	220 ± 15	20 ± 3
E - masculine	2	37	13 ± 8	2

be observed. This series of experiments demonstrate the necessary involvement at close range of female chemical compounds to induce the courter's wing vibrations. These experiments also suggest that the pheromone system is a necessary but not sufficient stimulus for courtship behavior. The presence of a second fly may be necessary to induce courtship behavior.

The dependence of $\overline{\Sigma T}$ on stimulant concentration is shown in Figure 1. A clear maximum is observed for the 1:100 dilution of the original extract. At higher concentrations there is an inhibition producing both a lower $\overline{\Sigma T}$ and a longer latency time which may be attributable to receptor adaptation.

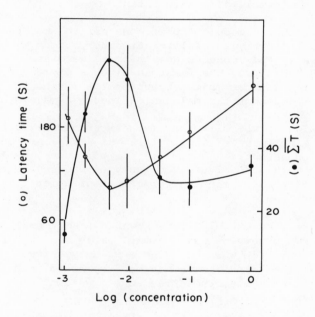

Figure 1: Dependence of the average wing vibration parameters (cumulative vibration time and latency time) on the concentration of female-specific stimulant. The concentration of the original extract was arbitrarily defined as one unit. The original extract was prepared as described in the text by washing a vial which had contained 100 flies for 1 hour with 0.5 ml hexane.

Preliminary gas chromatographic analysis of volatile substances from whole flies or hexane extract shows a shoulder which is specific to females. It was never observed from male materials, even when male courtship-stimulating young males were studied which suggests that the young male sex-appeal involves different compounds. The female peak migrates very close to cis-9 tricosene, a well known component of the *Musca domestica* pheromone system. When isolated by preparative gas chromatography (but always contaminated by a large overlapping peak), this fraction induced a sustained wing vibration of male pairs toward each other. Thus, this fraction probably corresponds to a component of the sex-stimulant of the female *Drosophila melanogaster* pheromone system.

Now I shall discuss preliminary observations obtained with smell-defective mutants.[7] Behavioral responses for odorants involve a complex pathway in the nervous system. A defect in any part of the system could lead to a modification or elimination of the olfactory response.[10] In order to discriminate the different compounds involved in the male wing vibration response, we have studied different strains of smell-defective mutants.

We have used the model of pheromone control proposed by Averhoff and Richardson[11] to try to explain our observations with these mutants. Basically this model suggests that there are two steps in the courtship behavior. The first involves a pheromone emitted by the female which induces male courtship; the second step involves two factors produced by the male which induce female acceptance.

Our results on the olfactory mutants are summarized in Table 2. In these experiments a single male and female are put in the observation chamber. When wild-type were used, we found that 90% copulated within 10 minutes but only 4% had copulated within 5 minutes. So we chose 5 minutes for our observation time in the remaining experiments.

When we studied the *olfC* mutant which cannot smell the ethyl acetate odor, we observed that the *olfC* males show a 2-fold increase in latency time when courting a wild-type female. This suggests that the female pheromone may be a compound with an acetate function. Alternatively, there may be a synergist of the female pheromone which contains an acetate group.

When we examined two different alleles of *olfA* (X1 and X8), a mutant which cannot detect benzaldehyde,[7] we observed a higher rate of attempted copulation in couples consisting of *olfA* females and wild-type males (*olfA$^+$*). This could mean that *olfA* females are unable to detect a compound normally emitted by males which interferes with the male pheromone action.

We have also examined two other classes of mutants: *olfB* which has a temperature-sensitive defect for benzaldehyde detection and *olfD* which is defective for both benzaldehyde and ethyl acetate detection. With both of these mutants we have observed only normal

Table 2. Effect of olfactory mutants on wing vibration parameters.

♀*	♂*	Number of couples Observed	% showing wing vibration	Latency time (s)	$\overline{\Sigma T}$ (s)	Attempted Copulation % showing behavior	Attempted Copulation Time to Copulation
wild	wild	40	90	40 ± 13	150 ± 15	90%	Within 600 s
						4%	Within 300 s
							(Ave.360 ± 30 s)
olfC (X10 allele)	wild	40	100	36 ± 11	153 ± 25	5%	Within 300 s
wild	olfC (X10 allele)	40	95	70 ± 18	110 ± 26	5%	Within 300 s
olfC (X10 allele)	olfC (X10 allele)	10	90	50 ± 24	152 ± 20	5%	Within 300 s
olfA (X8 allele)	wild	29	95	36 ± 13	70 ± 40	100%	Ave.155 ± 58
		11	100	43 ± 11	114 ± 40	None	Within 300 s
wild	olfA (X8 allele)	40	95	32 ± 14	120 ± 22	5%	Within 300 s
olfA (X8 allele)	olfA (X8 allele)	10	90	26 ± 7	81 ± 17	67%	Ave.167 ± 38
olfA (X1 allele)	wild	24	100	28 ± 11	76 ± 27	100%	Ave.174 ± 43
		16	95	38 ± 17	133 ± 30	None	Within 300 s
wild	olfA (X1 allele)	40	90	29 ± 13	155 ± 33	5%	Within 300 s
olfA (X1 allele)	olfA (X1 allele)	10	90	46 ± 17	63 ± 29	62%	Ave.180 ± 50

* All flies were 4 days old and were isolated from flies of the opposite sex within 12 hours after eclosion.

behavior with respect to latency time and $\overline{\Sigma T}$ for the male wing vibration.

The long distance attractant activity of males, females and hexane extract was examined using a horizontal air flow system illustrated in Figure 2. It is a simple Y-maze and consists of 3 parts: a starting vial, a Y-maze and 2 small cups where the odor is confined. Air is withdrawn at the junction causing a controlled and symmetrical in-flow into the cups and the Y-maze. The flies studied were segregated by sex within 12 hours after eclosion. They were deprived of food one day before the experiment and fed only with a 1% sucrose solution. The attractant odor (1, 10 or 25 flies) was collected by confining the flies for 2 hours in the cup without aeration.

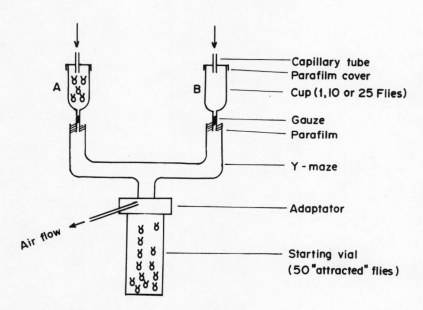

Figure 2: The olfactometer for testing long range attractive odors. Flies distribute in the arms of the Y-maze in response to an odor cue from the cups. Starting vial is 2.5 cm in diameter and 7 cm in length. The Y-maze is 1 cm by 12 cm and the cups are 1.5 cm by 2 cm.

The attracted flies (50 flies) are placed in the starting vial just before the experiment. When the air-flow is opened, they can walk freely and distribute themselves in the two arms. After five minutes, the distribution of flies is counted; this is repeated 6 times with the same flies. All these experiments are carried out at 24°C and in complete darkness. The response of flies was evaluated by the attractibility index (A.I.):[12]

$$A.I. = \frac{\text{number of flies in odor arm - number of flies in control arm}}{\text{total number of flies}}$$

We have used this index in our experiments because we always have a large proportion of flies remaining in the starting vial (30-50%). In the attractibility index, these unresponsive flies are neglected.

As summarized in Figure 3A with females in the odor cup, neither males nor females were attracted to the cup. Similarly, with a hexane extract from females, we have never observed males to be attracted to the cup. It may be that this female odor works only over a short distance or requires contact action in order to play a role in the courtship behavior.

In contrast, when males are in the odor cup (Figure 3B), females are attracted and males are weakly attracted. In Figure 3C we compare directly the attractiveness of two cups: one containing males (A) and the other containing females (B). Under this condition we observe an even higher attraction of the females to the males.

These results are preliminary since in the construction of our olfactometer, some problems remain to be resolved. For example, 30 to 50% of the flies tested are unresponsive to the flow of odors. One possible problem is that our running time of 5 minutes was too short. As suggested in Figure 4, longer running times of up to ~ 10 minutes appear to give a higher attractibility index. Another potential problem could be that pheromone antagonists are present in the air flow from the cups.

Table 3. Experimental design for studies in Figure 3.

Cup A	Cup B	Starting vial	Experiment
1, 10 or 25 females	control:air	(a) 50 females (b) 50 males	A
1, 10 or 25 males	control:air	(a) 50 females (b) 50 males	B
10 or 25 males	10 or 25 females	(a) 50 females (b) 50 males	C

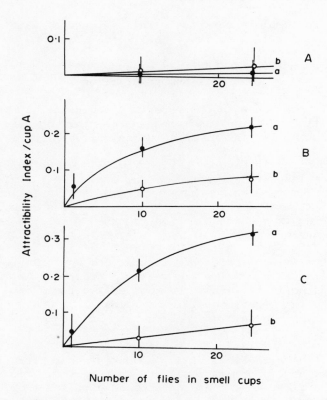

Figure 3: Olfactometer studies on the dependence of the attractibility index on the odor concentration. Experimental design is summarized in Table 3.

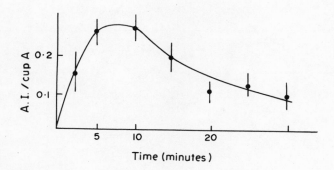

Figure 4: Kinetics of the female response to odors. Twenty five males were placed in cup A; 25 females in cup B. Fifty female "responders" were place in the starting vial and the experiment was run as in Figure 3C (a) except that various running times were tested.

We are interested in studying further the interactions between the molecular stimulant (pheromone) and its receptor. To accomplish this aim the following experiments are in progress: purification of the various active fractions; localization and analysis of the receptor using electrophysiological and biochemical techniques.

REFERENCES

1. E.O. Wilson, Chemical communication within animal species, in: "Chemical Ecology," E. Sondheimer and J.B. Simeone, eds., Academic Press, New York (1970).
2. K.E. Kaissling, Insect olfaction, in: "Handbook of Sensory Physiology", L. Beidler, ed., Springer-Verlag, New York (1971).
3. A.M. Sturtevant, Experiments on sex recognition and the problem of sexual selection in Drosophila, J. Anim. Behav. 5:351 (1915).
4. M. Bastock and A. Manning, The courtship of Drosophila melanogaster, Behaviour 8:85 (1955).
5. H.T. Spieth, Courtship behavior in Drosophila, Annual Rev. Entom. 19:385 (1974).
6. J.M. Jallon and Y. Hotta, Genetic and behavioral studies of female sex-appeal in Drosophila, Behav. Genet. 9:256 (1979).
7. V. Rodrigues and O. Siddiqi, Genetic analysis of chemosensory pathway, Proc. Indian Acad. Sci. 27B:147 (1978).
8. J.C. Hall, Courtship among males due to a male-sterile mutation in Drosophila melanogaster, Behav. Genet. 8:125 (1978).
9. D.A. Carlson, M.S. Mayer, D.L. Silhaeck, D.P. James, M. Beroza and B.A. Bierle, Sex attractant pheromone of the housefly: isolation, identification and syntheses, Science 174:76 (1971).
10. T. Kikuchi, Genetic alterations of olfactory functions in Drosophila melanogaster, Japan. J. Genetics 48:105 (1973).
11. W.W. Averhoff and R.J. Richardson, Multiple pheromone system controlling mating in Drosophila melanogaster, Proc. Natl. Acad. Sci. 73:591 (1976).
12. Y. Fuyama, Behavioral genetics of olfactory responses in Drosophila, I - Olfactory and strain differences in Drosophila melanogaster, Behav. Genet. 6:407 (1976).

A REVIEW OF THE BEHAVIOR AND BIOCHEMISTRY OF *dunce*, A MUTATION OF

LEARNING IN *Drosophila*

        Duncan Byers

        European Molecular Biology Laboratory
        Heidelberg
        Federal Republic of Germany

When it became apparent in the late nineteenth century that the accumulated experience of an adult animal does not pass to its progeny through the germ cells, biologists took up the problem of how animals learn from experience. The work of C.L. Morgan, E.L. Thorndike, I.P. Pavlov and others revealed that much of learning appears to occur by the formation and strengthening of connections between external situations and the behavior of the animal. The strength of the connections is primarily under the control of certain biologically significant stimuli, including heat, cold, food, water, pinches, bites, poisons, sour or bitter tastes, and related agents. Collectively these are termed reinforcers or rewards. To understand learning further, one wishes to know what neural and molecular changes occur during learning, how reinforcement brings these changes about, and how nerve cells use these changes for generating changed behavior.

A stimulus that elicits learning, such as a morsel of food (for a hungry animal), a pinch, or an electric shock, strengthens numerous connections. Associations form between the stimulus and the color, odor, sounds or other features of the environment, so that when these sensory cues later recur alone, they trigger appropriate behavior. This process is associative learning. In addition, a reinforcing stimulus strengthens some reflex connections whether or not they are active when the reinforcing stimulus occurs. This form of plasticity is named sensitization and is one type of nonassociative learning. An example of sensitization common in human behavior is that a painful stimulus enhances the strength of limb retraction reflexes elicitable by mild tactile stimuli.

The smallest dissected parts of brains found still capable of associative learning contain hundreds of neurons, and consequently the neural analysis of associative learning is primitive. Sensitization, however, occurs in some simple circuits. In the best studied case, the changes underlying sensitization of gill retraction in a mollusc are found to occur at certain synapses of the central nervous system. The changes mediating the sensitization appear to be the release of serotonin in the brain, causing synthesis of cyclic AMP in the synapses being sensitized, the cyclic AMP in turn stimulating a prolonged enhancement of neurotransmitter release.[1] Work with locust,[2] lobster,[3] and rat[4] indicates that mechanisms for synaptic modification by cyclic AMP may occur generally. Whether they are important in associative learning is unknown. These findings with cyclic AMP are described here because the subject of this review, a *Drosophila* mutant deficient in learning, has led by a different route to indications that cyclic AMP is important in learning.

The fruitfly *Drosophila* is competent at both sensitization and associative learning, and probing of the underlying mechanisms with mutants should therefore be possible. In this review, I will describe recent progress in this direction.

The best studied example of sensitization in any fly concerns the proboscis extension reflex of the blowfly *Phormia*. In a hungry fly, extension of the proboscis is elicited by contact of leg taste receptors with a sugar solution. The response to water or dilute sugar is enhanced (sensitized) by a prior stimulation with concentrated sugar solution. This reflex and its sensitization were studied for many years by Dethier and his colleagues, who have referred to the sensitization as the 'central excitatory state'.[5] Duerr and Quinn have shown that the proboscis extension reflex of *Drosophila* sensitizes similarly (unpublished). They find that some mutants deficient in associative learning are also abnormal in sensitization.

A demonstration of associative learning must include control experiments that are sufficient to distinguish it from sensitization. (As implied in the second paragraph of this review, the proper control is temporal separation of the reinforcement from the sensory cues, since by definition associative learning depends critically on their near simultaneity.) The first adequate demonstration of associative learning in any fly appeared in 1971. This was performed by Nelson, also using the proboscis extension reflex of *Phormia*.[6] In 1974, Quinn, Harris, and Benzer demonstrated olfactory learning in *Drosophila*, using a method suitable for selection of mutants with altered learning.[7] Three additional examples of associative learning in *Drosophila* using different sensory modalities are provided by Medioni and Vaysse,[8] Menne and Spatz,[9] and Booker and Quinn (unpublished). Sensory cues that *Drosophila* can learn include color, odor, taste, and limb position. The effective reinforcers known are electric shock, quinine powder (as a bitter taste), and strong vibration. Learning ability in *Drosophila* has proven to be more robust and general than was previously suggested, and one wonders how

learning might serve the fly in nature. One possibility is suggested
by the recent finding of Siegal and Hall that some features of normal
male courtship are modifiable by experience.[10]

In the olfactory learning experiment of Quinn, Harris and Benzer,
flies are trained and tested using metal grids (of printed circuit
material) spread with odorant solutions. Illumination is arranged so
that the flies' natural phototactic drive attracts them on to the
grids. For training, the flies are induced to run alternately on to
two grids with different odorants, with one of the grids electrified.
Learning is tested on two similar grids without shock. Normal flies
tend to avoid the electrified grid during training, and later, during
testing, they selectively avoid the odorant that was coupled with
shock. The flies selectively avoid whichever odorant was associated
with shock, indicating that the avoidance is due to associative
learning, not sensitization or inherent odor bias. After a training
of three 15-second periods with shock, the flies remember for three
to six hours.[11,12] With more training, the memory lasts longer.[7]

Screening for mutations of learning began in 1974 at the California Insitute of Technology by Dudai, Jan, Byers, Quinn, and Benzer
using this olfactory learning test.[13] Two mutants were found among
1500 mutagen-treated X-chromosomes. These have proven to be alleles
of one gene and are named $dunce^1$ and $dunce^2$. At Princeton University,
Sziber, Booker, Aceves-Piña, and Quinn have found five mutants with
aberrant learning.[14,15] These complement the *dunce* mutants and each
other in all combinations and therefore probably define five additional genetic loci. Only the *dunce* mutants are described in the
following paragraphs.

The *dunce* flies do not learn successfully in the olfactory
learning test; after the training, they do not selectively avoid the
shock-associated odorant. Sensory or motor incapacities that could
account for poor learning have not been found in *dunce*. During
training, *dunce* flies are attracted onto the odorant grids in the normal way by the light and are deterred by the charged grid, showing
that they can sense the electric shock. They are able to sense the
odorants used, because they demonstrate normal odor preferences in
a simple choice situation that does not require learning.[13,16]
(Dudai[16] and Dudai and Bicker[17] referred to the alleles $dunce^1$ and
$dunce^2$ as $dunce^{DB38}$ and $dunce^{DB276}$ respectively.)

With a modified form of the olfactory learning test, Dudai found
that *dunce* flies can form and express an unstable short-lived association of odorant and shock.[16] It appears that in *dunce* an association of shock with one odorant is quickly masked or erased when the
flies are exposed to a different odorant. In normal flies the
strength of an association does not appear to be influenced by other
odorants.[16] In the learning test used for isolating the *dunce* mutant, exposure to a control odorant occurred always between training

and testing, explaining why the unstable association in *dunce* was not previously detected. These findings suggest that *dunce* may be blocked in an early step of memory consolidation.

dunce flies have been observed in other learning tests. Duerr and Quinn find that they do habituate and sensitize, but are abnormal in the rates of decay of habituation and sensitization (unpublished) Aceves-Piña and Quinn have found that *dunce* larvae fail at olfactory learning also, though they can sense the odorants and electric shock used.[15] Booker and Quinn find that *dunce* flies are not as proficient as normal flies in learning to hold defined leg positions, though they are partially successful in this type of learning (unpublished). All these findings indicate that the normal *dunce* gene product may participate in various types of behavioral plasticity. Dudai and Bicker, however, found that *dunce* and the normal strain from which it was derived are equally competent in one form of visual learning.[1] This appears to demonstrate that the *dunce* gene product is not universally required for all learning in *Drosophila*. Present knowledge is insufficient to resolve whether the normal *dunce* gene product participates directly in learning mechanisms or serves some auxiliary function necessary for normal learning.

In other ways the *dunce* flies are remarkably normal. By external morphology they cannot be distinguished from normal flies. They seem healthy and vigorous; and tests of phototaxis, geotaxis, flight, locomotor activity, and sexual courtship reveal no defects. The electroretinogram and synaptic transmission at the larval neuromuscular junction are normal.

The two *dunce* mutants share one prominent trait in addition to poor learning. Both are poorly fertile as females, though to different degrees. After a single fertilization, individual normal females average 181 adult offspring, *dunce^1* females average 116 offspring (significantly different, $p<0.001$), and *dunce^2* females averaging 1.6 offspring each. Genetic mapping and complementation testing show that the infertility traits of *dunce^1* and *dunce^2* are allelic and map with the learning trait between *yellow* and *chocolate* near the tip of the X-chromosome.[18]

Classical and current estimates suggest that *Drosophila* has about 6,000 genes. Since at least 2,000 of these are known individually, one might expect that different geneticists are sometimes independently studying one gene. This has occurred with *dunce*, which was found independently by Mohler in a search for X-linked female-sterile mutants, and independently by Kiger and Golanty in a search for the genes of the cyclic AMP phosphodiesterase enzymes of *Drosophila*.

Cyclic AMP (cyclic adenosine monophosphate) was mentioned in the third paragraph of this account as a probable effector of synaptic

A LEARNING MUTANT

modification. It is also known as a trigger and regulator of other cellular functions (for a careful review see 19). Cyclic AMP is synthesized from ATP (adenosine triphosphate) by adenyl cyclase and hydrolyzed to AMP (adenosine monophosphate) by cyclic AMP phosphodiesterase.

The cyclic AMP phosphodiesterases of *Drosophila* have been studied by Kiger, Golanty, and Davis of the University of California at Davis.[20-24] In normal *Drosophila* they find two <u>soluble</u> forms of cyclic AMP phosphodiesterase activity. (It must be emphasized that the <u>insoluble</u> subcellular fractions also contain cyclic AMP phosphodiesterase activity, not yet characterized, that may yield further forms.) The two identified forms differ in molecular weight, substrate specificity, and rate of inactivation by heat. They probably represent independent cyclic AMP phosphodiesterase enzymes, though this is not certain. A gene controlling one of the forms was found using the methods of segmental aneuploidy and mapped within region 3D3 and 3D4 of the X-chromosome. Using overlapping chromosomal deletions characterized by Lefevre,[25] it was found that females with homozygous deficiency of five adjacent chromosomal bands including 3D3 and 3D4 survive to adulthood and are normal in appearance, but are sterile. To test whether the infertility of cyclic AMP metabolism are associated with the same genetic locus in this region, Davis and Kiger examined some of the 225 X-linked female-sterile mutants isolated by Mohler.[26,27] Two of these female-sterile mutants proved to map in region 3D3 and 34D and to lack the form of cyclic AMP phosphodiesterase that is associated with this region, demonstrating that the infertility and aberrant cyclic AMP metabolism are expressions of one gene. It is unknown whether this is the structural gene or a regulatory gene of the cyclic AMP phosphodiesterase.

News of this work reached me in 1978. The possible involvement of cyclic AMP in learning, the coincidence of female infertility, and an approximate coincidence of map position suggested that *dunce* and the gene affecting cyclic AMP phosphodiesterase could be identical. The truth of this hypothesis is demonstrated by the following evidence.[28]

1. The two mutants from Mohler learn poorly, like *dunce*. They do not complement *dunce*^{1} or *dunce*^{2} in learning or reproduction, and therefore are alleles of the *dunce* gene.

2. Mapping by complementation with various duplications and deficiencies places the original *dunce*^{1} and *dunce*^{2} mutants in region 3D3 and 3D4.

3. Females with homozygous deficiency of region 3D3 and 3D4 are viable and vigorous, but fail at learning, like *dunce*, and are sterile.

4. The mutants $dunce^1$ and $dunce^2$ have reduced levels (15% of normal) of the soluble cyclic AMP phosphodiesterase form that is undetectable in the two mutants of Mohler. Conformably, the concentration of cyclic AMP itself is elevated 1.6-fold in $dunce^1$ and $dunce^2$ and 6-fold in the two mutants of Mohler.

These findings show that the normal *dunce* gene product has roles in behavior, reproduction, and cyclic AMP metabolism. Moreover, the good viability of flies with homozygous deletion of *dunce* indicates that the normal *dunce* gene does not specify any essential vital function.

The current work with *dunce*, in several laboratories, is aimed at determining whether *dunce* is the structural gene of the cyclic AMP phosphodiesterase, where the *dunce* gene product is located in tissues and cells, and whether the poor learning in *dunce* is a direct consequence of aberrant cyclic AMP metabolism. We hope that the mutant will be useful in exploring the mechanisms of normal learning.

ACKNOWLEDGEMENTS

It is a pleasure to thank Seymour Benzer for many contributions to the work described here. The author was a Fellow of the Gordon Ross Medical Foundation at the Division of Biology, California Institute of Technology from 1974 to 1979 and is now a Research Fellow of the European Molecular Biology Laboratory.

REFERENCES

1. M. Klein and E.F. Kandel, Presynaptic modulation of voltage-dependent Ca^{2+} current: Mechanism for behavioral sensitization in *Aplysia californica*, Proc. Nat. Acad. Sci. U.S.A. 75:3512 (1978).
2. M. O'Shea and P.D. Evans, Potentiation of neuromuscular transmission by an octopaminergic neuron in the locust, J. Exp. Biol. 79:169 (1979).
3. E.A. Kravitz, P.D. Evans, B.R. Talamo, B.G. Wallace and B.A. Battelle, Octopamine neurons in lobsters: location, morphology, release of octopamine and possible physiological role, Cold Spring Harbor Symp.Quant. Biol. 40:127 (1975).
4. M.O. Miyamoto and B. McL. Breckenridge, A cyclic adenosine monophosphate link in the catecholamine enhancement of transmitter release at the neuromuscular junction, J. Gen. Physiol. 63: 609 (1974).
5. V.G. Dethier, R.L. Solomon and L.H. Turner, Sensory input and central excitation and inhibition in the blowfly, J. Comp. Physiol. Psychol. 60:303 (1965).
6. M.C. Nelson, Classical conditioning in the blowfly (*Phormia regina*): associative and excitatory factors, J. Comp. Physiol. Psychol. 77:353 (1971).

7. W.G. Quinn, W.A. Harris and S. Benzer, Conditioned behavior in *Drosophila melanogaster*, Proc. Nat. Acad. Sci. U.S.A. 71:708 (1974).
8. J. Medioni and G. Vaysse, Suppression conditionelle d'un reflexe chez la Drosophile (*Drosophila melanogaster*): acquisition et extinction, Comptes Rendus Soc. Biol. 169:1386 (1975).
9. D. Menne and H.C. Spatz, Color learning in *Drosophila*, J. Comp. Physiol. 114:301 (1977).
10. R.W. Siegel and J.C. Hall, Conditioned responses in courtship behavior of normal and mutant *Drosophila*, Proc. Nat. Acad. Sci. U.S.A. 76:3430 (1979).
11. Y. Dudai, Properties of learning and memory in *Drosophila melanogaster*, J. Comp. Physiol. 114:69 (1977).
12. W.G. Quinn and Y. Dudai, Memory phases in *Drosophila*, Nature 262:576 (1976).
13. Y. Dudai, Y.-N. Jan, D. Byers, W.G. Quinn and S. Benzer, *dunce* a mutant of *Drosophila* deficient in learning, Proc. Nat. Acad. Sci. U.S.A. 73:1684 (1976).
14. W.G. Quinn, P.P. Sziber and R. Booker, The *Drosophila* memory mutant *amnesiac*, Nature 277:212 (1979).
15. E.O. Aceves-Piña and W.G. Quinn, Learning in normal and mutant *Drosophila* larvae, Science 206:93 (1979).
16. Y. Dudai, Behavioral plasticity in a *Drosophila* mutant, *dunce*^{DB276} J. Comp. Physiol. 130:271 (1979).
17. Y. Dudai and G. Bicker, Comparison of visual and olfactory learning in *Drosophila*, Naturwissenschaften 65:494 (1978).
18. D. Byers, Studies on learning and cyclic AMP phosphodiesterase of the *dunce* mutant of *Drosophila melanogaster*, Ph.D. Thesis, California Institute of Technology (1979).
19. H. Rasmussen and D.B.P. Goodman, Relationships between calcium and cyclic nucleotides in cell activation, Physiol. Reviews 57:421 (1977).
20. J.A. Kiger Jr. and E. Golanty, A cytogenetic analysis of cyclic nucleotide phosphodiesterase activities in *Drosophila*, Genetics 85:609 (1977).
21. J.A. Kiger Jr., The consequences of nullosomy for a chromosomal region affecting cyclic AMP phosphodiesterase in *Drosophila*, Genetics 85:623 (1977).
22. J.A. Kiger Jr. and E. Golanty, A genetically distinct form of cyclic AMP phosphodiesterase associated with chromomere 3D4 in *Drosophila*, Genetics 91:521 (1979).
23. R.L. Davis and J.A. Kiger Jr., Genetic manipulation of cyclic AMP levels in *Drosophila melanogaster*, Biochem. Biophys. Res. Comm. 81:1180 (1978).
24. R.L. Davis, Biochemical characterization and genetic dissection of cyclic nucleotide phosphodiesterases of *Drosophila melanogaster*, Ph.D. Thesis, University of California at Davis (1979).
25. G. Lefevre, Jr., Dros. Inform. Serv., in press (1980).
26. D. Mohler, Female-sterile mutants in *Drosophila melanogaster*, Genetics 74:s184 (1973).

27. D. Mohler, Developmental genetics of the *Drosophila* egg. I. Identification of 59 sex-linked cistrons with maternal effects on embryonic development, Genetics 85:259 (1977).
28. D. Byers, R.L. Davis, and J.A. Kiger Jr., A defect of cyclic AMP metabolism in the *dunce* mutant of *Drosophila melanogaster*, in preparation (1980).

DISCUSSION

A. Fodor: Can you block learning with drugs?

D. Byers: The drugs caffeine and theophylline are known to block phosphodiesterase. It has been shown that feeding the flies caffeine blocks learning.

S.C. Lakhotia: Is development normal in the dunce mutant? You might expect cAMP levels to affect many processes.

D. Byers: As far as we can tell, development is normal in this mutant.

A. Garcia-Bellido: Can you study the dunce mutation using gynandromorphs?

D. Byers: It is difficult to test single flies in the learning paradigm, so gynandromorph analysis would be difficult.

Y. Hotta: Jeffrey Hall described the effect of a memory mutation on sexual behavior. Couldn't this be used in a single fly assay?

D. Byers: Yes, that could be used for mosaic analysis.

A. Garen: What is the basis for your classification of dunce as a behavioral mutant?

D. Byers: A behavioral mutant is one that affects behavior.

A. Garen: I was concerned about the fact that a mutation such as dunce may cause general defects and modify behavior as a secondary consequence.

CONCLUDING REMARKS

Linda M. Hall and Jeffrey C. Hall

This meeting brought together many investigators who are using *Drosophila* to study a wide-ranging array of biological problems. Because the problems being studied are so diverse, it might seem artificial to focus a meeting and a publication on one organism. However, there are many connections between the *Drosophila* topics presented here. For example, nearly all of these studies recognize the importance of genetics as a technique for studying biological questions. Both the technical and conceptual details of genetic analysis can be readily communicated among investigators who speak the same language, and an experimental approach which is used to solve a problem in one area of biology can stimulate investigators to approach problems in other areas.

The papers in this volume are arranged in a sequence beginning with the relatively basic genetic and molecular genetic topics, then moving to more complex developmental issues, and finally ranging to problems that involve higher functions of the central nervous system. This progression also ties the different areas together.

One of the threads that runs among all of the topics involves genetic mosaics. The use of genetically "mixed" animals emerges as an example of a versatile technique that finds wide-spread applications. In this volume we see genetic mosaics applied to problems in development, physiology, and behavior. The sophisticated genetic technology for generating mosaics was an important underpinning in all of these experiments. One feature of this mosaic technology exemplified in this volume involves the emergence of genetic and biochemical techniques to allow analysis of mosaicism in internal tissues as well as on the external covering of the animal.

Examples of the use of mosaicism in this volume involve experiments on lethal phenotypes, experiments on maternal inheritance and early development, analysis of pattern formation in a variety of tissues, and dissection of the role of the central and peripheral nervous systems in the development of normal neuronal wiring patterns. Mosaics are used to investigate such disparate problems as the question of autonomy of non-pupariating mutants and in the mapping of regions of the brain responsible for complex behaviors. Thus, we see a common technique tying together studies on development and neurobiology. Another example of genetics as a common conceptual tool is its application as a surgical tool to dissect a complicated process into its component parts. In this volume we see that mutants are used to determine the number and sequence of genes acting in the development of a particular tissue such as the muscle. They are also used to dissect complicated physiological processes such as the chemosensory and visual systems as well as the complex behaviors such as circadian rhythm, courtship and learning.

Although the purely genetic approaches continue to make important contributions, molecular approaches are rapidly being integrated into the various studies of *Drosophila* biology and many examples of this approach are included in this volume. Concepts developed in the analysis of molecular aspects of the expression of a particular gene are likely to be useful in developing models of cell differentiation and pattern formation, as well as in considering regulation of molecules important to neuronal function. Some molecular approaches such as "cloning" genes will provide a powerful way to analyze genes of interest to neuronal function as well as to analyze products of developmentally fascinating genes which to date have been studied on the basis of their morphogenetic abnormalities. Molecular approaches have also been used to define neurotransmitter receptors, ionic channels, more general features of cell membranes and also enzymes that control the metabolism of neurotransmitters. We are now beginning to see a fusion of these molecular approaches with genetic analysis and we see that mutants with impairments of brain morphology, brain chemistry, pheromone production or detection, and even learning can be useful in the analysis of the neurobiological underpinnings of higher behaviors.

Thus, by juxtaposing recent studies in *Drosophila* of development and neurobiology, we see that there are common conceptual approaches to these two complex areas of investigation. *Drosophila* is shown to be amenable to a wide variety of the analytical tools currently used in biology including: genetic manipulation and mosaic technology, monoclonal antibodies, biochemical investigations including those related to hormonal and neurochemical studies, quantitative physiological and behavioral recordings, and histological techniques ranging from those that allow neural pathways to be traced at the light microscopic level to ultrastructural investigations. Although the complexities of both development and nervous system functioning make these problems seem intractable in certain systems, we see that they are not necessarily so formidable in the tiny fruitfly. This is in part because of the strength of genetic analysis which allows one to dissect the roles of various parts of the whole system, using gene mutations that alter parts of the system, one by one.

International Conference on
Development and Behavior of *Drosophila*
Bombay, 19th-22nd December, 1979

List of Participants

P. Babu
Tata Institute of
   Fundamental Research
Bombay, India

S. Bhat
Tata Institute of
   Fundamental Research
Bombay, India

D. Byers
European Molecular
   Biology Laboratory
Heidelberg, West Germany

S. Balakrishnan
Indian Agricultural
   Research Institute
New Delhi, India

P.N. Bhavsar
Tata Institute of
   Fundamental Research
Bombay, India

P.J. Bryant
University of California
Irvine, CA., U.S.A.

J.A. Campos-Ortega
Universitat Freiburg
Freiburg, West Germany

M.K. Chandrashekaran
Madurai University
Madurai, India

A. Chovnick
The University of Connecticut
Storrs, CT., U.S.A.

B. Cotton
Universitat Zurich
Zurich, Switzerland

N.B. Dev
Tata Institute of
   Fundamental Research
Bombay, India

D.A. Desai (deceased)
Tata Institute of
   Fundamental Research
Bombay, India

S.U. Donde
Foundation for Medical Research
Bombay, India

A.K. Duttagupta
University of Calcutta
Calcutta, India

A. Fodor
Tata Institute of
   Fundamental Research
Bombay, India

A. Garcia-Bellido
Universidad Autonoma de Madrid
Madrid, Spain

PARTICIPANTS

A. Garen
Yale University
New Haven, CT., U.S.A.

N. Gautam
Tata Institute of
 Fundamental Research
Bombay, India

A. Ghysen
University of Brussels
Brussels, Belgium

K.S. Gill
Punjab Agricultural University
Ludhiana, India

J.C. Hall
Brandeis University
Waltham, MA., U.S.A.

L.M. Hall
Albert Einstein College
 of Medicine
Bronx, N.Y., U.S.A.

M. Heisenberg
Universitat Wurzburg
Wurzburg, Germany

Y. Hotta
University of Tokyo
Tokyo, Japan

S. Iyer
Tata Institute of
 Fundamental Research
Bombay, India

I. Kiss
Biological Resarch Center
Szeged, Hungary

K.S. Krishnan
Tata Institute of
 Fundamental Research
Bombay, India

S.J. Kulkarni
Tata Institute of
 Fundamental Research
Bombay, India

K. Kumar
University of Gorakhpur
Gorakhpur, India

H. Lamfrom
The Retreat
Ahmedabad, India

S.C. Lakhotia
Banaras Hindu University
Varanasi, India

P.A. Lawrence
MRC Laboratory of
 Molecular Biology
Cambridge, England

D. Majumdar
University of Calcutta
Calcutta, India

G. Morata
Universidad Autonoma de Madrid
Madrid, Spain

T. Mukherjee
Banaras Hindu University
Varanasi, India

A.S. Mukherjee
University of Calcutta
Calcutta, India

V. Nanjundiah
Indian Institute of Science
Bangalore, India

D. Nassel
European Molecular Biology
 Laboratory
Heidelberg, West Germany

D.L. Nasser
National Science Foundation
Washington, D.C., U.S.A.

A.G. Padhyde
Tata Institute of
 Fundamental Research
Bombay, India

PARTICIPANTS

W.L. Pak
Purdue University
West Lafayette, IN., U.S.A.

J. Palka
University of Washington
Seattle, WA., U.S.A.

M.L. Pardue
Massachusetts Institute
  of Technology
Cambridge, MA., U.S.A.

J.C. Parikh
Physical Research Laboratory
Ahmedabad, India

H.A. Ranganath
University of Mysore
Mysore, India

A.R. Reddy
University of Hyderabad
Hyderabad, India

V. Rodrigues
Tata Institute of
  Fundamental Research
Bombay, India

A. Sarabhai
The Retreat
Ahmedabad, India

H. Sharat Chandra
Indian Institute of Science
Bangalore, India

R.P. Sharma
Indian Argicultural
  Research Institute
New Delhi, India

A. Shearn
The Johns Hopkins University
Baltimore, MD., U.S.A.

O. Siddiqi
Tata Institute of
  Fundamental Research
Bombay, India

S.S. Siddiqui
Max-Planck Institut fur
  Experimentelle Medizin
Gottingen, West Germany

P. Simpson
Centre de Genetique Moleculaire
Gif-sur-Yvette, France

R.N. Singh
Tata Institute of
  Fundamental Research
Bombay, India

S. Singh
Tata Institute of
  Fundamental Research
Bombay, India

G. Sreerama Reddy
University of Mysore
Mysore, India

N.J. Strausfeld
European Molecular
  Biology Laboratory
Heidelberg, West Germany

J. Szabad
Biological Research Centre
Szeged, Hungary

P. Thammana
Tata Institute of
  Fundamental Research
Bombay, India

G. Thor
Tata Institute of
  Fundamental Research
Bombay, India

T.R. Venkatesh
Tata Institute of
  Fundamental Research
Bombay, India

V.G. Vaidya
Poona
India

R. Venard
Tata Institute of
  Fundamental Research
Bombay, India

K. Vijay Raghaven
Tata Institute of
  Fundamental Research
Bombay, India

K. White
Brandeis University
Waltham, MA., U.S.A.

E. Wieschaus
European Molecular Biology
  Laboratory
Heidelberg, West Germany

M. Wilcox
MRC Laboratory of Molecular Biology
Cambridge, England

S. Zingde
Cancer Research Institute
Bombay, India

PARTICIPANTS

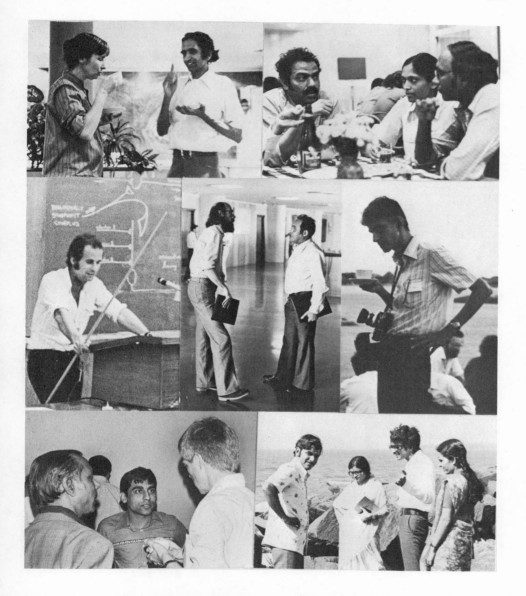

Top row: M.L. Pardue, K.S. Krishnan, M.K. Chandrashekaran, V. Rodrigues, P. Thammana.

Middle row: P.A. Lawrence, A. Ghysen, A. Garcia-Bellido, O. Siddiqi.

Bottom row: A.S. Mukherjee, S.S. Siddiqui, M. Heisenberg, S.J. Kulkarni, S. Zingde, T.R. Venkatesh, A.G. Padhye.

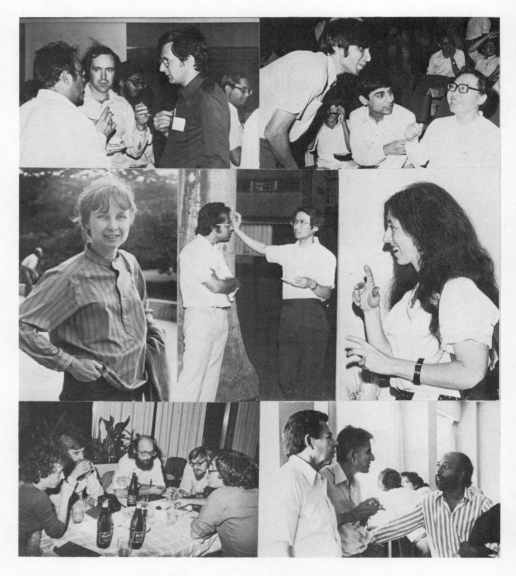

Top row: A. Garcia-Bellido, E. Wieschaus, G. Thor, J. Szabad, R.P. Sharma, P. Babu, S.S. Siddiqui, Y. Hotta.

Middle row: M.L. Pardue, N.B. Dev, W.L. Pak, L.M. Hall.

Bottom row: P. Simpson, P.J. Bryant, A. Ghysen, M. Wilcox, G. Morata, A. Chovnick, O. Siddiqi, K. Mukherjee.

PARTICIPANTS

Top row: P. Babu, K. White, M.L. Pardue, P.K. Maitra, V. Nanjundiah, R.N. Singh, P.J. Bryant.

Middle row: A. Shearn, K. White, L.M. Hall, S. Balakrishnan, N. Gautam, I. Kiss.

Bottom row: J. Palka, A. Garen, R. Venard, J.C. Hall.

Top row: N.B. Dev, J.A. Campos-Ortega, Mrs. Campos-Ortega, P. Thammana, K. S. Krishnan, A. Sarabhai.

Middle row: M. Heisenberg, R. Narasimhan, D. Byers, O. Siddiqi.

Bottom row: N.J. Strausfeld, A. Ghysen, G. Sreerama Reddy, E. Wieschaus, K.S. Gill.

INDEX

A

Acetylcholine, 295,442-444
  agonists and antagonists of, 255
  control of behavior by, 449
Acetylcholinesterase, 305-307, 310,315-316
  aggregates of, 309
  of electric eel, 309
  mosaics of, 442-446,449, 454
  mutants of, 442-443,445-446,449
  subunits of, 308
Acid phosphatase mutant
  as nerve cell marker, 436, 449-450
Alkaline phosphatase, 315-316
Amnesiac mutant, 433,435
Antenna
  ganglion innervated by, 374,383
  sensory axons from, 279-280,283
Antennapedia mutant, 242
Antibodies
  antigens binding with, 198
  in anti-sera, 195
  binding of, 196-197
  monoclonal, 193
Aristapedia mutant, 242
ATPase, 315,328
Avoidance behavior, 364,387
Axons
  from bristles, 237,241
  of campaniform sensilla, 238,256
  in *Daphnia*, 224
  pioneer fibers of, 226, 258
  program of outgrowth of, 257
  projections of, 225,228, 241,255
  of retina, 224-225
  sensory, 223,226-227,242, 247,272

B

Bar-eye mutant, 286-287
Behavioral mutants, 248,301,347, 355-357,366-369,397-398, 409-410,425,441,460-461, 467,469-472,474
Benzaldehyde, responses to, 362-363,367
Biogenic amines, 375
Bithorax locus, 141,157,227,242 252,267
  bithorax mutants of, 142-145, 147-148,150-151,239,247, 275
  bithoraxoid mutant of, 144
  contrabithorax mutant of, 239
  deletions involving, 143-144 147
  giant neurons in, 283
  interactions with white locus, 36
  triple mutants (*bx pbx/Ubx*) of, 229,233,237,239,253,256, 258,277-278,283,286-287
  ultrabithorax mutants of, 142-143,147-148
Blind mutants, 334-335,345-346, 397,398
  in courtship, 426-427,435,457
Blowfly, 468
Brain, 215,279
  deuterocerebrum of, 280,290
  glycoproteins from, 318
  involved in courtship, 436,443
  membranes from, 319
  mushroom bodies in, 373-374, 376
  mutants of, 373,378-387
  protocerebrum of, 290
Bristles, 228
  abnormalities of, 25-26,96,104
  neurons of, 251,291
Brute force, 299
α-bungarotoxin, 294,297
  binding activity of, 296
  binding of radioactive, 295,299
α-bungarotoxin binding component
  pharmacological specificity

of, 296
purification of, 297-298
Buridan's paradigm, 397-399
Butyrylcholine, 307

C

Caffeine, 73,474
Calcium ions,
 chelation of, 325
 effect on protein phosphorylation, 328
Calliphora
 nervous system of, 283
 opsin concentrations in rhabdomeric membranes of, 340
Cell competition, 129-131
Cell death, 116,119,211,213, 217,262-263
Cell interactions, 130-131
Cell-lethal mutations
 temperature sensitive alleles of, 117,119, 121-122,124-125
Cell line of *Drosophila*, 198
Cell surfaces, 193,225
 interactions between, 227
Cerci, 225-226
Cerebrosides, 317
Chemicals
 olfactory responses to, 366,371
Chemoreception, 347,358,361,384
Chemoreceptors, 347,364-365,370
 electrophysiological responses of 350-354
 L cells in, 351-353,354
 S cells in, 354,356
 W cells in, 350-352,356
Choline acetyltransferase, 442
 courtship in mutants of, 444,446-447
 mosaics of, 444,447-449,
 temperature-sensitive mutants of, 444,446, 448
Choice behavior, 398,400
Chordotonal organs, 229,274

Chromocenters, 63
Circadian rhythms
 in courtship, 438-442,449
 eclosion in, 410,413,417-419
 ion gradient model of, 412, 414-415
 locomoter activity in, 409-411
 membrane model of, 413-415
 mutants of, 409-410
 2-oscillator models of, 417,420
 oscillators in, 414,418
 phase-responses in, 412,418-420
 in plants, 422
 slave (driven oscillators in) 414,417
 subjective day and night in, 410,413-414
 temperature and light induced, 418-419,421
Clonal analysis, 153,159
Cobalt ions
 filling of axons with, 229,249, 268,269-275,279-280,287
 trans-synaptic fills by, 273
Cobratoxin, 299
Cockroach
 effect of α-bungarotoxin on, 297
Comatose mutant, 301
Compartments, 124,132-133,159,251, 262,290
 neurons of, 252
Complementation units, 28,30,170-172,409
Conconavalin A, 318-320
Contact guidance, 258
Courtship behavior, 375,383,390, 425-426,437-438,444,449-450,453-454,457-458,463, 474
 communication in, 457
 female immobility in, 427
 index for, 426,428-430,446, 454,458,460
 learning and memory in, 433-435,449,469
 mosaic analysis of, 402,428-429,437,441,443,447-448, 450

odor cues in, 386,447,457-461,463
rejection behaviors in, 431
relation to circadian rhythms, 438-442,449
visual control of, 401,427,447
wing display and vibration in, 436,457,459,461
young males in, 430
Courtship song, 434-435,438,442-443,445
effects of acetylcholinesterase mutants on 445-446
effects of periodicity mutants on, 438-441
fluctuations in interpulse intervals of, 438-442
interpulse interval in, 437,446,454
mosaic analysis of, 436,445,450
mutant of, 436-437
Cricket
cercal system of, 225
tympanic organ of, 224
Cyclic AMP, 324-326
involved in learning, 468,470-472,474
Cyclic AMP phosphodiesterase
inhibitor of, 324,470-472,474
Cyclic GMP, 327

D

Daphnia
glial cells, 226
nerves of, 224-225
Degeneration staining, 249
Dendrites, 269
Deafferentation, 233-234,237
Deficiency-of-silver (*Df(1)svr*) chromosomal region
effects on embryonic development, 211-219,221
Deletions, chromosomal, 143-144,147,172,202,213

affecting embryonic neural development, 206,208-211,215,221
of parts of X chromosome, 204-205,219,409,471
Deoxyglucose
mapping of behavior-specific neural activity by, 402-404
Developmental pathways, 160,247
"switches" involving, 160-161
Determination
states of, 157
Diiosopropylfluorophosphate, 305
Distal outgrowth, 112,115,117-119,122
DNA
autoradiography of, 61,63,67,77
inhibition of synthesis of, 71
replication of, 57,60-68
replicons of, 58,66,73-75
time of replication of, 63,65-66,69
Drosophila hydei, 72,428
Drosophila pseudoobscura, 391,414,418-420
Drosophila simulans, 428,438
Drosophila subobscura, 427
Duplications, chromosomal, 172,204,213
Dunce mutants, 467,474
effects on cyclic AMP metabolism, 470-472, 474
defective learning in, 469-470
development of, 474
poor female fertility in, 470-471

E

Ecdysone, 134,165,168,170,179
Electric shocks
in learning experiments, 468
Electroretinogram, 335-336
Embryo, 203
blastoderm of, 198
denticle belts of, 217
development of, 86,91,206
epidermis of, 198
genetic abnormalities of, 205
head segments of, 211,217
Keilin organs of, 211,217

lethal phenotype of, 202
mechanoreceptors of, 217
nervous system of,203,211,216
pharynx of, 211,217
ventral pits of, 211,217
whole mounts of, 201
Ethyl methane sulfonate
mutations induced by, 96,
164,188,469
Ethyl acetate, responses to,
362-363,367
Expressivity, 143-144
effects of temperature
on, 151,153
Eye
digitonin extracts of, 337,
339
freeze fracture of, 341

F

Female-sterile mutations, 90,
92-93,96,98-105,470-
471
Fine structure recombination
mapping, 7,16,31
Flight behavior, 399
Flightless mutations, 186-190
Foci, of mutant gene action
of courtship behavior, 402,
436,443
involving lethality, 176-
179
of muscle defects, 186
Freeze fracture electron microscopy of ninaA eye, 341
Fruitless mutant, 431
pheromones from, 432
Fuchsin staining of embryos,
202-203,205,208,214

G

Gene dosage, manipulation of,
4-5,301,341,342,471
Gene expression
sequential activation,109
stage specific regula-
tion of, 163
Geotaxis, 387

Gastrulation, 87
"Giant" flies, 136
Glass-eye mutant, 427,432,435
Glycoproteins, 318-320
Golgi impregnation, of neurons,
280,283
Guidance
avoidance of, 397,399
visual, 396
Gustatory mutants, 347,355-356
behavioral responses of, 357
physiological responses of,
356-357
temperature-sensitive, 356
Gynandromorphs, 97,135,164-165,167,
176-180,184,428,436,441,
443,447-449,474

H

Haltere, 145,150,238-239,255,259,
268,275,278
nerves of, 263,272
projection of axons from, 229-
231,233-234,236,242,255,
263,271,273,275
ventral axons from, 276
Haplo-insufficient gene loci, 219
Heads of flies, 313-314,323
enzymes in, 315
phospholipid and cholesterol
contents of, 316
protein pattern of, 317,320
Heat shocks, 41-43
RNA transcription induced by,
43,45,48
salivary gland chromosome puffing induced by, 42
studied by nucleic acid hybridization, 44-47
Homeosis
direct, 157,162
indirect, 157-158,160,162
Homeotic mutants, 141,155,157,227,
232,237,247-249,252,262,
269
lethal alleles of, 158-160
projection of sensory axons
in, 229-231,233-235,241,
255,257,271,273,288

INDEX

re-routed nerves in, 236, 238,263
transformations involving, 142,144-145,147,152, 156,159,160,239
Horse radish peroxidase
filling of axons with, 229, 249,253-254,272-273
Housefly, 309,377-378,460
Hoverfly, 399,401
Hybridomas, 194
Hypoderm
of embryo, 209,216-217
Hypomorphic mutation, 145

I

Imaginal discs
development of, 155-156
genetic abnormalities of, 134,136,158,165
injection of, 159
small disc phenotype of, 158
surfaces of, 193
transplantation of, 186
Immunofluorescence, 194
Ion detectors, 352
In situ hybridization, 44-45
Ion channels, 293,297
Isoelectric focusing, 300

K

K10 mutant, 90
Krüppel mutant, 91,93

L

Labellum
chemoreceptors of, 348,359
physiological responses
of hairs on, 351,354, 357
Larvae, 313
chemoreceptors of, 370
development of, 88,157,382
glycoproteins from, 319
"marker" enzymes in, 315
membrane fractions of, 320

muscles of, 183-184
olfactory responses of, 362-365,368
taxis of, 365
Laser lesions, 226
Learning, 374,386,467-469,472
in courtship, 433-435,449
mutants of, 467,469-472,474
olfactory cues in, 469
reinforcers of, 468
Lectin-affinity column, 299
Legs
discs of, 118,168,175
genetic abnormalities of, 110
nerves of, 263
supernumery, 120
Lethal mutations, 91,95,135,172
affecting pupal stage, 170-179
developmental characteristics of, 173
embryonic, 201
lethal phase in, 171
Ligands, cholinergic
muscarinic, 295
nicotinic, 295
Locomotion of adult flies, 391,409
Locust
cerci of, 226
grafting experiments in, 242
nervous system of, 250,258

M

Maroonlike mutant, 100-101
Mated females
courtship effects of, 386, 433-435
Maternal effect mutations, 95
Mechanoreceptors, 217,228,250
Membranes, 313,323,328
from brains, 319
fragments of, 306,309-310
glycoproteins in, 318
isolation of, 314
from larvae, 320
from mitochondria, 324
of photoreceptors, 331
proteins in, 293,306,314,324-325

Memory, 469
    consolidation of, 470
    in courtship, 433-435,449,
        474
    mutant of, 433-435,474
Metamorphosis, 163-164,174
    of nervous system, 223-224
Metarhodopsin, 332,335
    conversion from rhodopsin
        333
Mesothorax, 142,147-148,152-153
    axons of, 256
    dorsal nerves of, 263
    sensory structures of, 249
Metathorax, 142,148
    axon branches of, 286
    leg nerves of, 263
Mice, immunized, 194
Minute mutations, 25-26,130,233
    compartments and, 132
    complementation map of,
        28,30
    fine structure map of, 31
    phenotypic effects of, 30
    salivary gland map of, 27
Mitochondria
    enzymes from, 316
    membranes from, 324
Mitotic recombination, 90,93,
        101-103,105,131,133,
        160,232
Mosaics, 85,89,90,92,105,233,
        474
    clones in, 131-133,153
    gynandromorphs, 97,135,
        164-165,167,176-180,
        184,428,436,441,443,
        447-449
    induced by mitotic recom-
        bination, 90,93,101-
        103,105,131,133,160,
        232
Moth
    antennal lobes of, 224
    metamorphosing nervous
        system in, 224
Movement detectors in nervous
        system, 391,402,404
    wiring scheme of, 393

Muscles
    actin in, 183
    adult, 184,188
    fibrillar, 184,186-187
    indirect flight, 184
    larval, 183,187
    myosin in, 183
    neuronal projections to, 291
    precursors of, 186-187
    proteins in, 190
    sarcomeres of, 190
    tergal depressor of tro-
        chanter, 184
    tubular, 184
    Z-bands of, 190
Mushroom bodies, 373-377
    Kenyon cells of, 374-377,380,
        382
    lobes of, 374-377,380-382
    mutants of, 378-387
Mushroom bodies-deranged mutant
        379-390
    behavior of, 383-386
    development of, 380,382
    larvae of, 382

N

NADH diaphorase, 315
Neither-inactivation-nor-after-
        potential
    ninaA mutant, 334,336-338,340-
        341,345-346
    rhodopsin concentrations in,
        338
Nervous System
    anlage of, 213
    cellular cortex of, 203,445-446
    CNS projections in, 252,284-286
    condensation of, 212
    developmental analysis of, 247-
        248
    genes controlling development
        of, 202,220,444
    genetic defects in, 204,207-
        211,220,444
    movement detectors in, 393
    neurogenesis of, 201,205,257-
        259
    neuromeric commissures in, 209

INDEX   493

neuropile of ganglia in,
    203,208,270,279,445-
    446
pathways in, 247,258
protocerebral lobes of, 215
subesophageal ganglion in,
    210,214-215
toxin binding in, 295
ventral portion of in
    embryo, 208-211,214
visually triggered activity
    in, 403
Neurons
    choice of pathways by, 247,
        252-253
    conduction of impulses with-
        in, 301
    developmental history of,
        250
    giant, 283,286
    homeotic, 252-253,255
    interganglionic, 277
    metathoracic sensory, 253
    misrouting of, 263-264
    segmental identity of, 249
    sensory, 223,226-227,242,
        247,250
Neurosecretion, 382
Neurotoxins, 293-294,301
    binding properties of, 294
Neurotransmitters
    receptors for, 293,295
Nicotine, 295,299-300
    drug resistant mutants to,
        299-300
No-action-potential mutant, 301
No-object-fixation mutant, 397-
    398
No-receptor potential mutant
    (*norpA*), 334-335,345-
    346
    interactions with *rdgB*
        mutants, 336
    in courtship, 427,432,435
Notch locus
    affect on embryonic neural
        development, 206,214,
        216,220
    deletions of, 205-206

Notum, 153,262
    bristles of, 259
    duplications of, 262-264
Non-pupariating (*npr*) mutants,
    165-166,170,179
    autonomy of, 167
    imaginal discs of, 168
Nucleotidase, 316

O

Olfactometer, 462-464
Olfactory mutants, 361,366-369
    in courtship, 429-430,433-
        435, 460-461
Olfactory responses, 361-362,375
    384-387
    in courtship, 428-433,447,457-
        461
    of females, 464
    indexes for, 362-363,463-464
    in learning, 469
Oocytes, 85,89,103-104
    chorion of, 86-87,92
    follicle cells associated
        with 88-91,103-198
    nurse cells associated with,
        103
    oogenesis of, 85
    polarity of, 86,88-89,91
Opsin, 339-342
Optomotor responses, 391,394
    control system of, 392-393,404
    mutant of, 397-398,428
    positive feedback in, 395
Ovaries, 99,197
    development of egg chambers in
        99,103
    transplantation of, 99-101

P

Paralytic-temperature-sensitive
    mutants, 301,427,434,446
Pattern
    deficiencies and duplications
        of, 113-117,119
    genetic abnormalities of, 110-
        111

models for formation, 111, 124
positional values of, 113
pre-patterns of, 109
triplications of, 111,120-121
Periodicity mutants, 409,412, 415,454
  eclosion in, 410-413,438
  effects on courtship song, 438-441
  locomotor activity in, 410-411
Pharynx, embryonic, 211,217
  genetic defects of, 218-219
Pheno-effective period of mutation, 206,219
Pheromones
  affecting courtship, 428-431,435,454,457-459
  chromatography of, 431-432,459-460
  general, 462-464
  of housefly, 460
  varied concentrations of, 429,459,464
Phosphatidylcholine, 317
Phosphatidylethanolamine, 317
Phosphatidylserine, 317
Photo-activatable cross-linking procedure, 300
Phosphorylation reactions, 324-328
Phototaxis, 384
Photoreceptors, 331,333
  desensitization of, 337
  potentials of, 332-333,337
  quantum bumps in, 332,335
  rhabdomeric membranes of, 338,340
Phototransduction, 331,336
  genetic defects in, 335
Plants
  circadian rhythms in, 422
Pleura, 262
Pole cells, 89
Polyteny, 198
Position effect variegation, 17-18,20
Positional information, 113,156

Proboscis extension response, 347-350
  in gustatory mutants, 355,357
  inhibition of, 349-350
  in learning tests, 468
Prolonged-depolarizing after potentials, 332-334,337
  mutants defective in, 334,336-338,340
Proteins,
  electrophoretic gels of, 49-51, 300,306-308,317,319,327
  of membranes, 320,324
  of muscles, 183,190
  phosphorylation of, 323-328
  in vitro translation of, 48-52
Pseudocholinesterase, 307-308,311
Pseudo-search behavior, 395-396,399
Pupae, 175
Pupariation, 134,136
  genetically induced delay of, 165
  non-pupariating phenotype of, 164
Puromycin, 72
Pyridine, responses to, 363

Q

Quinine, responses to, 349,355,468,
  mutants of, 355

R

Receptor-degeneration mutant, 334-336
  interactions with norpA mutants, 336
  suppressors of, 336
Regeneration, 111,117-118,262-264
Reproduction, 426,450
Rhodopsin, 331-333,335,337,339-341
  conversion to metarhodopsin, 332
  genetically altered levels of, 338
Ring gland, 167,169
RNA transcription, 43,45,48
"Rocket" electrophoresis, 9,19

Rosy gene, 3
  genetic dissection of, 6-9, 16
  polytene chromosome map of, 4
  regulatory variants of, 11, 14-15

S

Salicylaldehyde, responses to, 362-363
Salivary gland, 197
Salivary gland chromosomes, 4,27,169
  autoradiograph of, 44
  puffs of, 42,60,169-170, 179
  2B1-10 region of, 169-173
  1B "X-chromosomal" region of, 211-219
  3B "X-chromosomal" region of, 409
  3D "X-chromosomal" region of, 471
Saxitoxin, 294
  radioactive, 297
Segmental aneuploidy, 301,471
Selective medium, 194
Sensilla, 273
  sensilla campaniformia, 228-229,234-235,238-239,241,250,253,256, 259,274,276,279,290
  sensilla trichoidea, 274, 348
Sensitization, 468
Sephadex column, 309-310
Serial reconstruction of sections, 271
Shibire-temperature-sensitive mutant, 427,434
Silver impregnation of neurons, 272
Size control, 133
Smellblind mutant, 361,369, 429-430,433,435
Sodium chloride, responses to, 349,351,355,387
  mutants of, 355-356

Sorting out of cells, 150
Spectral sensitivity
  of visual function, 340
Sphingomyelin, 317
Sterility
  of females, 86,90,96,104,470-471
  of males, 431
Succinate dehydrogenase, 315-316
Sucrose gradient, 314
Sugars, responses to, 349
  mutants of, 355-356
  physiological responses to, 352

T

Taste responses, 347-348,358
  mutants of, 355-357
Temperature
  effect of on nerve misrouting, 263-264
Temperature sensitive mutations, 117,119-122,124-125,301, 356,427,434,444,446-448
Thoracic ganglion (of adult), 233, 254,256,263,276,279-280, 284,288
  involved in courtship, 436,443
  supernumerary axons in, 287
Thorax
  cuticle of, 142
  duplications of, 233,262
  muscles in, 188
Transfer RNA, 26
Transformer mutant, 437,439,443, 445,450
Transient receptor potential mutant, 334-335
Translation of mRNA's 48,52
  cell free systems of, 48,50-51
Transdetermination, 157-158
Transvection, 36
Tread compensator, 399-400
Trichogen cell, 239
Two-dimensional polyacrylamide gels, 50,317,319

U

Uridine
  incorporation of radioactive, 43-45,170

V

Visual behavior, 396,398,401, 403,
  in courtship, 426-428,447
Vitamin A
  flies deprived of, 339-340

W

White locus, 35,336,409
  interactions with bithorax locus, 36
  interactions with zeste gene, 37-39
  various alleles of, 35
Whole mounts
  of embryos, 201
  stained with fuchsin, 202-203
Wing, 135,228,238-239,255-256, 259,268,275,278,291
  abnormal positions of, 188
  basal lamina in disc of, 196
  binding of antibodies to disc cells of, 196-197
  compartments of, 134,148,262
  disc of, 116,118,134,168,195, 263-264
  dorsal axons from, 274
  epithelium in disc of, 195-197
  genetic abnormalities of, 115-116,235-237
  homeotic, projection of axons from, 229,231, 233,235-236,254,256-257,271-273,275-278
  peripodial membrane in disc of, 195-197
  projections of axons from, 229-236,242,255,272
  removal of, 233-235
  trichomes of, 150
  triple bristle row of, 148,150, 153,228
  ventral axons from, 276
Wing mutations, 115-116,235-237
Wingless mutant, 115,117,233,235-237,255-256,262-263
Wound-healing, 112

X

Xanthine dehydrogenase, 3-4,10
  electrophoretic mobility variants of, 5
  expression during development, 10-11
  variation in activity of, 9

Y

Y-maze
  in olfactory experiments, 462

Z

Zeste gene, 35,38,409
  interactions with white locus, 37-39

DATE DUE

	DISCHARGED		DISCHARGED
05. 06. 83			
JUN 0 7 '88			
	DISCHARGED		
AUG 0 9 1995	2 9 1995		
Pcla #301254			
da 3/1/05			

DEMCO NO. 38-298